PERSPECTIVES ON NATURE CONSERVATION – PATTERNS, PRESSURES AND PROSPECTS

Edited by **John Tiefenbacher**

Perspectives on Nature Conservation – Patterns, Pressures and Prospects
Edited by John Tiefenbacher

Published by InTech

Janeza Trdine 9, 51000 Rijeka, Croatia

Publishing Process Manager Romana Vukelic

Technical Editor Teodora Smiljanic

Cover Designer InTech Design Team

Additional hard copies can be obtained from orders@intechopen.com

Perspectives on Nature Conservation – Patterns, Pressures and Prospects,
Edited by John Tiefenbacher
p. cm.
ISBN 978-953-51-0033-1

For EU product safety concerns:
IN TECH d.o.o., Prolaz Marije Krucifikse Kozulić 3, 51000 Rijeka, Croatia,
info@intechopen.com or visit our website at intechopen.com.

Meet the editor

Dr John P. Tiefenbacher (BS from Carroll College, Wisconsin, MS from University of Idaho, and PhD in Geography from Rutgers University) is currently Professor and Director of the James and Marilyn Lovell Center for Environmental Geography and Hazards Research in the Department of Geography at Texas State University in San Marcos, Texas. Dr Tiefenbacher's research has focused on Environmental Geography or the relationship between people and their natural environments. Dr Tiefenbacher has published on a diverse set of topics that examine perception and behaviors of people with regard to natural and technological hazards, environmental problems, and personal preconceptions about the places they inhabit. He has undertaken studies throughout the United States and in Mexico. His current research pertains to the challenges of urban hazards and evacuating cities, adaptation to climate change, the responses of agricultural industries (particularly wine growing) and climate changes in several regions (principally South America, the U.S. and Europe) of the world.

Contents

Preface

Perspectives on Nature Conservation is a collection of chapters that demonstrate the diversity of information and viewpoints that are critical for appreciating the developing gaps and weaknesses in local, regional and hemispheric ecologies, and also for understanding the limitations and barriers to accomplishing critical conservation projects. The organization of this book is intended to emphasize, through these reports of original research by an array of international scholars, the linkages between the geographic foci of conservation projects (i.e. they are focused in specific locations or in certain types of places, and often require management of spatial behaviors and processes) and the biological substances (flows biotic and abiotic materials, chemicals and energy through time) that we conceptualize as "nature". I have organized the chapters into five sections that take the reader through perspectives of diminishing spatial scales (that is to say, from smaller to larger landscapes, covering increasingly larger portions of the surface of the Earth).

First, a chapter by Krzysztof Rostański provides a conceptual perspective for understanding the place of nature (not only the biological forms, but the geometric forms and dimensions that nature forms) in our society, the inseparable thread that it is in all modern societies (indeed, there can be no culture without nature and, dialectically, to define nature defines the definer of it) and its conservation. Rostański emphasizes that even "incomplete" natural landscapes (or disturbed ecosystems) can serve the nature needs of society, and that logic is demonstrated in the long-practiced approaches to the "greening" of cities and the tenets of landscape architecture.

Nature conservation is founded on the notions of loss and change. Substantiation of the species that represent nature and natural conditions requires the expertise and discipline of biologists who catalogue and track the composition of the resident communities of the present and the past. Several chapters in the second section ("Understanding Patterns, Change, and Conservation Needs") exemplify the methods of those who study the evolution of species and communities vis-à-vis changing environments, in order to predicate conservation on rational expectations of success. Coiffard, Gomez and Daviero-Gomez discuss their examination of the speed and timing of the emergence of landscape ecology of angiosperms (during the Cretaceous) and their eventual domination of terrestrial environments. The analytical methodology of clustering is one commonly used by others, defining the assemblages of species that cohabitate.

Genetics is fundamental to the evolutionary process, and the scientific capacity to discriminate and track genetic changes itself is rapidly evolving. Akhalkatsi, Ekhvaia and Asanidze detail the extraordinary hearth of diversity in the Caucasus Mountains of central Asia that provided the seeds of modern agricultural cultivars. Their study of the remnant wild relatives of cereals, legumes, grapevines and other crops in Georgia demonstrates that development and economic preferences has reduced the fortitude and diversity of the genetic pool in their original source region. The authors discuss the implications of the trend for conservation efforts in the region.

The third chapter, by Baumbach, examines the distribution and conservation status of a set of plant species that either tolerate or prefer soils that are high in metal compounds. The plants themselves may not attract a large following of devotees (or detractors for that matter); the environments within which they flourish are not normally desirable. Heavy metal contamination from industrial emissions and mining in Germany (where this study is conducted) are usually related to undesirable, "brown" or derelict sites, which usually attract remediation that suggests elimination of the environments in which they thrive. This would seem to make conservation of these assemblages more challenging.

The third section of this book features landscape metrics and their use in conservation. The first chapter in this group is a review of the use of measures of the characteristics of landscapes and ecosystems in nature conservation studies. Geographical concepts are applied to ecological circumstances and examined for their role in the perpetuation or devolution of ecosystems. Csorba and Szabó discuss the patterning of landscapes and the roles of fragmentation, patch development, size and connectedness on habitats, and the application of these concepts and relationships to practical projects in conservation.

Eggers and Köhler present an example of such an application. Their study of the fragmentation of woodland throughout Germany on avian and mammalian habitats demonstrates that landscape metrics can be employed to evaluate the impact of fragmentation on habitat suitability for fragmentation-sensitive species. The incorporation of analyses of fragmentation of potential habitat at different scales (scales appropriate to the species in question) can illuminate needs for landscape patch connectivity to facilitate conservation and land use management.

Cardozo, Naretto, Zak and Chiaraviglio face the conservation of species from a very different angle. While the metrics of landscape are used to evaluate the suitability of habitat, the authors seek to better understand the role of landscape in both the presence and absence of "sister" species (in this case species of lizards in north central Argentina that normally occupy separate but adjacent habitats) and the differential use by the sexes of the two species in the zone of contact between them. They find that the ecological interaction between the species within the contact zone is vital to their survival, and that understanding the landscape characteristics that foster this interaction is vital to species and ecosystem conservation efforts.

The fourth section of this book regards regional conservation perspectives. Three case studies demonstrate approaches that consider the regional implications and constraints on efforts that strive for regional landscape regimes. First, Makhzoumi, Talhouk, Zurayk and Sadek discuss an overlooked aspect of nature conservation: bio-cultural diversity in rural portions of long-settled landscapes. They argue that not only do the species of woodland patches and scrubland maquis of Lebanon deserve conservation to perpetuate their existence, but also that their conservation would provide the foundation for tourism and development opportunities to establish a vital economic base. To support this notion, they provide a landscape design approach that integrates local needs with the resources of the past and present, and links the region to national and international objects for a peaceful, self-reliant, local economic development through rural landscape heritage.

Politics and economics become increasingly more relevant to conservation projects as the spatial extent of the project increases. Kalikhman, in an examination of the challenges posed by the bureaucracy of the Russian Federation and sub-national regional organization of the Lake Baikal region after the collapse of the Soviet Union. While the legal definitions of the control of space within the basin of the inland sea created a pastiche of political control, approaches to conservation based upon either natural boundaries or administrative units seem to fall short of what is needed for effective and satisfying conservation of the region. The author provides an integrated plan as an alternative strategy to mitigate the weaknesses of the traditional approaches to conservation. The case exemplifies the perception among some that conservation projects can have negative implications for economic activities, and this creates a barrier to establishing consensus support for such projects. The integrated approach proposed by Kalikhman strives to navigate the political and economic barriers to an effective conservation plan.

A third case, by Atik, Sayan, Karagüzel, and Yildirim, regards the challenge of saving nature from the nature lovers. In Termessos National Park in Turkey's Mediterranean region, the visitors are degrading the vegetation along trails in some parts of the park. While recreation has long been regarded to be a friend of conservation, particularly recreation by those trying to engage with nature in the way that Rostański discusses in the first chapter of this book, its popularity has demonstrated that even a good thing can go too far. The experiences of park managers in the United States, for instance, had begun to see the prospects of "loving nature to death" by the 1960s, as national park visitation skyrocketed. It wasn't until then that ecologists had become acutely aware of the fragility of nature, particularly in more extreme environments. As the (at least relatively) wealthy are now able to visit the national parks of the less developed countries of the world with greater frequency and duration, similar impacts are becoming clear. The trade-off is difficult to control because visitation brings economic activity and it's wealth that many regions cannot afford to forego. Carrying capacity is the conceptual approach used to determine an objective rate of tourism, but it should be determined by the ecology, not the economy, which is often the proxy chosen. It is the ecological-economic tension that often generates the greatest challenges to effective regional conservation planning.

Conservation of larger areas involves greater complexity and sets of issues that confront the conservationist change and become more difficult. The global perspective is indeed the youngest conservation approach (in an endeavor that is less than a century old itself); the challenges that face global conservation are still murky, the strategies developed to achieve it are relatively undeveloped, and the international agreements to enable it are insufficient. A chapter by Restrepo-Aristizábal, Heggestad and Acuña-Rodríguez concludes this volume. It elaborates on the use of landscape ecology and its principals in the development of a policy for nature conservation throughout the Neotropics (of Latin America) in the context of global (particularly global climate) change. Conservation, particularly conservation of intact forests and other carbon-sequestering environments, is now of critical importance to not just saving species, but more importantly to providing the breaks to slow the warming of the Earth's climate. Conservation can be a vital tool to increase storage of atmospheric carbon and could be (partially) effective even without the limitation of greenhouse gas emissions that is being bargained for at the United Nations Committee of the Parties meetings that are following up on the Kyoto Protocol. The will to make the necessary changes to combat global warming head-on may elude the world's diplomatic communities, but a regional conservation plan could serve as a stopgap measure, or at least a preliminary step toward regaining a balance in the world system.

The chapters in this book hold many more valuable insights to a broad assortment of issues and debates in conservation specifically, and in science in general. Each study was undertaken with its own purposes and scholarship in mind. While not intending to put words into the mouths of each of the authors of these chapters, I have tried to highlight some meaningful issues that hold this text together. The chapters speak for themselves (or their authors) and this one speaks for me. It is up to each scholar to interpret the literature and find its meaning. I expect that the readers of this volume will find many meaningful lessons herein.

Dr John P. Tiefenbacher
Professor
Director, James and Marilyn Lovell Center for
Environmental Geography and Hazards Research
Department of Geography
Texas State University-San Marcos
USA

Part 1

The Place of Nature Conservation in Modern Society

1

Modelling Nature in Ecologically Oriented Urban Context

Krzysztof M. Rostański
Silesian University of Technology
Poland

1. Introduction

Although every model implies simplification, it also indicates the most vital elements of a given structure to follow, which is the basic assumption of the present article. Modelling could also be understood as the shaping of an existing natural object. The question remains how this could be implemented in harmony with nature. Greenery makes human-ordered space clearer and more complete. When formed after the aesthetic norms of the related architecture, the greenery usually acquires geometrical shapes. However, when natural systems are followed closely, the structures thus obtained with their free form and ecological values emphasised, are able to improve our living conditions. The present work attempts to prove that even if we try to be in keeping with nature only in form, both the aesthetic and the ecologic aims are possible to achieve. The quality of designed greenery cannot be ignored, considering the need to preserve nature in our neighbourhood, and to provide the maximal biological efficiency. The main aim of the present work is to show the importance of nature in our surroundings while discussing various aspects of its role in the urban greenery. The use of the local plants with their particular features could make the architecture look exceptional, adding a strong local identity. On the other hand, using foreign ornamental plants without proper consideration leads to the formal homogeneity of modern cities, despite all their aesthetic value. The integration of built up space and local natural landscape implements the goals of The Convention on Biological Diversity and The European Landscape Convention, being also a principle of rational naturalism. Moreover, natural and naturalistic greenery, functioning beyond our consciousness at the level of biological reaction, is conducive to the proper functioning of living organisms, including human beings.

The thesis of the work could be expressed by the statement that the human being is part of his local nature together with the products of civilisation created by himself. However, a more suitable habitat for him should be simple, emerging from the natural surroundings organically, yet still being human-made. This very idea is supported by naturalism.

Naturalism is understood as the idea of following natural processes and creating naturalistic relations modelled on them. It supports our healthy lifestyle and provides us with tools to implement its principles. Naturalistic features could be used for land developing in varying degrees. It is hard to say that the naturalistic form is always the best choice, as also geometry has its advantages and in some situations seems to be even better. Naturalism, however, has many advantages worth noticing. Some naturalistic elements are so flexible that they could be used even in the conditions under human pressure.

2. Subject and methods

The present work covers the moderate climate zone of the Northern Hemisphere. The time period discussed goes back to the 17th century but the last hundred years is considered to be the most significant. The subject of the research refers to the naturalistic elements of landscape architectural forms and the designed space components.

The issue of modelling nature in the urban context is described by means of aspects considered by the present author as vital. The adopted research method resembles the phenomenological attitude by describing firstly the particular aspects, and thus revealing the characteristic features, then providing a summary, and finally formulating a general view. The aspects considered involve the evolution of the naturalistic idea in design and the respective law regulations, the functional naturalism, the formal naturalism and the elements of perception and semiotics.

The first aspect to be considered is naturalism in the history of landscape architecture, which starts in the Far East gardens, inspiring the European journalism and, resulting in naturalistic elements of landscape parks and gardens. The present ideas of the ecological approach to design are supported by legal codifications and international agreements, charts and directives. The historical overview proves the significance of naturalistic greenery. The functional naturalism shows the influence of natural greenery on our habitat, human needs and health. The formal naturalism presents selected features of naturalistic elements in greenery compositions, their formal patterns and structures abstracted from the chosen natural communities. These factors constitute the basic framework of a naturalistic composition. In order to cause certain human reactions, the factors should be supplemented with the ideological content and the structural principles of a composition, which relates to the problem of semiotics and perception. Each aspect mentioned above leads to the conclusions describing the most important principles and targets of the naturalistic design, called models. The final effect of the reasoning is the heuristic model of the naturalistic design used by the author to demonstrate a selection of design styles based on the varying degrees of their correspondence to nature.

3. Development of the naturalistic idea in the historical context

3.1 The beginnings

One of the oldest examples of nature-inspired composed greenery comes from the Chinese gardens, where the landscape modelled on real places was built with plants and animals characteristic of the area. Its form was mostly free, and yet it was full of religious symbolism. The following words from the *Tao Te King,* Book 51, probably strongly influenced the Eastern gardening - "All things (…) receive their forms according to the nature of each, and are completed according to the circumstances of their condition. (…) A cart is more than the sum of its parts"(McDowell & Stewart, 1988).

The Japanese gardens were strongly dominated by metaphysical symbolism. The *Sakuteiki,* an 11th cent. gardening manual, stresses that gardens should not be a true copy of nature but rather its interpretation referring to the religious and literary content. (Takei & Keane, 2008). The Japanese garden was in harmony with local nature, which manifested itself in the use of the local plants exclusively and their arrangement after the natural habitat.

In Europe the religious symbolism was hard to find in naturalistic ideas in gardening. Generally, the ideas were expressed by means of the formal patterns e.g. the sublime style, imitating the dynamic and expressive natural phenomena, which seemed to be quite popular.

3.2 The picturesque or the natural

One of the first European manifestations of the naturalistic ideas was a statement by Henry Wotton (1568-1639), of 1624 - "First, I must note a certain contrariety between building & gardening: For as Fabrics should be regular, so Gardens should be irregular, or at least cast into a very wild Regularity"(Wotton, 1624). Following this way of thinking, William Temple (1628-1699) designed a waving path across the meadow in Moor Park (1680-1690), which set the trend for landscape parks. The beginnings, however, were not easy and for many years, gardening was dominated by geometry and the architectural way of thinking. Throughout the 18th and the 19th centuries, the reverse direction of inspiration, namely from nature to architecture, was gradually approached by W. Kent (1685–1748), L. Brown (1716–1783) and other designers. Humphry Repton (1752-1818) combined two different standpoints, those of an architect and of a gardener. J.C. Loudon's (1783-1843) attempts to copy the natural dispersion of trees in gardens were also quite significant (Turner, 2005) .

The subject of naturalistic ideas was raised by journalists and philosophers, among whom a great impact was exerted by Jean-Jacques Rousseau (1712-1778), who was convinced that "sanctity lay in unadorned nature"(Shepard, 2002). The beauty of nature, however, was not generally valued. Greenery was often overwhelmed by the architecture of pavilions to such an extent that Archibald Alison (1757-1839) suggested founding parks without them so that the greenery was more visible (Alison, 1853). Improving nature was generally criticised, e.g. Friedrich Schiller did not want the designed greenery to stand out from the natural one (Kruft et al, 1994).

The approach encouraging to model nature artistically finally led to the cosmopolitan gardens with plants from all over the world, which is visible in the gardens of the picturesque and the gardenesque styles. Also, Central Park in New York (1858), was initially designed as a garden with many plants from different parts of the world (Treib, 1999). Only very few have still survived, with some of them invasive and hard to control. It is in the human nature to have a collecting instinct and a will to constantly enrich gardens. People still perceive nature, as Joan I. Nassauer writes, from the angle of a picturesque convention and not an ecological value (Nassauer 1995).

The natural aesthetics was widely adapted in the national style, very popular in England and Germany in the 19th century. In the works of Alexander Humboldt (1769-1859) and Ernst Haeckel (1834-1919), the natural aesthetics encompassed the ecological aspects of local nature, which helped to arouse ecological awareness and the appreciation of real natural simplicity.

3.3 The geometric or the free line style

From the very beginning greenery arrangement had two parallel styles: the free form and the geometric style. The latter was preferred by those accustomed to the baroque forms, but it also had its critics, e.g. Hermann Pückler-Muskau (1785-1871), a landscape planner. His opinion was supported by the Laputians, the crazy philosophers from Swift's *Gulliver's Travels*, who "conceived the idea of regulating vegetation by geometry, and exhibiting

Euclid's diagrams on a plot of ground, they could not have devised any thing more preposterous than trees clipped into cubes and other mathematical figures, or disposed in formal rank and file" (Cochrane, 1835).

A considerable degree of refinement was noticeable in the gardens designed by Gertrude Jekyll (1843-1932), being sometimes free in their form, and sometimes making use of geometrical forms, with the geometrical flowerbed edgings which surrounded freely mixed groups of plants (Bisgrove, 1992). William Robinson (1838-1935) in turn, enchanted by a variety of plant arrangements in nature, fostered naturalistic gardens of ornamental plants (Robinson, 2009). Similarly, J.P. Thijsse (1865-1945), a Dutch designer, showed the possibility of creating naturally-looking arrangements of mixing local plants with ornamental ones. The examples above prove that geometry and free form are often hard to separate.

3.4 In harmony with ecology

The issue of native plants in the garden and the similarity of their arrangement to the natural plant communities appeared in the works of Elisabeth Holden (1871-1920) in England in 1906 (Andrews, 2003). At the same time a similar idea of imitating nature in park designing was developed in the United States by Jens Jensen (1860-1951) (Hobhouse, 2005). He composed naturalistic gardens and is best-known for his prairie gardens, proving that it is possible to build attractive natural-looking gardens. Jensen's achievements undermine the validity of the statement made in 1909 by one of the garden city designers, Raymond Unwin (1863-1940) that "any attempt to copy nature must be futile, for the conditions of natural growth are so complex as to be quite beyond the power of the gardener to understand or reproduce. He can only hope at best to parody, and is much more likely to caricature"(Unwin, 1994). At the turn of the 20th century the International Style was developed which brought examples of fully geometric gardens, totally alien to nature, such as the garden round the Noailles Villa at Hyéres, France (1927) designed by Gabriel Guevrekian (1900-1970) (Turner, 2005).

In the first half of the 20th century the most fully integrated designs of buildings and their natural surroundings are represented by the selected designs of F.L. Wright (1867-1959). His prairie houses adjusted to the landscape and the native greenery, especially his Fallingwater, Penn. (1937), which despite, probably, being damp, shaded and cold, shows an absolutely exceptional idea of creating a building in the form corresponding to the human way of thinking i.e. the Euclidean coordinates, and blending into the natural wood. The house is a true house, the wood around is natural and everything seems to be linked organically. A similar idea can be found in Dessau, Germany, where The Bauhaus Master Houses (1926) stand only among pine trees, which are characteristic of the surrounding local woods. The examples of using native plants supported somehow the modernistic idea of preserving the natural features of the material used.

The ideas included in the plans made by Ian McHarg (1920-2001) were inspiring for the city authorities planning their ecological policy. He claims that using the natural conditions will always give better and more permanent solutions than transforming everything when following the arbitral decisions dictated by the artistic vision (McHarg, 1992). He defined the adapted environment as "requiring the least work of adaptation". On the other hand, when analyzing people's needs, we have to acknowledge their rights to culture and even to fashion. This is probably when the idea of the sustainable development was born.

3.5 The time after modernism

The end of the 20th century brought many ideas, some of them popular and controversial. In the works of Martha Schwarz (1950-), the art reaches a state where the greenery is sometimes made of plastic, e.g. Splice Garden in Whitehead, Mass. (1986). Whereas in modernism the popular slogan was "form follows function", Bernard Tschumi (1944-) states that now "form follows form" (Nesbitt, 1996), which results from the increasing domination of the artificial world. The La Vilette Park, Paris, France (1982-1993) designed by Tschumi – "the largest discontinuous building in the world" was planned as an aesthetic pattern (Turner, 2005). The art is delightful and attractive but the real life follows its own way, gathering people together under the canopy of trees which form an ordinary little wood in the eastern part, while the unusual interior of the park remains empty. Frederick Gibberd (1908-1984) had an idea of the park space which should surprise people rather than lead them, resembling the way nature is perceived by a nature lover whose walk in the woods often looks like the Brownian motion. Such space, however, will not be clear to the majority of people because it requires the appropriate knowledge and interest. The greenery of the Lanxmeer district in Culemborg, Holland (1993-) was developed as both a creation of semi-natural habitats and a limited aesthetic design of ornamental plants. Michel Desvigne (1958-) and Christine Dalnoky (1956-) designed a park imitating the form of a riverine forest in the post-industrial land near Millenium Village (1999) in London (Donadieu, 2006). Although it looks very natural, it is noticeable that the trees are planted in regular spaces. The naturalistic ideas are common in the postindustrial areas. The size of those areas prevents their full and accurate management, thus creating room for common plants and their free succession, which happens very often. The existence of the "Berlin Wall" is the origin of the Nature-Park Südgelände, Berlin (2000), where common and alien plants form a bird sanctuary there. The place is of low value for botanists, but its local function is considerable. The experience of the Ruhr region parks in Germany shows that the highly adaptable alien plants can make communities of long durability. It is argued what influence they will have on the natural plant communities and what is the meaning of the term natural in this context. It is a fact that expansive non-native plants have been spreading, which will lead to the catastrophic uniformity of greenery, with the only differences resulting from the climate zones, and this is only a question of time. Some hope, however, is in the growth of the ecological awareness. Owing to them the features of the post-modern urbanism are beginning to include those referring to naturalism: respect for the regional values, the ecological approach to urbanisation and spontaneity (Bańka 2002). Those feature could be found in the four gardens designed by Richard Haag at Bloedel Reserve (1978) near Seattle, WA (Cooper, 2003). He preserved the growing trees and initiated the undergrowth reconstruction and development in its natural form. He also placed the logs of cut tree stumps, leaving them to decay naturally. Particularly, objects of the Land-Art type use the natural context for the background of their installation. Following a similar principle, Susan F. Child built a garden in Grande Island, VT, by placing footbridges and other small architecture elements in the natural area (Cooper, 2003).

3.6 The historical model

When summarizing the development of the naturalistic tendencies one can notice two approaches to the relation between built-up areas and greenery. The first one treats greenery as totally different from human creations, the second one views nature and culture as related to each other, both being elements of the human environment. Culture is seen here as the evolutionary consequence of the world development.

	Composition	Greenery	Aesthetic value	Example
1.	Formal	geometry; sometimes lack of greenery	designed, modelled elements	Unwin, Schwartz
2.		natural greenery outside or the filling of geometric elements like boskets; buildings and greenery create a compact system of interiors		Augustan style
3.	Sublime	naturalistic greenery with expressive accents; natural greenery penetrating the composition;, buildings in contrast to greenery; connection with ornamental green elements		sublime style
4.	Buildings placed in the greenery	hardly designed	contrast between architecture and nature	Le Corbusier

Table 1. Greenery as an environment different from human creations

	Composition	Greenery	Aesthetic value of greenery and architecture	Example
1.	Formal	geometric, treated as a building element of permanent formal features; architecture and greenery penetrate and complete each other	coherent	Guevrekian
2.		modelled after natural patterns; ornamental plants; architecture and greenery penetrate each other		Jekyll
3.	Free form	modelled after natural patterns; ornamental and native plants; plant succession accepted; harmony between buildings and greenery	to a large extent independent	Robinson
4.		resembling natural patterns; ornamental and native plants; architecture and greenery of similar form	coherent	The Güell Park
5.	Buildings and greenery as an organic whole	minimal intervention	synergic	The Fallingwater

Table 2. Nature and culture as complementary elements of human habitat

4. The naturalistic idea in law

The idea of sustainable development is considered as significant in the urban area planning. Surprisingly enough, the term is understood differently by various specialists. It seems important to show its significance, especially the treatment of native plants, in three aspects: urban conventions, general principles of sustainable development, and finally biodiversity protection. The most important legal acts concerning the issue of native plants in the urban area planning are presented below.

4.1 Urban conventions

The first to recognize recreational areas and their greenery as an important element of a municipal structure was the Athens Card of 1933, which was followed by general legal acts. At the Vancouver conference (Habitat I) of 1978, ecological aspects related to human settlements were acknowledged to be essential for their proper development. In order to counteract any symptoms of unsustainable development, the Habitat agenda was accepted at the Istanbul conference (Habitat II) of 1996. The degradation of native plants results from the lack of balance, although it was not explicit in the document. The European Landscape Convention of 2000 emphasised the necessity of protecting the local landscape values both human-created and natural, which implies native plants protection. A similar idea is conveyed by The New Charter of Athens 2003, which recommends that cultural environment and natural environment be joined in harmony. The biodiversity protection seems to be understood as only protecting existing objects. Although greenery is given an important role, it is treated as vague mass. No attention is paid to its quality resulting from its correspondence with local nature.

4.2 Sustainable development

According to the Declaration of the UN Conference on Human Habitat (Stockholm 1972), human environment is to be created in cooperation with nature, which implies the necessity of noticing the processes occurring in nature, strengthening their positive manifestations and limiting the negative ones. A similar idea was expressed at the *Paris Convention on the Protection of the World Cultural and Natural Heritage* of 1972, which discussed the natural areas of value but omitted the significance of the areas transformed by man, e.g. wasteland and its big potential for the free plant succession. The turning point was the *Report of The World Commission on Environment and Development* of 1987 compiled by Gro Harlem Brundtland, which introduced the definition of sustainable development. Its consequence was the Rio de Janeiro Conference of 1992 and its Agenda 21, which discusses the issue of synergy in the civilisation development, i.e. a full and effective development requires mutual respect of its elements. Nature, meant as greenery and the natural processes, is as important as economy, society and space order. The duration of this system requires the possibility of self-regeneration, which entails allowing the process of natural succession. The aim of the eco-development is to maintain biodiversity and human health, which is included in the *European Cities Card for Eco-Development* passed in Aalborg in 1994, and also confirmed by the *United Nations Millennium Declaration* of 2000. One of the main remarks of the *Local Agenda 21* of 2002 is the threatening of natural environment by the spreading of invasive alien species, which will lead to biodiversity degradation. Unfortunately, the invasive species are often treated initially as ornamental plants or ones very effective in land reclamation. The problem is that nature and biodiversity are understood differently by various specialists.

4.3 Biodiversity

Starting with the urban acts regulating space in its broad meaning, through the principles of sustainable development, which try to link all the activities conducted in the environment, one encounters the issue of biodiversity, i.e. the role of the native species and their communities in the human environment, including the urban areas. The *Council Directive 79/409/EEC on the conservation of wild birds,* known as the 'bird directive' suggests that the attention should be paid to the state of biodiversity in the areas not covered by legal protection. *The Bern Convention* of 1979 on the protection of European wild flora and fauna and their habitats claims that human action is not an element of the natural processes and must be controlled. Introducing alien species to native nature is not desirable, especially when they are expansive species which are able to transform natural communities destructively. For the *Council Directive 92/43/EEC on the conservation of natural habitats and of wild flora and fauna,* known as the 'habitat directive' it is important to protect and reconstruct the habitats significant for the EU. Its aim is to provide habitats for the more and more rare animal species, which could also apply to the greenery in the urban areas. This directive points out to the species under strict protection. Their appearance even in the urban areas should result in determining preservation zones. The problem is with its practical implementation, as such species sometimes occur in the municipal wastelands, which are usually treated as areas excluded from protection. This is the case, for instance, in the post-industrial areas of Upper Silesia, Poland. According to the *Convention on Biological Diversity,* ratified by the EU in 1993, counteracting the threats to the environment by various forms of protection and sustainable use of biodiversity is to be implemented also in the areas beyond the protected zones. The Convention formulates the following protection targets: natural ecosystems, trends in preserving wild species, the local supply of genes of wild and domesticated species, including the local varieties of crops and farm animals. Introducing species modified genetically and expansive alien species to the environment poses a threat to biodiversity. The announcement of the Committee for the European Council and the European Parliament referring to the thematic strategy on the urban environment of 2006 postulates limiting the loss of natural habitats and biodiversity in the urban areas. The 3rd Meeting of the Cartagena Protocol Parties on biologic safety, held in Curitiba, Brazil, in 2006, points to the existence of the direct link between the maintaining of biodiversity and the profit derived from nature, i.e. food production, clean water, nutrients circulation and climate regulation. More detailed conclusions from the research on the changes in the state of biodiversity could be found in the cyclic publications of the CONTUREC conferences. National legal acts sometimes make it obligatory not only to protect but also to actively act in order to support biodiversity. For instance, the Polish Environment Protection Law, of 2009, mentions e.g. reconstructing plant habitats, which should be implemented in land development projects. It also points to the necessity of forestation, afforestation, or concentrations of plants. This should be all done to protect the soil, animals, to shape the climate or to meet other needs related to providing biodiversity, natural balance and recreational needs of people.

4.3 The legal model

A number of international legal acts stress the cooperation between man and greenery without any confrontation. They express their appreciation of the local identity and protest against the global space homogeneity. The impact of human activities in the natural processes is especially clear in the case of synanthropic plants, colonising the areas transformed by man. The native species of this kind seem to play a positive role in the urban areas. However, ruderal plants are omitted in the legal acts and conventions, and the justification for their appreciation could be derived only indirectly. That may result from their absence in communities regarded as valuable naturally, or from their peculiar aesthetics they represent, or finally from their defiance of human activities. People have a peculiar idea of cleanliness and order, thus fighting whatever is accidental and destroys the order. But is it really desirable? Without the natural succession there is no evolution of communities registered in the natural processes. Human intervention is necessary, however, to control the spreading of alien species in local nature.

	Principles	Actions	Effects
1.	sustainable development	combining cultural and natural aims	synergy
2.	using resources in a way assuring conditions for their recreation	protecting and actively supporting the natural processes	recovering the lost values
3.	identity as the basic culture value, biodiversity as the essential biologic value	combining cultural and natural uniqueness	value diversity
4.	control of human arbitrary decisions	contextual planning in the field of culture and nature	limitation of space homogenization and the spread of alien species
5.	monitoring the state of the environment in the protected and not protected areas because they both affect the quality of the natural system	wastelands treated as an important source of genetic material in the urban areas; monitoring the state of greenery and possible threats	limiting the expansion of alien species; evolutionary recreation of natural habitats
6.	durability and desired ecological value of habitats	using native plants and their varieties; keeping the sources of natural succession	a high quality and a high ecological potential of greenery in the urban areas

Table 3. The legal model

5. Functional naturalism

The real functional value of an urban structure manifests itself in harmonizing civilisation with nature. When separated, both are hostile environments for people, when combined, they provide living conditions. Any simplification will result in a faulty solution, which refers both to 'the world of technology' and 'the world of false nature'. Discussing the function of naturalistic greenery in human environment requires distinguishing at least the basic principles governing nature as a system. It is important to show its usefulness values resulting from those principles. The building of such structures requires, in practice, conducting the appropriate land evaluation and accepting the suitable implementation principles. Discussing those issues leads to defining the natural and functional model, i.e. collecting all the basic features of naturalistic greenery in its functional aspect.

5.1 Nature as a system

A system is a collection of elements and relations among them. A natural system is a hierarchic arrangement: organism - population - biocenosis/biotope - ecosystem - biosphere. Each layer is related to the others and cannot be eliminated from the system. The ecosystem is the most essential unit for the designed implementation scales. The borders among the ecosystems are not distinct, being established statistically. Henryk Zimny compares an ecosystem to an orchestra, whose individual elements may act independently, but it is in the tuned up team that they form a unit being more than their sum (Zimny, 2002). Ecosystems evolve, depending on the changes in their environment. The mature ones have a dynamic balance of biocenoses and a stability of the system. Greenery elements in the urban area create a natural system. All its elements generate relations among each other, which results in the influence of their location in the structure upon their development (Alexander, 2002).The natural value of the area depends on the number of species (McHarg, 1992). Each species plays a role, so when two of them happen to have the same role, one is always eliminated. Taking it into consideration, it is possible to claim that the number of species corresponds to the number of roles played in the area, i.e. it determines the measure of complexity of processes occurring there. The stable state is alive, not stagnant. It is kept owing to the natural processes, which are permanent, whereas the form or composition are not permanent. Thus nature teaches us that the essential features are freedom, changeability, usefulness and competence.

One can distinguish a few fundamental principles for creating natural structures. First, the holistic principle resulting from the metapopulation theory, according to which it is impossible to predict any changes of a given population without identifying changes of all the populations of the same species within a region. In order to get a full picture of all the changes in a given population, one has to consider not only the habitat occupied, but also other habitats potentially possible to settle (Mitka, 2004). Second, the islands theory, according to which the number of species depends clearly on the size of the island. It is applied successfully to the so-called habitat islands, e.g. forest enclaves (Kornaś & Medwecka-Kornaś, 2002). Third, the seasonal rhythmic one, based on the clear seasonal changes in communities dominated by deciduous plants, which is characteristic of the moderate climate. It is related to the seasonal changes in the lighting conditions, the composition of the undergrowth and the microclimatic conditions. By contrast, coniferous forests are almost uniform all year round. Fourth, the form in nature results from the function and is an information carrier. Nothing is accidental. Following this principle, even the size has its optimum and is limited. Fifth, a characteristic feature of natural forests is possessing trees in different ages, which helps to make them durable and able to revive naturally. Natural forests have different layers: tree canopies, young trees, bushes and forest floor. A variety of species helps to protect forests from pest, frost, pathogenic mould and crunching by animals because each species has a different resistance to those factors (Wika (ed.), 1999). Municipal parks usually have a simplified structure and trees of a similar age, which makes them grow old and lose their values due to illnesses.

5.2 Usefulness value

In the contemporary world an important issue is whether greenery, natural or naturalistic, is useful to people. Civilisation complements nature but requires harmony. We are impressed with geometric forms. Naturalistic greenery, which is generally admired, does not impose on the observer. Thus, it is not surprising that our holiday places differ from our working or living places. We feel good in nature because, among other things, the natural sounds do not require any response. A city dweller may get away from everyday activities in a place different from his usual environment. Natural or naturalistic greenery offers such a possibility. According to sociological research, the closeness of green areas increases the feeling of satisfaction and is conducive to mental health. When staying in a forest park, the feeling of stress and headache disappears, and the speed of the process depends on the time spent there and the physical effort made. (Henderson et al, 2009). It seems important that naturalistic greenery should be modelled in a way providing visual penetration, which is essential for the feeling of safety. In this aspect it is difficult to accept the full freedom of development. An important issue is always the cost of the land maintenance. Naturalistic forms, designed as possibly permanent solid structures, are very expensive to preserve their form and layout, e.g. typical Japanese stroll gardens. A naturalistic garden being a kind of a wild garden allows spontaneous succession and lush growth without frequent intervention, which is not possible in the case of a geometric garden. It could be left without any care for a season or two and it will not be destroyed. On the other hand, trees and bushes grow, taking up more and more space. The moment comes, however, when the original function of the area may lose its value due to too much shade, dampness or undergrowth. A natural forest with trees in different ages has always the same function, whose value will not decrease despite the age of the oldest trees.

5.3 Assessment of the area receptivity to naturalistic development

Sherwin Greene determined four factors essential in an urban layout assessment (Greene, 1992). They could all well be easily applied to naturalistic greenery.

Criterion	Aim	Feature
function	comfort	continuity, safety, comfort, variety
space order	orientation and comprehension	cohesion, clarity, continuity, balance
identity	surrounding image, its unique features	focusing attention, unity, character, form, particular features
charm and attraction	willingness to stay	scale, suitability, matching, right choice, vitality, harmony

Table 4. Assessment criteria by Sherwin Greene

Criterion	Aim	Feature
levels of scale	a large diversity of the wholeness at various levels of scale	diversity degree; centre accentuation, combining with the surroundings through rhythms and similarities; intensification of the strengths of centres through the order of sizes and scales
strong centres	distinct centres control the surroundings	existence of centres; their strength as centres of attention
boundaries	forming a field of visual forces clarifying the centre	distinctiveness; separation power and its need
alternating repetitions	alternating repetitions and rhythms strengthening centres, intensifying one another	degree of interlock of interior and exterior forms
positive space	positive spaces of organized character and function with a dominant centre	space cohesion and its complexity degree
good shape	harmonious, often symmetrical, dense, with a well-marked centre, standing out from the surroundings	biomorphic, geometrical, with a degree of mutual penetration
local symmetries	strengthening local centres with symmetry	harmony with accentuation and power of domination
deep interlock and ambiguity	local centres related strongly to the surroundings	relations clarity; degree of diversity and distinctiveness
contrast	distinct centres and variety	diversity degree; accentuation
gradients	gradual transition of structures	harmony; reference
roughness	lack of live objects identity; authenticity, morphological roughness	form perfection; uniqueness
echoes	rhythms of similar elements repetitions	repeatability; similarity
the void	the most profound centres as a perfect wholeness with a kind of void inside	balance of background and accent; diversity of accents
simplicity and inner calm	live objects forming a wholeness simple and pure in form	form adequacy; a degree of domination and complexity
not-separateness	many layers of centres turning smoothly into the surroundings	a degree of monotony and binding with the surroundings; accentuation

Table 5. Assessment criteria by Christopher Alexander

Another useful idea is that of Christopher Alexander, the author of fifteen fundamental characteristics (Alexander, 2002). Although, originally described as the features of the degree of form vitality, they could be also used to describe the degree of structure organic development. Alexander creates the theory of natural structures based on the idea of 'wholeness', expressing an arrangement with a degree of 'life'. "The wholeness of any portion of the world is this system of larger and smaller centres, in their connection and overlap" (ibid.). All the elements tend to cooperate in synergy generated by the configuration as a whole.

Associating architectural objects with naturalistic greenery requires creating elements of greenery whose features are similar to natural communities with a similar species composition and located in similar habitats. To assess the possibilities of creating such habitats, Firbank suggests six criteria (Firbank et al, 1993)

Criterion	Aim	Feature
site selection	optimal localization	physical characteristics, local natural attributes, management history, current techniques, time and means available
consequences of other developments	optimal target	
implementation assessment	plan correction	implementation characteristics, soil quality, water conditions, species richness, presence of indicator plants, synanthropic and invasive plants, habitat structure
implications of different management options	optimal maintenance method	degradation during implementation, compensatory actions and those resulting from maintenance, implementation time, expenses expected and unexpected
predicting the off-site impacts	preparing protection from possible inconveniences	habitat conducive to pest, conditions for predators

Table 6. Assessment criteria by L.G. Firbank

Creating naturalistic space requires a pre-assessment of the area. The most essential criteria for deciding whether to provide aid to the existing biodiversity by introducing missing native plants are given below. The decision depends on the present biological value of an object and on its other features (Rostański, 2007).

Factor	Criterion	Feature
ecological	role in natural structure	local nature reserves, element of ecological corridor, additional element
	contact with natural structure	effective contacts, island elements
	habitat diversity	high or low, depending on the area
	land form	dense; with developed shoreline; segmented
	size of the area (Supuka 1998)	above 2 ha; 0.5-2 ha; below 0.5 ha
	soil conditions	acceptable; requiring partial change; requiring full change
	water conditions	acceptable, requiring retention increase

Factor	Criterion	Feature
floristic	habitat potential	potential natural flora, synanthropic flora, alien ornamental species
	floristic value	protected and rare plants, natural and degraded plant communities, native plants, synanthropic flora
	tendencies	increasing role of native plants, synanthropic or expansive alien ones
faunal	biodiversity	existence of mammals and other animals, existence of birds
cultural	role in urban structure	ordering elements, aesthetic value, natural value

Table 7. Assessment criteria by Krzysztof M. Rostański

5.3 Greenery layout

Naturalistic greenery, planned to counteract nature degradation or as a place of a particular aesthetic role, requires appropriate implementation methods in its developed form. The important factors here are the area size, the afforestation density, the percentage of the area not covered by greenery and the number of people to penetrate the area.

Nature degradation is counteracted by the habitat restoration, protection of spontaneous re-naturalization processes, habitat transportation and habitat creation (Tokarska-Guzik, 2001) When discussing the principles of the habitat creation, Trueman (2006) states that "a spontaneous community is likely to be more valuable than a created community". It is essential to adjust the designed species and the target community type to the habitat without strictly imitating a species composition found in books and changing the habitat conditions, which is usually very expensive. And finally, it is important to have the support of the local community for the created object (Trueman, 2006). According to J. Supuka (Supuka 1998), an important role in the natural system is played by greenery areas of more than 2 hectares.

According to the research conducted in 1995 by the present author in two districts of Katowice, Poland - Koszutka and Ligota - the border maximum value of tree canopy density is about 30% of the open area. Higher values cause too much shading, which is inconvenient for the inhabitants. It is equivalent to approximately 120 trees/ hectare. Natural forests have a definite recreational absorbency. For pine forests the value is 2 - 10 people/hectare/day. For oak-hornbeam forests and floodplain forests the value is 15 - 17 people/hectare/day. The highest absorbency is of meadows, pastures and ruderal plant areas. Their value is up to a 100 people/hectare/day (Krzymowska-Kostrowicka, 1991).

5.4 The functional model

This model describes general functional features of naturalistic greenery. It has been determined on the basis of the analysis of the functioning of the essential natural elements, the assessment of the possibility of using naturalistic greenery and the principles of its creation.

	Principles	Operations	Effects
1.	sustainable development	distinguishing natural and cultural processes	obtaining basis for design
		holism	
		conflict mitigation	optimizing solutions and lowering maintenance costs
		maintaining relations supporting essential processes	
		following natural processes	
2.	good environment functioning	maintaining hierarchic system	element synergy
		providing structural continuity	
		controlling autonomous and free succession	self-regulation ability
		providing structure receptiveness to changes	
		securing variety	
		preserving well-functioning elements	functionality
		transforming badly-functioning elements	
		securing harmony of form and function	
		creating socially acceptable aesthetics in places requiring accents	clarity
		counteracting element fragmentation	process durability
3.	human friendly environment	selecting plant communities with anticipated and recognized impacts	modification of the physical functions of an organism
		securing access to green areas	impact on well-being
		limiting psychic stimuli typical of urban environment	

Table 8. The functional model

6. Formal naturalism

Creating naturalistic greenery requires identifying typical natural forms by imitating natural communities, matching species with habitat conditions, or just following natural dispersion of accents. One can assume that formal naturalism is imitating aesthetic features of natural communities in a greenery arrangement. The degree of faithfulness may vary from the exact copy to the accent placement method discussed above. Designers may go further by shaping greenery after inanimate nature elements, plant or animal forms. Their arrangements of ornamental plants are even shaped after particularly attractive communities.

6.1 Accent geometric dispersion pattern

Natural plant dispersion in green areas usually seems chaotic and accidental. It turns out, however, that certain geometric structures, often known for many years, are able to describe such spatial structures. The chaos reveals to be a highly organised order, and not pure accident. The basic mathematical construction, exciting people since the ancient times, is the golden ratio, called by Euclid the "extreme and mean ratio" at the turn of IV and III cent. B.C. (Livio, 2003). In fact, as supported by examples, this ratio when applied to complex spatial structures uses up the space maximally. As this ratio occurs in nature, it could be assumed that it bridges the gap between nature and technology. For instance, the flowers of Biting Stonecrop (*Sedum acre*) reflect the golden ratio structure based on the pentagram.

Another example of the golden ratio in nature is the phenomenon of phyllotaxis. The succeeding leaves occur on the stem with a place shift angle whose ratio to the full angle corresponds to the following numbers of the Fibonacci sequence. The larger sequence numbers are taken, the closer their proportion to the golden ratio is. It turns out that phyllotaxis represents the state of minimal energy needed to maintain the form. In the sunflower it enables to pack the seeds maximally (Livio, 2003).

Fig. 1. The point layout in the basic attractor of the 'dragon' fractal

In order to create a plane arrangement of elements with an impression of natural randomness, the most useful is the spiral construction of the golden ratio. It makes use of another mathematical construction, bringing us closer to the 'chaos' of nature, that of the fractal calculus. It is popular to build visual presentations of fractal formulas called fractal attractors. One of the fractals, called 'dragon', when built with the appropriate parameters (using GCFract software), has the spiral of the golden ratio as its attractor. The basic parameters of a typical 'dragon' cause its inside to be filled with points in a free way, visually very 'chaotic'. The point distribution reflects in a perfect way the habitat arrangement of a given species in the natural meadow. It also reflects the 'chaotic' distribution of trees in the forest grown in natural succession.

6.2 Patterns referring to the features of natural elements

Formal naturalism draws its forms from nature observation. In contrast to natural functionalism, the observation is based on the admiration of natural aesthetics, being close to an artistic view in its treatment of nature as inspiration rather than an immediate pattern. This is a transformation of the given reality into a reality emphasizing only selected features. The patterns are drawn form the features of inanimate nature, plants, animals or selected ecosystems. Inanimate nature inspires geomorphic patterns, which is seen, for instance, in the shapes of rocks, hills, waterfalls rendered by means of plants shaped appropriately or matched so that their species or variety features imitate the desired features. The tamarisk (*Tamarix sp.*) forms a green mist, when in bloom the mist is pink. In the Kenroku-en garden, Kanazawa, Japan, the azaleas are shaped into boulders surrounding the creak and the gravel bank has its equivalent in the group of irises. Plants create certain characteristic forms naturally, which is an inspiration for phytomorphic patterns, e.g. the motives of corridors along streams or animal paths with the canopy of trees forming a vault seen at a distance. Interesting phytomorphic patterns are created by Patricia Johanson, who enlarges the form sizes of authentic flowers or their shoots and fills in the resultant patterns with plants, e.g.

her design of 2004 called The Ellis Creek Water Recycling Facility - 'Morning Glory Pools'. Zoomorphic patterns, inspired by the shapes of animals, are popular particularly as a way of hedge modeling. Forms closer to naturalism, though, are single tufts of grass looking like curled up cats. The shapes resembling animals could sometimes be achieved by delicate pruning, supporting a branch, and, finally appropriate lighting up.

The formal reference to selected features of natural plant communities could be called as drawing from ecomorphic patterns. The most popular example is rock gardens. Ecomorphic patterns are very close to the real habitat creation.

6.3 Plant communities as a pattern of form

The form closest to nature is a greenery arrangement patterned after natural plant communities. Apart from the form itself, an important role is also played by the native species composition. Those areas are not suitable for intensive recreation, but they may constitute, instead, the background or the ecological accent in the park. It may be necessary to thin out the forest floor and the shrub layer near buildings, as too dense greenery is not healthy for people. Environmental conditions determine the possibility of plant growth to a large extent. The species best adjusted win, owing to the mutual competence, thus similar habitats have similar plants. In nature this order manifests itself in natural plant communities. They have also a characteristic artistic expression consisting of the plastic features typical of the plants, the tree stand density degree, the division into layers, the density and diversity of the forest floor, the spread of greenery, leafage, conifer needles, the uncovering of the land surface (uncovered soil, sand, other rock material), the share of water area, the degree of shading, the colour, the changeability of the aspects in time. Each region has its characteristic communities. Even each location has its particular aspect, a particular species set and domination. Thus, it is difficult to create a standard table of species composition and their percentage share. A botanist should pay attention to important elements and to assess a given place thoroughly every time. It is also suggested that in order to preserve the clarity of the community origin, and its artificial creation in particular, rare and indicator species of the natural origin feature should not be used in designs. One of the most frequent forest communities in Poland is the oak-hornbeam forest.

6.4 Oak-hornbeam forests

The forests have a clear division into two layers of tree levels. The upper level is formed by oaks (*Quercus sp.*), linden trees (*Tilia sp.*) and partly by ashes (*Fraxinus excelsior*). A few metres below, the forest is dominated by hornbeams (*Carpinus betulus*) and Norway maples (*Acer platanoides*). The shrub level is diverse but not dense, due to the considerable overshadowing. The most popular trees of the lower level are hazels (*Corylus avellana*), spindles (*Euonymus sp.*), dogwoods (*Cornus sanguinea*), guilder roses (*Viburnum opulus*) and honeysuckles (*Lonicera sp.*). The forest floor plants usually occur in patches forming distinct textures. In the spring, before the trees come into leaf, oak-hornbeam forests are sunny and the undergrowth comes into bloom luxuriantly. The area is dominated by white, yellow and blue flowers against the vivid greenery. In the autumn, the green forest floor forms the background for the white and beige leaves of Solomon's-seal (*Polygonatum multiflorum*), gradually enriched with the yellow and beige colours of the shed tree leaves.

Fig. 2. The characteristic features of the oak-hornbeam forest are the tree trunks, the clearly visible shrub level and the forest floor divided into patches of different colour and texture.

Oak-hornbeam forests in Poland occur in rich soil with a high oxygen production. The groundwater level depends on the habitat. The air humidity varies from 40-80%, and is visibly higher at night. The shrubs and the forest floor provide a high germicidal ability. The air circulation is minimal when the undergrowth is thick but the impact on the human body is positive when it is looser. Then it is stimulating, by improving our natural immunity, blood circulation and pressure. The recreational absorbency varies from 15 people/hectare/day in oak-hornbeam forests in barren soil to 6 people/hectare/day in rich soil (Krzymowska-Kostrowicka 1991).

Fig. 3. A scheme of the oak-hornbeam forest arrangement from Fig. 2

6.5 The formal model

In the process of designing naturalistic greenery it is not only the beautiful and unique phenomena that are essential but also the structural patterns which decide the aesthetic form of plant arrangements as a whole and their connections with the surroundings.

	Principles	Actions	Effects
1.	Structural	in distance relations, making reference to the ratio between numbers in the Fibonacci sequence, incl. the golden ratio	habitat conditions diversified, self-regulation abilities facilitated, arrangement imperfection covered up
		dispersing accents in imitation of the fractal structure	
		using a hierarchic order of centres-accents	space clarity and functionality
2.	Formal	free distribution of elements	visual naturalness of structure
		imitating a variety of patterns: geomorphic, fitomorphic, zoomorphic and ecomorphic ones	aesthetic creation
		grading faithfulness to nature	different degree of durability and self-regulation
			adjusting to different habitats
		referring to natural plant communities	predictable bio-climate
			local uniqueness

Table 9. The formal model

7. Nature modelled in the aspect of spatial perception

The way in which greenery is observed is marked with the quality of our perception. The ability to perceive depends on the structural features of the object and the observer's sensitivity. For a greenery designer, it is the basic tool of creating a form.

7.1 Perceiving elements of a naturalistic arrangement

It is essential to perceive the following elements: forces binding the arrangement, its dynamism, harmony with the surroundings and the hierarchy conducting the observer. Arnheim uses the term of the field of visual forces, which fills the arrangement and in which particular elements attract or repel one another (Arnheim, 1977). The balance of the composition, like in nature, consists in minimizing the energy of the tensions. The dynamic effect appears when the balance is disturbed. The power of the tensions depends on the distinctiveness of the objects, their sizes, location and similarity. Solids which are hung higher appear heavier. Thus, trees with high crowns seem heavier than trees whose crowns reach the ground. The greenery covering an area seems to mainly raise its level. For the dynamism to appear in the composition, it is essential to provide a stable arrangement pattern as its frame of reference. The impression of stability is created by healthy, straight plants and by harmony with the natural composition of the habitat, whose perception is accessible only to observers with some knowledge of nature.

Although in natural communities plants are arranged freely, there are visible repetitions and rhythms e.g. tree trunks, tufts of grass. The arrangement dynamism could be achieved by introducing rhythm disturbances. Thus, tilted trunks, undergrowth locally concentrated, or deformed elements make distinct and dynamic accents as long as the number of accents is not too high. Their excess, however, introduces real chaos and blurs the composition. A whole is more than the sum of its components. Distinct groups of objects create interactions where not only particular objects have weight but also their common outline. One tends to look for accent points by following the range of looks with one's eyes. When creating a composition, it is important to consciously locate those points. The natural order consists in reducing any potential tensions, which makes it very similar to the principles of a balance creation in a composition. Distinct contours, being against nature, are eliminated spontaneously. Any geometrizing of greenery elements increases the tensions of visual forces.

7.2 Naturalistic greenery perception in the sphere of reaction

Walking in the forest produces impressions often difficult to describe. Their sources are as abstract as those related to music. They, probably, have a similar mechanism of creating aesthetic feelings. Following the concept of creating feelings in music (Lissa, 2008), we could try to describe the process of greenery perception. At first, the observer notices the image of particular objects (flowers, leaves, branches), sounds (insects, leaves), smells of plants, which builds characteristic creations forming representations and as a result structures, e.g. meadows and forests. They are associated with the impressions stored in our memory by recalling past events or stirring imagination, which is followed by reactions producing aesthetic feelings. If the thinking process leading to greenery assessment could be presented in this way, its critical element would be comparison with one's own experience, whose lack or association with unpleasant situations, will stifle our ability to assess positively. Such assessment mechanisms are associated with e.g. perception filters described by Amos Rapoport (Rapoport, 1977). According to him, the perceived world consists of the real world elements which remain in the observer's mind after filtering information through the filters of one's cultural and one's personal experiences. Following the Maslow needs pyramid (Koster, 2005), the green area should provide the feeling of security in the first place.

An important aspect of naturalistic greenery, which should be mentioned, is catering for people's sense of direction. In the Amager Commons in Copenhagen, each intersection of the alleys is planted with different native trees or shrubs, which provides people who have natural knowledge or are observant enough with clear directions. Points of reference are always needed and they could be created by means of naturalistic greenery. Natural space is perceived in pictures with hidden archetypes directing our reactions subconsciously. Our reaction to some of them is biologically conditioned regardless of our experience with native nature, but others require formation by our early contact with nature. However, without suitable education stressing ecological values and contact with native nature, the functional naturalism will not be accepted by society, no matter whether our reactions are conscious or not.

7.3 Semiotic aspects of naturalism

People communicate by means of signs, so when getting to know their surroundings, they also look for meaningful elements. The ability to understand the semiotic layer in our surroundings is affected not only by our experience and cultural knowledge but also, in the case of greenery, by our ecological knowledge and consciousness. Nature creates places which we sometimes find nice, sometimes horrifying, exposing human weaknesses. One's experience of a place is more important than the physical objects in it, which is confirmed by Yi-Fu Tuan, for whom a place is created not only with material components but also with speech, gestures and meaningful objects (Bańka, 2002). A place full of meanings possesses its *genius loci*, or it is space with a soul, whose spirit is determined by the relations of five features-dimensions: things, order, character, light and time (Norberg-Schulz, 1980, as cited in Bańka, 2002). Their different combinations make different types of landscape. If referred to nature, it is possible to demonstrate a few examples formed accordingly. A greenery composition of native trees, e.g. birches, builds a relation with time through the spring youth, the vividness of green and also with light through the brightness of the trunks and leaves. If a trace of human order is introduced by planting the birches on an orthogonal grid, a particular character of coordinated space will be achieved. If the 'chaotic' natural order is preserved, as a result, the impression of freedom and vitality will be produced. By retaining the species composition typical of a birch wood, we will achieve the faithfulness to things, the natural hierarchy of space elements and the character in accordance with the birch wood archetype. The material dimension will be disturbed by introducing buildings, but a building placed here will be enriched with the particular character and colour of this type of wood.

The world of nature has always seemed to possess nonphysical meaning, which connects it closely with the culture of the area where it is found. The natural elements and phenomena have always been perceived as signs. In the created green space, the signs appear in a literal form, interpreted directly, e.g. thorns as a warning, tree shade as a promise of the cool. In naturalistic compositions, signs usually occur as archetypes deciphered beyond consciousness. Archetypes reach our consciousness in a complex as a symbol. It is not easy to create meanings conveying emotion, admiration or reflection. To do it indirectly, ambiguously and in a way hard to define requires great art. One could mention a few symbols related to naturalism: 'ancient forest' - a wild place, difficult to access, unfriendly to people; 'paradise' - a friendly place of bliss and peace. As Shepard put it, "scenery is not scenery without the right cultural baggage"(Shepard, 2002). An important role in spatial symbolism is played by quotations, which in naturalistic greenery is the area elements representing a specific habitat surrounded by the background with a different character. This quotation could be represented by the pine forest in the courtyard of the National Library in Paris.

When moving in a green area, we discover signs which form a kind of narration. Thus, when greenery is shaped with people - its users - in mind, it is important to introduce narrative elements. Just like in architecture. "A building in which nothing is designed for sequence is a depressing experience"(Arnheim, 1977). The consecutive elements of a set of places arranged in order create mood through the surprise and secret of the next step.

A park which is simplified and predictable discourages us from taking a walk. Ambiguity stimulates meditation, whereas unambiguous objects escape our conscious attention. It should be pointed out that the richness of narration increases if all the senses are engaged.

7.4 The perception model

Although a natural forest may seem to be devoid of any meanings, in fact it does have many. When shaping naturalistic greenery, its meanings could be enriched by those which will be clear in a given culture and will convey meanings worthy of deciphering. The meanings could be developed by comparing the design ideas with the model given below.

	Principles	Actions	Effects
1.	a high degree of complexity close to information noise	minimization of visual tensions	lack of enforced reactions
			reducing outlines of buildings
2.	lack of frame separation	constant completing a composition with all elements of the visual field, incl. the background with a particular weight	full unification with the surroundings
3.	arrangement hierarchy	stressing a wholeness focusing visual forces at different specification levels	depth of form
		characteristic size	efficiency
		ordering	minimal energy to maintain stability
		using the golden ratio and the fractal accent dispersion	a compromise between the stability impression and preserving the live visual tensions
4.	structure clarity	placing clear accents	securing Cartesian coordinates
		popularizing natural knowledge	ability to appreciate nature
		moderate use of semiotic elements	decreasing the observer's involvement
		harmony with the human needs hierarchy	functionality
		combining processes characteristic of nature and culture	organic design
5.	dynamism	creating static balance disturbance	raising visual tension
		intentional lack of closing a form	
		form repetition disturbance	
		lack of hierarchy	
		natural form deformation	
		seasonal changeability, growth, movement caused by wind	
		introducing symmetry as an alien form	

	Principles	Actions	Effects
6.	aesthetic experience of nature	sensual perception of images, smells and sounds - creating impressions - associating structures - developing feelings	ability to appreciate natural aesthetics
		assessment: of the spatial layer - physical features, of the time layer - recollections and anticipations, of the emotional layer - mood and feelings	experiencing to the full
7.	semiotic aspect	sensitivity to archetypes	behaviour optimization, behaviour management
		conveying cultural meanings through space narration and symbolism	stirring imagination
		form ambiguity	
		using 'quotations' from local nature	giving a unique character

Table 10. The perception model

8. Holism in nature modeling

The considerations presented above lead to the conclusion that naturalism developing in Europe since the 17th century, appearing sporadically even earlier, is still important, at least in greenery modeling. Its origin is the growing social consciousness that the human being part of nature when opposing nature destroys also himself. Unfortunately, those tendencies, commonly called ecological, are often marred with the individuality of the point of view. Nature is sometimes protected in one aspect only with means degrading it in another. It is always essential to consider problems as holistically as possible. The problem is too complex for us to be able to fully control the relations between the human and surrounding nature. The human-caused damage could be minimized if the existence of the variety of aspects and their importance are realized. An illustration of this holistic attitude is the final model. The principle of nature modeling is the same for each model. However, it manifests itself in many ways which sometimes repeat themselves, and are sometimes visible only from the perspective of a definite aspect.

When analyzing the relation between modeling nature and the naturalness degree of formed structures, it is possible to distinguish four styles corresponding to the successive degrees of faithfulness to natural patterns and the presented holistic model. These are: the natural style, the organic style, the picturesque style and the geometric style.

The natural style is characterized by preserving the original natural elements to the largest degree possible. Architectural objects and plastic installations are placed in this context, using the natural values as the arrangement base. Designed greenery is a copy of the original and its aim is to mask the changes caused by the process of erecting buildings.

The organic style is represented by objects transformed by designers with an attempt to preserve their natural aspect. The basic principle is sustaining and stimulating natural processes together with developing the aesthetic layer by adding distinct human-formed elements, e.g. re-introduction of species or stimulation of natural plant succession.

The ecological aspect of greenery compositions designed in this style does not really lose its significance despite the geometry of native plants and a limited use of non-expansive ornamental plants. The style also tends to maintain a high level of building elements and tended greenery in contrast to spontaneous flora.

The picturesque style is characterized by combining ornamental and native plants. The final effect is to give deep aesthetic impressions. The spatial composition is free, close to the fractal geometry. The picturesque style uses a species diversity, which guarantees conformity with the habitat conditions and resistance to the periodic pest invasions.

The geometric style dominates spaces highly urbanized, e.g. roof gardens or vertical gardens. It is a form of art forced onto space, a sculptor's creation, often far from the local ecological aspects. Certain naturalism traces could be found in the form of a free accent dispersion, but only in the geometrically limited fields. There is no room for chance, everything being designed and given in its final form.

	Principles	Actions	Effects
1.	Naturalism as an idea	searching for harmony between nature and culture	influence on legal regulations
		promoting the living world	formal creation manifestations treated as expression of artificiality and stagnation
		imitating natural laws	building structures based on nature and accepting it
		searching for justification in nature of human actions as a negative idea	treating human actions as nature manifestation, leading to robbing nature of its resources
2.	Naturalism manifestations in culture	juxtaposing nature with culture	greenery closer to natural, not clearly bound to buildings
		creating structures combining nature and culture	structures binding greenery formally to buildings, tamed nature
			buildings as an organic consequence of natural forms, buildings placed in natural greenery
		creating naturalistic built-up area	considering natural processes in development
			preference for complexity and adjustment
			natural elements adaptation
			creating structures referring to natural ones
			minimizing energy needed for construction and maintaining
			creating optimal conditions for functioning based on self-regulation
			development of nature protection active methods
			promoting native flora
			using wastelands for reconstructing local natural values

	Principles	Actions	Effects
			referring to natural plant communities stressing the local uniqueness and identity
			using the predictable psychosomatic influence of natural communities
		sustainable development	synergy of building elements and greenery
			easing conflicts
			sustaining natural processes
			hierarchy and diversity
			development of legal regulations
			education in distinguishing and appreciating natural communities
3.	Naturalistic structure	building harmony of hierarchic wholeness	structure with a diverse force of accents at various levels of scale
		optimal size	functionality
		free composition	problem of spatial coordination
			naturalness impression
		spatial relations based on the golden ratio and fractal structures	using space optimally
			adjusting to human perception
		added geometrical elements	stressing important places
			artistic content message
		using material and intuitive elements	message affecting various senses
			narration
		securing the organic character of relations	harmony with habitat conditions, changeability in time, transience, complexity, free arrangement, border fluidity, form dynamism, subtle accents
4.	Perception of naturalistic structures	the field of observation is formed by a streak of looks, with the scene determined by the accent, and not by the frame	composition built with spatial objects, their context and background
		minimal visual tensions	lack of uncomfortable sense involvement, dissimilarity from everyday effort habitat, sense of freedom
		vague, implicit and complex forms	
		affecting all senses	causing and controlling reactions
		using meanings associated with archetypes or symbols	narration with references to native nature
		following the hierarchy of human needs	optimal functionality

Table 11. The holistic model

9. Conclusions

The richness of natural greenery, its complexity, and its unsurpassable potential are values not to be despised. Native plants could be included in greenery compositions even in city centers. If the habitat conditions do not allow to create semi-natural communities, it is, at least, possible to adopt the species diversity degree or the plant placement, which brings notable ecologic and useful benefits. In the urbanized area, an important role is played by the aesthetic and useful factors. The question of being organic does not mean the so-called naturalness or the anthropogenic character of elements filling the space, but their connection with the processes generated by its users and its wider spatial context. The key here is naturalistic greenery. In order to secure the positive social attitude to naturalistic greenery, it is important to popularize its real values in the various aspects: faithfulness to natural processes, spatial pattern, changeability in time and the emotions related. The holistic model presented above is designed as the basis of the naturalistic approach to creating landscape enclosures. It is not necessarily the main design model, but recommended as one of its layers. The direction of the designed solution will show, at the successive stage of elaboration, which of the mentioned-above models could be useful in developing the essential themes of the design. The fundamental issue is getting to know the local conditions well and using them to the full, which will determine the style of the object. The environment, the human being and his needs constitute a system requiring harmony and mutual respect and it has to be realized that the synergy beneficial to everybody is at stake.

10. References

Alexander Ch. (2002). *The Nature of Order. Book 1, The phenomenon of life.* The Center for Environmental Structure, ISBN 978-0-9726529-1-9, Berkeley, USA

Alison A. (1853). *Essay on the Nature and Principles of Taste.* Harper & Brothers Publishers, New York, USA

Andrews J. (2003). *The Country Diary Book of Creating a Wild Flower Garden.* Chancellor Press, ISBN 0-7537-0774-8,London, UK

Arnheim R. (1977). *The dynamics of architectural form.* University of California Press, ISBN 978-0-520-03551-5, Berkeley, USA

Bańka, A. (2002). *Społeczna psychologia środowiskowa.* [A social environment psychology]. Wydawnictwo Naukowe Scholar, ISBN 83-7369-005-0,Warszawa, Poland

Bisgrove R. (1992). *The Gardens of Gertrude Jekyll.* Frances Lincoln Limited, ISBN 10: 0-7112-0746-1, London, UK

Cochrane J. G. (1836). *The Foreign Quarterly Review.* Vol. XVI, Adolphus Richter & Co. London, UK

Cooper P. (2003). *Gardens without boundaries.* Mitchell Beazley, ISBN 1-84000-738-9, Londyn, UK

Firbank L.G., Arnold H.R., Eversham B.C., Mountford J.O., Radford G.L., Telfer M.G., Treweek J.R., Webb N.R., Wells T.C.E. (1993). *Managing Set-Aside Land for Wildlife.* Institute of Terrestrial Ecology, ISBN 0-11-701568-7, London, UK

Greene S. (1992). Cityshape: Communicating and Evaluating Community Design, *Journal of Amarican Planning Association,* Vol.58, No.2, pp. 177-189, ISSN 0194-4363

Henderson J., Irving D., Gibb E., Procter J. (2009).*Geenspace for a more successful and sustainable Scotland.* Geenspace Scotland, ISBN 978-0-9550921-1-4, Stirling, UK

Hobhouse P. (2005) *Historia ogrodów*. [The story of gardening]. Arkady, ISBN 83-213-4375-9, Warszawa, Poland

Kornaś J. & Medwecka-Kornaś A. (2002). *Geografia roślin*. [A geography of plants]. PWN, ISBN 83-01-13782-7, Warszawa, Poland

Koster A. (2005). Urban Green Space as reflection of people's needs. in: *Conference proceedings Quality and Significance of Green Urban Spaces*. pp.11-17, Van Hall Larenstein, Holland, April 14-15, 2005.

Kruft H.-W. (1994). *A history of architectural theory: from Vitruvius to the present*. Princeton Architectural Press, ISBN 1-56898-001-9, New York, USA

Krzymowska-Kostrowicka A. (1991). *Zarys geoekologii rekreacji*. [An outline of geo-ecology recreation]. Uniwersytet Warszawski, ISBN 83-900143-0-0, Warszawa, Poland

Lissa Z. (2008). *Wybór pism estetycznych*. [A selection of aesthetic writings]. Universitas, ISBN 97883-242-0883-8, Kraków, Poland

Livio M. (2003). *The Golden Ratio: The Story of Phi, The World's Most Astonishing Number*. Broadway Books, ISBN 0-7679-0816-3, New York, USA

McDowell J. & Stewart D. (1988). *Concise guide to today's religions*. Scripture Press, ISBN 0-946515-55-7, Amersham-on-the-Hill, UK

McHarg I. (1992) *Design with Nature*. John Wiley & Sons, Inc. ISBN 0-471-11460-X, New York, USA

Mitka J. (2004). Znaczenie teorii metapopulacji w ochronie gatunkowej. [The role of metapopulation theory in species conservation]. *Roczniki Bieszczadzkie* [The Bieszczady annals] Vol.12 (2004), pp.149-170. ISSN: 1233-1910

Nesbitt K. (1996). *Theorizing a new agenda for architecture*. Princeton Architectural Press, ISBN 1-56898-054-X, New York, USA

Rapoport A. (1977). *Human Aspects of Urban Form*, Pergamon Press, ISBN 0-08-017974-6 Oxford, UK

Robinson W. (2009). *The Wild Garden*. Expanded edition, Timber Press, ISBN 13: 978-0-88192-955-3, Portland, USA

Rostański K. M. (2007). The area evaluation with the scope of abilities for designed plant succession. in: *Landscape Assessment - From Theory to Practice: Applications in Planning and Design. 18Th International Annual ECLAS Conference*. pp.335-341, Faculty of Forestry, ISBN 978-86-7299-137-6, Belgrade, Serbia, October 10 - 14, 2007

Shepard P. (2002). *Man in the Landscape. A Historic View of the Esthetics of Nature*. The University of Georgia Press, ISBN 0-8203-2440-X, Athens, USA

Supuka J. (1998). Importance of urban vegetation for ecological stability of towns. *Ecology /Bratyslava*, supplement 1 (1998), pp.110-117, ISSN 1335-342X

Takei J., Keane M. P. (2008). *Sakuteiki. Visions of the Japanese Garden*. Tuttle Publishing, ISBN 978-0-8048-3968-6, Tokyo, Japan

Tokarska-Guzik B. (2001). Przyrodnicze zagospodarowanie terenów górniczych. [The natural development of miting areas] in: *Przywracanie wartości użytkowych terenom górniczym*. [Restoring the usefulness value to mining areas] Pilecka E. (Ed.) pp.209-222, PAN, WARSZTATY, ISBN 83-87854-03-4, Kraków, Poland,

Trueman I. (2006). *Habitat creation*, University of Wolverhampton, UK, a presentation at Silesian University, Katowice, Poland, May 2006

Turner T. (2005). *Garden History, Philosophy and Design 2000 BC – 2000 AD*. Spon Press, ISBN 0-415-31749-5, London, UK

Unwin R. (1994). *Town Planning in Practice. An introduction to the art of designing cities and suburbs*. Princeton Architectural Press, ISBN 1-56898-004-3, New York, USA

Wika S. (ed.) (1999). *Lasy Województwa Śląskiego*. [The forests of the Silesian Province]. Wydawnictwo Kubajak, ISBN 83-87971-02-2, Krzeszowice, Poland

Wotton H. (1624). *The Elements of Architecture*. Guy Kirkham F.A.I.A., Springfield, UK

Zimny H. (2002). *Ekologia ogólna*. [General ecology]. Agencja Reklamowo-Wydawnicza Arkadiusz Grzegorczyk, ISBN 83-88823-61-2, Warszawa, Poland

Part 2

Understanding Patterns, Change and Nature Conservation Needs

2

Deciphering Early Angiosperm Landscape Ecology Using a Clustering Method on Cretaceous Plant Assemblages

Clément Coiffard[1], Bernard Gomez[2] and Véronique Daviero-Gomez[2]
[1]*Museum für Naturkunde, Berlin,*
[2]*CNRS-UMR 5276 Terre, Planètes, Environment,*
Université Lyon 1 (Claude Bernard), Villeurbanne,
[1]*Germany*
[2]*France*

1. Introduction

Fossils open up windows on past lives and evolutionary mechanisms inaccessible to other fields of biology. Palaeontology has highlighted five moments of major extinctions in the life history on Earth: Ordovician/Silurian, Late Devonian, Permian/Triassic, Triassic/Jurassic and Cretaceous/Tertiary (Sepkoski, 1986). These five crises shaped life (apparitions versus extinctions of species), as well as diversity, ecology, and landscape. So fossils are not objects only useful for taxonomy and systematics, but their ecological traits can be deciphered to reveal macro-evolutionary processes and landscape ecology.

Today, flowering plants or angiosperms include the greatest terrestrial plant diversity consisting of over 250,000 species. They constitute the most abundant and diverse plant group, and notably provide food and medicine to present-day animals and human beings. The sudden appearance of angiosperms in the fossil record, their mostly cryptic evolution during the Early Cretaceous (-145 to -100Ma), and their subsequent rise to dominance in all terrestrial habitats during the Late Cretaceous (-100 to -65Ma) remain one of the most puzzling mysteries in evolutionary biology. These events are often quoted in the literature as "Darwin's abominable mystery". The ascent to dominance by the new plant groups (i.e. angiosperms versus ferns and conifers) affected the whole terrestrial ecosystems, food chains, soil chemistry and atmosphere composition.

In order to better understand the chronology and mechanisms behind the considerable turnover from gymnosperm-dominated vegetation to angiosperm-dominated vegetation and their consequences, we have studied megafossil plant remains from the Barremian-Campanian of Europe. Because of the fossil record often provides incomplete and biased data, we have used the Wagner's Parsimony Method (WPM). It was developed in 1961 by Wagner who worked on the systematics of ferns. The WPM consists of a system of hierarchical classification; it is a classical clustering method now used in phylogeny, however it can be applied to fields other than cladistics (Masselot et al., 1997). In synecology, it can be used for classifying localities or samplings as a function of biological content. They can also be related based on environmental conditions, so as to provide evidence of the relationship between palaeo-biocoenoses (known from fossils) and the palaeo-biotopes (known from sedimentology). Modern ecology concepts can be applied to a reconstruction of vegetation, and it is possible to relate ecological strategy and evolutionary history. We have applied the WPM for grouping localities by their megafossil plant content. For this study, we have run three successive WPM analyses using three time intervals: (1) Barremian-Albian, (2) Albian-Cenomanian and (3) Turonian-Campanian. The clusters obtained consist of seven types of depositional environments: (1) estuary mouth, (2) salt marsh, (3) freshwater swamp, (4) floodplain, (5) levee, (6) channel and (7) lacustrine deposits. They are each compared to independent sedimentological and palaeoenvironmental settings (e.g. crevasse splay deposits and lithographic limestones).

2. Geological setting

During the Cretaceous, oceanic accretion increased and wide volcanic provinces formed (such as the traps of Deccan) around the Cretaceous/Tertiary (K/T) boundary. The volcanic activity was associated with the releases of large volumes of CO_2 into the atmosphere, and consequently provoked greenhouse episodes in particular during the Cenomanian, which is considered to be the most recent warmest episodes of the Phanerozoic (Veizer et al., 2000). Also major perturbations in the carbon cycle occurred in the oceans. There were several oceanic anoxic events (OAEs) during which large quantities of organic-matter-rich black shales were deposited. These organic accumulations can have two origins (Leckie et al., 2002): (1) an increase in the concentrations of CO_2 or nutrients may have led to an increase in the marine production: production oceanic anoxic events (POAE); (2) an increase in rainfall precipitation may have led to an increase in detritic inputs: detritic oceanic anoxic events (DOAE).

In the last decades, palaeoclimate estimates were obtained from several marine and terrestrial proxies. For instance, oxygen isotopes were widely used to study palaeotemperature variations from marine animal shales (e.g. Bowen, 1961), while clay mineralogy provided information on conditions during formation of clays and therefore helped to reveal precipitation patterns (e.g. Ruffell & Batten, 1990). Plants were also used to infer palaeoclimates using leaf morphology (e.g., Climate Leaf Analysis Multivariate Program, CLAMP, Wolfe, 1993; Spicer & Herman, 2010 and reference therein) or pCO_2 (Retallack, 2001). On the basis of the data available for Western Europe, the time interval from the Barremian to the Santonian is divided into five climate stages: (1) Barremian - Early Aptian; (2) Late Aptian - Early Albian; (3) Albian; (4) Cenomanian - Early Turonian; (5) Late Turonian - Santonian.

2.1 Barremian - early Aptian arid phase

Ruffell & Batten (1990) identified an arid phase from the Barremian to the middle Aptian of Western Europe based on a review of climate indicators (e.g. evaporites and clay mineralogy). During the same time interval, Haywood et al. (2004), based on a climate computer simulation, suggested that England had mean annual temperature of 22-26°C and precipitation of about 1400-3000mm with a very marked drought season. A first OAE 1a (Fig. 1), with thin deposits of black shales occurred during the Early Aptian. It may be explained by changes (1) in the oceanic circulation, (2) in the hydrologic cycle with an increase of precipitation and erosion in a context of CO_2–induced global warming, and (3) in the oceanic fertilisation with an increase in iron of volcanic origin leading to higher oceanic productivity and carbon burying (Leckie et al., 2002). Also there were an abrupt diminution of $^{87}Sr/^{86}Sr$ and a negative excursion of $\delta^{13}C$ that suggest a relation with the super panache (plume) of Otong-Java eruption (Fig. 1). These pikes may also be explained by volcanic light-isotope carbon inputs or by dissociation of methane hydrates (Leckie et al., 2002). After OAE 1a, the values of $\delta^{13}C$ values continued to decrease, suggesting that the carbon burying lasted during the middle Aptian in parallel with the sea-level rise.

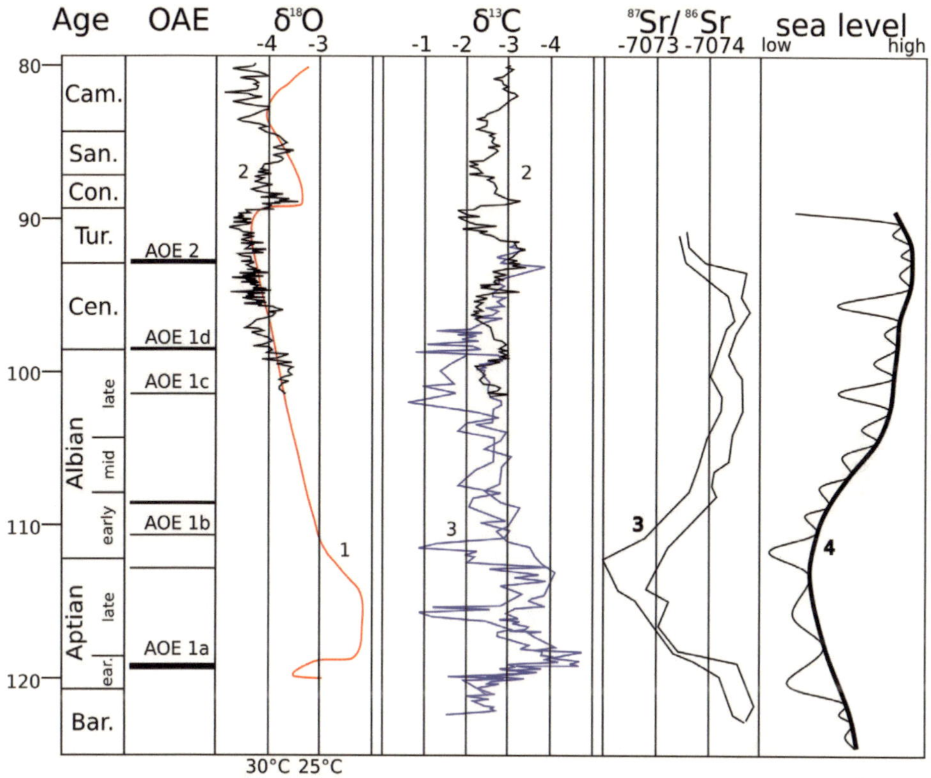

Fig. 1. Main proxies to estimate the Middle Cretaceous global changes (1, Steuber et al., 2005; Rudistes; 2, Stoll & Schrag, 2000, Foraminiferes; 3, Leckie et al., 2002; 4, Haq et al., 1988).

2.2 Late Aptian - earliest Albian humid phase

Negatives excursions of $\delta^{13}C$ happened again during the upper Aptian and the Aptian/Albian (OAE 1b) (Fig. 1). Also, in Europe, there was a transition from an arid phase (Ruffell & Batten, 1990) to a humid phase as indicated by the deposits of detritic quartz at the bottom of the Vocontian trough (Wortmann et al., 2004). These changes were attributed to a positive retroaction, the carbon-burying during the OAE 1a and the level "Jacob" of OAE 1b resulting in lower atmospheric CO_2 concentration (Weissert & Lini, 1991). They are associated with a global cooling of about 5°C, a growth of polar caps, and a sea-level fall (Steuber, 2005). According to Erbacher et al. (1998), the black shales of level "Jacob" of OAE 1b may have formed because of the inputs of detritic organic matter (DOAE) related to an increase in the precipitation. In contrast, the level "Paquier" of OAE 1b is considered as a phase of warming (8°C according to Price et al., 2003) and wetting (Heerle et al., 2003).

2.3 Albian warming

During the Albian, the oceanic accretion and the sea levels continuously increased (Fig. 1). Thus, more CO_2 entered into the atmosphere and Europe became an archipelago of islands (Fig. 2; Haq, 1988). This new input of CO_2 provoked a global warming phase with the temperatures rising from 12-14°C to 22-26°C in the North Atlantic (Fig. 1) (Leckie et al., 2002; Price et al., 2002). Also, clay mineralogy of sediments in northwestern Germany indicates a change to semi-arid conditions in Europe characterised by seasonal precipitations (Mutterlose et al., 2003).

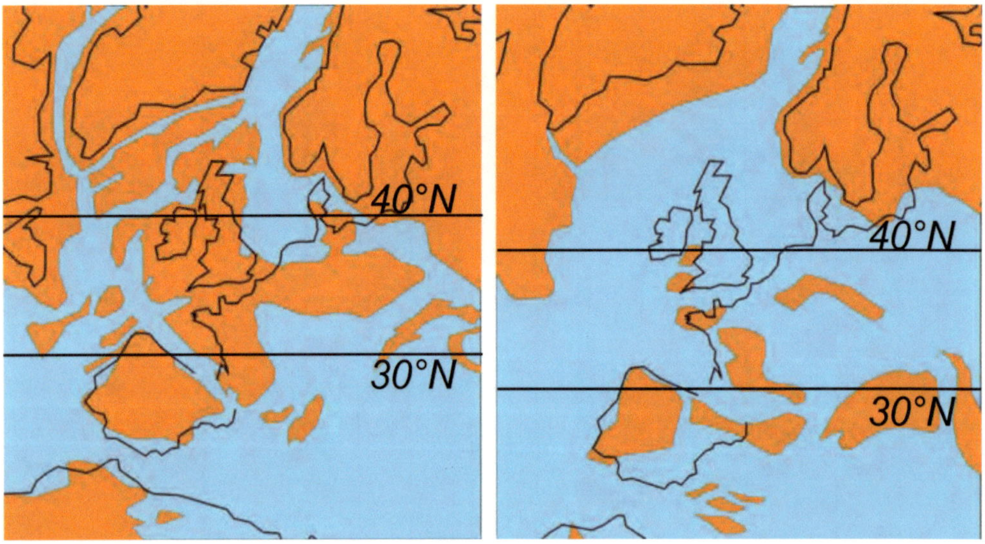

Fig. 2. Palaeogeography of Europe during low (Berriasian-Barremian) and high (Late Cenomanian) sea levels (modified from Ziegler, 1988).

During the lower Upper Albian, a new phase of oceanic anoxic event (OAE 1c) was produced by inputs of detritic organic matter (Leckie et al., 2002) and can be compared with the humid interval described by Fenner (2001) from clay mineralogy. The OAE 1c is followed by a warming of about 5°C (Fig. 1) (Stoll & Shrag, 2000) and seasonal precipitation (Fenner, 2001).

The Albian ended by another phase of oceanic anoxic event (OAE 1d; Leckie et al., 2002) related to input of detritic organic matter. The cyclicity of these OAEs might be due to a periodic change in the oceanic circulation favouring the upwelling of nutrient-rich waters and increasing the productivity (Leckie et al., 2002). Bornemann et al. (2005) also suggested oceanic circulation changes might be controlled by Monsoon-type climates. During OAE 1d, a new regression connected again to other islands in the European archipelago (Fig. 1-2) (Haq, 1988).

2.4 Cenomanian - early Turonian thermal maximum

The Cenomanian is considered the most recent warm time-interval of the Phanerozoic (Veizer et al., 2000). The Middle Cenomanian is, however, marked by a slight cooling and a new regression (Fig. 1; Stoll & Shrag, 2000; Haq, 1988). CLAMP analysis of the megafossil leaf angiosperms from the middle Upper Cenomanian of Peruc-Koryčany indicated that the region of today's Czech Republic had a subtropical climate with a mean annual temperature of 17-20°C and precipitation of about 1400mm (Table 1; Kvaček, 2000).

Fossil flora	MAT (°C)	MAP (mm)	DSC (months)
Peruc-Koryčany Cenomanian	17-20	1400	9.5-10.5
Zliv, Klikov, Hluboká Senonian	14-17	1300	7.5-10

Table 1. CLAMPs from the Late Cretaceous of Europe (MAT, mean annual temperatures; MAP, mean annual precipitations; DSC, duration of the growth season).

At the end of the Cenomanian, a new phase of oceanic anoxic event (OAE 2) is associated with a marine regression. The spike of more negative $\delta^{13}C$ is related to the burying of oceanic organic matter (Keller et al., 2001), while the fall in $^{87}Sr/^{86}Sr$ is induced by the strong oceanic accretion and the super panache of Kerguelen.

2.5 Late Turonian - Santonian cooling

From the Late Turonian to the Early Santonian a cooling and a regression occurred, and were related to a glaciation (Fig. 1; Stoll & Shrag, 2000; Miller et al., 2003). CLAMP analysis of the megafossil leaf angiosperms from the Senonian of the region of Zliv, Czech Republic indicated that it was under a warm temperate climate, slightly cooler than that of the Cenomanian, with a mean annual temperature of 14-17°C and mean annual precipitation of about 1300 mm (Table 1; Hermann & Kvaček, 2002).

During the Cretaceous, climates varied greatly. In particular, during the Albian, the temperatures had reached plus 10°C in a few millions of years. These high climate variations came with wide fluctuations of climate zones and, subsequently, distant migrations of terrestrial plants. These global climate changes, in addition to eustatic oscillations (compartmenting terrestrial environments into isolated islands), created new niches and new evolutionary opportunities for the clades (e.g. development of extra-tropical megathermal forests).

3. Method

3.1 Synecological principle of the Wagner's Parsimony Method

The Wagner's Parsimony Method (WPM) consists of constructing one or more trees in which the objects (e.g. taxa in phylogeny, localities in synecology) are related to others based on their characters (e.g. nucleotides or morphology characters in phylogeny, presence/absence of a taxon in synecology). The tree(s) are dichotomous-branched trees (i.e. trees with each split forming two branches). The most parsimonious tree is that which minimises the number of changes in character states of each node on the whole tree. The total number of state changes of the whole character set for the same tree is called length of the tree. So this method consists of looking for the tree (or the trees) with the smallest length.

When several trees of minimal length (so-called most parsimonious) are created, it is possible to create a tree called the "strict consensus" that represents the existing groupings in all of the most parsimonious trees. When different solutions of grouping branches are possible, the consensus tree presents a polytomy that employs non-dichotomous branching. The most parsimonious trees are rooted with an external group to polarise the characters.

3.1.1 How are the data coded?

In the case of the palaeosynecology, the objects studied are the localities, and the characters used are the presence/absence of taxa, these data are obtained from the plant lists. With these data, we build a matrix with the localities as columns and the taxa as rows. Each box of the matrix takes the value 0 or 1 depending on whether the taxon is absent or present, respectively, in the locality. The taxa used must be identified at the species level and doubtful or uncertain data («cf.» or «aff.») can be coded by «?». The final treatment of these uncertain or missing data (or «?») can be made as in the classical cladistic analyses, by more parsimonious inference on the final trees. This treatment estimates whether or not the presence of particular taxa is consistent with the analysis.

3.1.2 Taphonomic biases and data collecting

Because taphonomic biases deeply constrain the contents of palaeontological assemblages, we compare the localities with similar biases. For instance, it would be absurd to compare insect assemblages of ambers with those of lacustrine lithographic limestones. Ideally, specimen collection would be verified to have been undertaken identically in all of the localities studied. The problem is the same and the solution is analogous to working on living organisms. The proportion of the (palaeo-)biological diversity collected must be estimated by establishing a curve of diversity per locality (Perochon et al., 2001). If the collection(s) are insufficient for one locality, then they must be completed in the field.

3.1.3 Differences between characters and attributes

In addition to the characters defined above, the localities studied include more information called attributes. The attributes are not used for building the parsimony trees. So, the trees obtained have no link to the attributes and the value of attributes can be tested. In our palaeosynecological case, the attributes most often consist of sedimentological and taphonomic data.

3.1.4 Why and how to choose the external group(s)?

To root a tree, we must define the states of primitive (plesiomorphic) and derived (apomorphic) characters (sensu Nel et al., 1998). Then, the characters constitute a so-called primary polarisation, since the latter is done before the WPM analysis.

According to Masselot et al. (1997) and Nel et al. (1998), the external group must be defined before the polarisation of the characters, that is before determining whether the state of a character is primitive or derived. The supposed primitive state of a character is thus that observed in the external group. This external group can correspond to one or more stations or hypothetic of real samplings. The use or real ecological units (i.e. stations and samplings) is preferable because they introduce fewer a priori hypotheses.

In the particular case of a palaeosynecological study on the evolution of ecosystems through time, the time interval or the oldest locality studied can be chosen as the external group. With this choice, the presence of an "ancient" taxon within the external group is considered as primarily plesiomorphic and the presence of a "more recent" taxon as primarily apomorphic. Several external groups (i.e. more ancient time intervals or localities than all other localities studied in the internal group) can be used alternatively or simultaneously and their impact on the analysis tested.

4. Results

The European Cretaceous plant record is well-constrained and dated stratigraphically and shows well-preserved angiosperms from the Barremian to the Maastrichtian. Furthermore, European palaeobotany provides many detailed studies of Cretaceous floras for analysis that encompass a full range of deposition environments. Three age intervals were studied in three successive WPM analyses: (1) Barremian-Albian, (2) Albian-Cenomanian and (3) Turonian-Campanian. The clusters obtained are compared to independent sedimentological and palaeoenvironmental settings for each age. The clusters consist of seven types of depositional environments: (1) estuary mouth, (2) salt marsh, (3) freshwater swamp, (4) floodplain, (5) levee, (6) channel and (7) lacustrine deposits (Coiffard et al., 2006, 2007, 2010). Plant taphonomy (i.e. the deductions of the characteristics of production, transport, burying, and fossilization based on the states of preservation of megafossil plant remains) strongly supports the close proximity between the living environments and the final depositional environments in all fossil assemblages retained for the WPM analyses. One may assume that plant communities living in similar biotopes fossilize in similar depositional environments. Thus the resulting cluster corresponds to communities that grew at or near the environment where they deposited and are recorded. The palaeo-vegetations or communities evolved in space and in time, and they may be interpreted in terms of landscape ecology. Thus each plant assemblage may reflect environmental mosaics juxtaposed in the same geomorphological unit (e.g. channel margins ranging from sandy banks to "hardwood" forests). Nowadays, such environmental mosaics take place within a few meters. All these factors together contribute to make a robust data set of the evolution of angiosperms from the Barremian to the Campanian that can be traced through various ecosystems and related to other plant groups occupying the same niches. These data sets produce a view of angiosperm radiation in three phases that reconciles the previous scenarios based on the North American record.

4.1 Phase 1: Barremian-Aptian freshwater lake-related wetlands

Worldwide, angiosperm megafossils from the Barremian are very rare, with the earliest records found in freshwater lake or wetland habitats of China and Spain (Sun et al., 1998, 2002; Gomez et al., 2006; Dilcher et al. 2007). In an aquatic community, these angiosperms competed with charophytes that dominated macrophytic associations since the Permian (Martín-Closas, 2003; Sun et al., 1998, 2002). In the Barremian of Europe, chloranthoid/*Afropollis* pollen indicates terrestrial angiosperms (Heimhofer et al., 2005), although matoniaceous fern thickets and open conifer woodlands in floodplains dominated the terrestrial vegetation. Such a vegetational physiognomy is consistent with the fact that

Western Europe underwent an arid phase from the Barremian to the Middle Aptian (Ruffell & Batten, 1990; Haywood et al., 2004). Aquatic angiosperm megafossil remains (Coiffard et al., 2007) and terrestrial chloranthoid, lauralean or magnolialean fossil pollen (Heimhofer et al., 2005) appear highly diversified in the Late Aptian. This diversification was accompanied by the closure of woodlands, conifers spreading over most environments, and near extinction of matoniaceous ferns. In contrast, terrestrial leaf megafossils are very rare except for *Quercophyllum* from the Aptian/Albian of Arnal (Portugal) that probably grew along freshwater lakes or pond margins. The Barremian-Aptian fresh waters and related wetlands are sketched at the bottom of Figure 3 (Fig. 3.1).

Fig. 3. Evolution of plant association through time and space. 1, Barremian; 2, Aptian/Albian; 3, Albian, 4, Cenomanian; 5, Campanian.

4.2 Phase 2: Aptian-Albian understory floodplains (excluding levees and backswamps)

The vegetation of Europe was highly diverse during the Albian (in floodplains, ca. 11 taxa per locality vs. 8 during the Barremian and 6 during the Aptian/Albian, Coiffard et al., 2007), probably forced by a warming period beginning in the early Albian (Leckie et al., 2002; Price & Hart, 2002). Angiosperms exhibited a wider ecological range during the Albian, occurring in significant numbers for the first time in floodplains and probably some became vines (e.g. *Menispermites* Lesquereux; Dobruskina, 1997). Also, angiosperms competed with Osmundaceae taxa (e.g. *Cladophlebis* Brongniart), which nearly disappeared during the Albian (see Fig. 3.2 and 3.3).

In contrast to the small-leafed, ramified angiosperm habit, ferns and gymnosperms that were replaced by angiosperms show a large-leafed, monocaul habit. The small-leafed, ramified habit is more resilient in case of limb fall: if the meristem of a monocauline plant is destroyed, it may die or recover slowly while a ramified plant can use many active meristems that will allow the plant to recover more easily. In closed forests of the floodplains, angiosperms occupied the understory, while conifers formed the canopy. In contrast, matoniaceous fern thickets were exclusive to open vegetation of floodplains. Most floodplain angiosperms belonged to core angiosperms (i.e. Eudicots and Magnoliids). The increase in Eudicots during the Albian could be due to a poleward migration driven by global warming. This migration is also supported by the poleward dispersion of tricolpate (Eudicots) pollen grains shown by Hickey & Doyle (1977).

4.3 Phase 3: Cenomanian-Campanian levees, backswamps and coastal swamps

At the beginning of the Cenomanian (Fig. 3.4), European angiosperms were already widespread, and inhabited most environments, except apparently estuary mouths (Coiffard et al., 2006). In backswamps, they replaced Cycadales, and to a lesser extent, matoniaceous and dicksoniaceous ferns. In coastal swamps, they competed with matoniaceous ferns (e.g. *Weichselia* Stiehler), and constitute the earliest record of halophyte angiosperms (Coiffard et al., 2006, 2009). Angiosperms exhibited clear tree habit, especially trees of Lauraceae and Platanaceae which inhabited disturbed channel margins. In such a niche, they may have displaced conifers due to a greater growth rate of seedlings in accordance with Bond's hypothesis (Bond, 1989; Becker, 2000; Coiffard et al., 2006). This may also have included the shorter life cycle of the angiosperms compared to the gymnosperms present in similar habitats.

From the Turonian (Fig. 3.4 and 3.5), various genera of the Platanaceae spread over more stable floodplain environments while Bennettitales and Dicksoniaceae decreased (Coiffard & Gomez, 2010). Palms competed with Cupressaceae (*Taxodium* affinities) in the backswamps. Thus, palm seedlings developed large leaf crowns that shade and impede weeds, whereas conifer seedlings must compete with weeds. During the later growth stage, palm monocaul habit represents a lower energetic cost than Cupressaceae trunks, resulting in either a faster growth or higher allocation to reproduction. Cupressaceae were maintained in cooler areas, while palms do poorly in cool and cold temperate climates.

The reconstruction of vegetation changes through the Cretaceous (Fig. 3) shows that angiosperm colonization happened in three stages: (1) during the Barremian-Aptian early angiosperms were restricted in water bodies of freshwater wetlands; (2) during the Aptian-

Albian angiosperms spread through understory floodplains (excluding levees and backswamps); and (3) during the Cenomanian-Turonian they finally colonized the natural levees, backswamps and coastal swamps.

These results show that the rise to dominance was not uniform and took place in space and time at different rates according to the environments.

5. Discussion

5.1 Usefulness of megafossil floras for palaeoecological studies

Plant taphonomy (i.e. the deductions of the characteristics of production, transport, burying, and fossilization based on the states of preservation of megafossil plant remains) strongly supports the close proximity between the living environments and the final depositional environments in all fossil assemblages analysed with the WPM. One may assume that plant communities living in similar biotopes fossilise in similar depositional environments. Thus the resulting cluster corresponds to communities that grew at or near the environment where they deposited and are recorded. This view is supported by the congruence between the clusters obtained with the WPM and the palaeoenvironmental data (i.e. each cluster corresponds to a single type of palaeoenvironment).

The palaeovegetations or communities evolved in space and time, and they may be interpreted in terms of landscape ecology. Thus each plant assemblage may reflect environmental mosaics juxtaposed in the same geomorphological unit (e.g. channel margins ranging from sandy banks to "hardwood" forests). Nowadays, such environmental mosaics take place within a few meters. Thus, if the palaeofloras chosen are parautochthonous (i.e. deposited close to the living place) the palaeovegetation can be reconstructed at the landscape scale.

5.2 Congruence with other northern mid-latitude record

A similar pattern of gradual rise to dominance of angiosperms is observed in the northern mid-latitudes of North America. However, gaps in the sedimentary record may be taken into account for explaining a few differences in the fossil record. The North American Potomac group succession was widely used to understand the early evolution of angiosperms. Doyle & Hickey (1976) and Hickey & Doyle (1977) suggested that fossil angiosperms from the Aptian-Lower Albian of Potomac Zone I were understory shrubs and grew on levees and in floodplains. Pinnately compound leafed platanoids (e.g. *Sapindopsis* Fontaine) from the Upper Albian of Potomac Zone IIB formed early successional shrubs in the same environments. These events may correspond to the phase 2 of our scenario. Platanoids (e.g. *Araliopsoides* Berry) evolved along disturbed channel margins from the Cenomanian of the Potomac Zone III and colonized floodplains from the Cenomanian-Turonian of the Potomac Zone IV. The latter two zones may be in agreement with our phase 3, mentioned above. However, Doyle & Hickey (1976) and Hickey & Doyle (1977) hypothesized that early angiosperms were riparian weeds before the Aptian, but had no fossil record to support their hypothesis.

Retallack & Dilcher (1981a) proposed a reconstruction of the Albian/Cenomanian Dakota Formation flora, and showed that angiosperms occupied a variety of environments.

Lauraceous trees/shrubs (e.g. *Prisca/Magnoliaephyllum*) occupied swampy woodlands and margins of coastal lagoons, angiospermous shrubs (e.g. "*Acerites*") colonized margins of tide-dominated deltas, platanoids were common around freshwater lakes, levees and swales of freshwater coastal stream sides (e.g. *Araliopsoides*), and conifers forested dry floodplains (Retallack & Dilcher 1981b). Closer inspection shows some differences with Europe (Kvaček & Dilcher, 2000) but overall the Dakota flora and their ancient environments fit well with the beginning of the third phase of our scenario mentioned above. The Magnoliales (e.g. *Liriophyllum* Lesquereux, *Didromophyllum* Upchurch et Dilcher) collected from North America were lacking in Europe, and the Laurales were represented by different taxa (e.g. *Pabiana* Upchurch et Dilcher in Northern America vs. *Cocculophyllum* Velenovský in Europe).

According to Crane (1987), North America apparently lacks early freshwater continental records compared to Europe. This situation led to the idea that early angiosperms were mainly present in disturbed floodplains such as the Potomac Zone I. However, we now know that European, Chinese, and North American records (Gomez et al., 2006; Sun et al., 1998, 2002; Wang & Dilcher, 2006) show that angiosperms occurred in aquatic environments from at least the Barremian. The occurrence of early aquatic angiosperms opens the way for reinterpretation of the North American records. The Aptian peltate leaves *Proteaephyllum reniforme* Fontaine was considered to be a step into aquatic environments (Hickey & Doyle, 1977), angiosperms having been temporarily helophytes that are flooded, before changing to become permanently flooded hydrophytes. However, it could be an early aquatic angiosperm that appeared in our phase 1.

5.3 Evolutionary implications

Darwin's "abominable mystery" about the sudden appearance of rather modern genera of flowering plants can now be understood. What most questioned Darwin was "the sudden appearance of so many extant taxa of flowering plants in the Upper Chalk (Darwin, 1875). At that time, he was using a very limited Cretaceous plant record compared to present knowledge. Thus, in the last decades numerous Lower Cretaceous localities containing earlier angiosperms have been described as extinct angiosperm leaf morphotypes. Our scenario supports the view by Darwin (1875) that "the presence of even one true angiosperm in the Lower Chalk makes inclined to conjecture that plant of this great division must have been largely developed in some isolated area, whence owing to geographical changes, they have last succeeded in escaping, and spread quickly over the world". So the rise to dominance of angiosperms was a process that lasted over more than twenty million years.

During the last twenty years, the development of phylogenies has led to two main hypotheses. A posteriori, these were, however, only supported by the fossil records of North America and especially the adequacies with Doyle and Hickey's scenario (Doyle & Hickey, 1976, Hickey & Doyle, 1977).

Taylor & Hickey (1996) suggested that early angiosperms were perennial rhizomatous plants resembling herbaceous Magnoliids such as Piperaceae and competed with ferns and sphenopsids in disturbed areas such as stream margins. These angiosperms then invaded the aquatic niche in the early Cretaceous thanks to their rhizomatous growth habit and efficient seedlings, thus creating a stable environment.

Feild et al. (2004) hypothesized a disturbed forest understory and/or streamside shrubby ecology similar to the ecology of extant basal angiosperm lineages (i.e. Amborellaceae, Nymphaeales, Austrobaileyales, and Chloranthaceae). Taylor and Hickey's "paleoherb" (Taylor & Hickey, 1996) and Feild et al.'s "dark, damp and disturbed" (Feild et al., 2004, 2009) hypotheses are supported by the occurrence of angiosperms in disturbed floodplain environments from the Aptian (e.g. lower zone I of the Potomac group, USA; Hickey & Doyle, 1977; Hochuli et al., 2006). This early angiosperm ecology may fit our phase 2, but is unlike our phase 1. However, Feild et al. (2004) suggest that the colonization of aquatic environment occurred early in angiosperm evolutionary history. If so, one may question whether earliest angiosperms were shade tolerant as proposed by the "dark, damp and disturbed" hypothesis (Feild et al, 2004).

In a recent study of the evolution of angiosperm vein density during the Cretaceous, Feild et al. (2011) showed that the leaf hydraulic capacities of angiosperms escalated several steps during the Cretaceous. This change occurred in three phases. During the first phase (Lower Cretaceous), angiosperms are not significantly different from gymnosperms. The second phase (Mid-Cretaceous) witnesses an increase above gymnosperms but still below living angiosperms. Finally, angiosperms reached their modern value during the Upper Cretaceous. Those can be paralleled with the three phases observed in the ecological widening of angiosperms.

However, angiosperms were not the only clade that experienced a diversification during the Cretaceous: core Leptosporangiate ferns, Pinaceae, Gnetaceae and Podocarpaceae also underwent extensive diversification (Crane & Upchurch, 1987; Kelch, 1998; Rydin et al., 2003, 2006; Schneider et al., 2004; Wang et al., 2000). So, there is an underestimation of the turnovers of the whole vegetation. Thus, most researchers have worked on innovative features in angiosperms, whereas when dealing with angiosperm evolution we should also think and work upon angiosperm evolution in time and space as they occupied changing environments through times.

6. Conclusion

The survey of the megafossil plant remains from the Barremian to the Campanian of Europe reveals that the rise to dominance of angiosperms in the environments was non-synchronous. Thus, the early ecological evolution of angiosperms that shaped the Cretaceous landscapes was a gradual diversification in space and time of taxa and of their ecology. In contrast to the famous quote by Darwin, the appearance of angiosperms was not sudden, but it lasted from the Barremian (or earlier) to the Cenomanian, that is at least 30 million years.

The evolution of angiosperms at the local scale of Europe mirrors the gradual spread at a global scale previously reported from low latitudes to high latitudes during the Early Cretaceous by Hickey & Doyle (1976). More recently, Coiffard & Gomez (2012) suggested that angiosperm rise to dominance was probably driven by the climates, and especially the rather continuous global warming from the Barremian to the Cenomanian favoured the spread of eudicots to high latitudes and the establishment of megathermal forests.

7. Acknowledgment

C.C. received financial support from the von Humbolt Foundation. B.G. and V.D. received financial support from the CNRS-UMR 5276 Terre, Planètes, Environnement, INSU NOVAMBRE of the Institut National des Sciences de l'Univers, and projects CGL2008-00809/BTE, CGL2008-00550/BTE, CGL2009-11838/BTE, and CGL2011-27869/BTE funded by the "Ministerio de Ciencia e Innovación" of the Spanish government, and project SGR2009-1451 funded by the Catalan government.

8. References

Becker, P. (2000). Competition in the regeneration niche between conifers and angiosperms: Bond's slow seedling hypothesis. *Functional Ecology*, Vol.14, No.4, (August 2000), 401-412, ISSN 0269-8463, doi:10.1046/j.1365-2435.2000.00455.x.

Bond, W.J. (1989). The tortoise and the hare: Ecology of angiosperm dominance and gymnosperm persistence. *Biological Journal of the Linnean Society*, Vol.36, No.3, (March 1989), pp. 227-249, ISSN 0024-4066, doi:10.1111/j.1095-8312.1989.tb00492.x.

Bornemann, A.; Pross, J.; Reichelt, K.; Herrle, J.O.; Hemleben, C. & Mutterlose, J. (2005). Reconstruction of short-term palaeoceanographic changes during the formation of the Late Albian 'Niveau Breistroffer' black shales (Oceanic Anoxic Event 1d, SE France). *Journal of the Geological Society*, Vol.162, No.4, (July 2005), 623-639, ISSN 0016-7649, doi:10.1144/0016-764903-171.

Bowen, R. (1961). Oxygen Isotope Paleotemperature Measurements on Cretaceous Belemnoidea from Europe, India and Japan. *Journal of Paleontology*, Vol.35, No.5, (September 1961), 1077-1084, ISSN 0022-3360.

Coiffard, C.; Gomez, B.; Kvaček, J. & Thévenard, F. (2006). Early angiosperm ecology, evidence from the Albian-Cenomanian of Europe. *Annals of Botany*, Vol.98, No.3, (September 2006), 495-502, ISSN 0305-7364, doi:10.1093/aob/mcl125.

Coiffard, C.; Gomez, B. & Thévenard, F. (2007). Early Cretaceous angiosperm invasion of Western Europe and major environmental changes. *Annals of Botany*, Vol.100, No.3, (September 2007), 545-553, ISSN 1095-8290, doi:10.1093/aob/mcm160.

Coiffard, C.; Gomez, B.; Thiébaut, M.; Kvaček, J.; Thévenard, F. & Néraudeau, D. (2009). *Eucalyptolaurus depreii*, gen. et sp. nov., intramarginal veined Lauraceae leaves from the Albian-Cenomanian of Charente-Maritime (Western France). *Palaeontology*, Vol.52, No.2, (March 2009), 323–336, ISSN 0031-0239, doi:10.1111/j.1475-4983.2009.00845 .x.

Coiffard, C. & Gomez, B. (2010). The rise to dominance of the angiosperm kingdom: dispersal, habitat widening and evolution during the Late Cretaceous of Europe. Lethaia, Vol.43, No.3, 164-169, ISSN 0024-1164, doi:10.1111/j.1502-3931.2009.00181.x.

Coiffard, C. & Gomez, B. (2012). Role of latitude and climate in dispersions, radiations and rise to dominance of early angiosperms through the Cretaceous. *Geologica Acta*, Vol.10, No.1, (March 2012), ISSN 1695-6133, doi:101344/105000001701.

Crane, P.R. 1987. Vegetational consequences of the angiosperm diversification. In: *The Origins of Angiosperms and Their Biological Consequences*, E.M. Friis, W.G. Chaloner & P.R. Crane (Eds.), pp.107-144, Cambridge Univ. Press, Cambridge, England, ISBN 978-0521323574.

Crane, P.R. & Upchurch, G.R. (1987). *Drewria potomacensis* gen. et sp. nov., an early Cretaceous member of Gnetales from the Potomac Group of Virginia. American Journal of Botany, Vol.74, No.11, (November 1987), 1722-1736, ISSN 0002-9122, doi:10.2307/2444143.

Darwin, F.; Seward, A.C. (eds.), 1903. More letters of Charles Darwin. A record of his work in a series of hitherto unpublished letters. London, John Murray, volume 1 and 2, 494pp. and 508pp.

Dilcher, D.L.; Sun, G.; Ji, Q. & Li, H. (2007). An early infructescence *Hyrcantha decussata* (comb. nov.) from the Yixian Formation in northeastern China. *Proceedings of the National Academy of Sciences of United States of America*, Vol.104, No.22, (29 May 2007), 9370-9374, ISSN 0027-8424, doi:10.1073/pnas.0703497104.

Dobruskina, I.A. (1997). Turonian plants from the southern Negev, Israel. *Cretaceous Research*, Vol.18, No.1, (February 1997), 87-107, ISSN 0195-6671, Doi: doi:10.1006/cres.1996.0051.

Doyle, J.A. & Hickey, L.J. (1976). Pollen and leaves from the mid-Cretaceous Potomac Group and their bearing on early angiosperm evolution, In: *Origin and early evolution of angiosperms*, C.B. Beck (Ed.), pp.139-206. Columbia University Press, Columbia, SC, ISBN 9780231038577.

Erbacher, J.; Gerth, W.; Schmiedl, G. & Hemleben, C. (1998). Benthic foraminiferal assemblages of late Aptian-early Albian black shale intervals in the Vocontian Basin, SE France. *Cretaceous Research*, Vol.19, No.6, (December 1998), 805-826, ISSN 0195-6671, doi:10.1006/cres.1998.0134.

Feild, T.S.; Arens, N.C.; Doyle. J.A.; Dawson, T.E. & Donoghue, M.J. (2004). Dark and disturbed: a new image of early angiosperm ecology. Paleobiology, Vol.30, No.1, (January 2004), 82-107, ISSN 0094-8373, doi:10.1666/0094-8373(2004)030<0082: DADANI>2.0.CO;2, .

Feild, T.S.; Chatelet, D.S.; Brodribb, T.J. (2009) Ancestral xerophobia: a hypothesis on the whole plant ecophysiology of early angiosperms. Geobiology, Vol.7, No.2, (March 2009, 237-264, ISSN 14724669, doi:10.1111/j.1472-4669.2009.00189.x.

Feild, T.S.; Brodribb, T.J.; Iglesias, A.; Chatelet D.S.; Baresch, A.; Upchurch, G.; Gomez, B.; Mohr, B.; Coiffard, C.; Kvaček, J. & Jaramilla, C. (2011). Fossil evidence for Cretaceous escalation in angiosperm leaf vein evolution. *Proceedings of the National Academy of Sciences of United States of America*, Vol.108, No.20, (17 May 2011), 8363-8366, ISSN 1091-6490, doi: 10.1073/pnas.1014456108.

Fenner, J. (2001). Middle and Late Albian geography, oceanography, and climate and the setting of the Kirchrode I and II borehole sites. *Palaeogeography, Palaeoclimatology, Palaeoecology*, Vol.174, No.1-3, (20 September 2001), 5-32, ISSN 0031-0182, doi:10.1016/S0031-0182(01)00285-1.

Gomez, B.; Daviero-Gomez, V.; Martín-Closas, C. & de la Fuente, M. 2006. *Montsechia vidalii*, an early aquatic angiosperm from the Barremian of Spain. *7th European Palaeobotany and Palynology Conference*, abstract volume, pp. 49, Prague, Czech Republic, September 6-11, 2006.

Haq, B.U.; Hardenbol, J. & Vail, P.R. (1988). Mesozoic and Cenozoic chronostratigraphy and eustatic cycles, In: *Sea-level changes: an integrated approach*, C.K. Wilgus, B.S. Hastings, H. Posamentier, J.V. Wagoner, C.A. Ross & C.G.S.C. Kendall (Eds.). Special Publication of the Society of Economic Paleontologists and Mineralogists, Tulsa, Oklahoma, No.42, 71-108, ISBN 978-0918985743.

Haywood, A.M.;Valdes, P.J. & Markwick, P.J. (2004). Cretaceous (Wealden) climates: a modeling perspective. *Cretaceous Research*, Vol.25, No.3, (June 2004), 303-311, ISSN 0195-6671, doi:10.1016/j.cretres.2004.01.005.

Herrle, J.O.; Pross, J.; Friedrich, O. & Hemleben, C. (2003). Short-term environmental changes in the Cretaceous Tethyan Ocean: micropalaeontological evidence from the Early Albian Oceanic Anoxic Event 1b. *Terra Nova*, Vol.15, No.1, (February 2003), 14-19, ISSN 0954-4879, doi: 10.1046/j.1365-3121.2003.00448.x.

Heimhofer, U.; Hochuli, P.A.; Burla, S.; Dinis, J. & Weissert, H. (2005). Timing of Early Cretaceous angiosperm diversification and possible links to major palaeoenvironmental change. *Geology*, Vol.33, No.2, (February 2005), 141-144, ISSN 0091-7613, doi: 10.1130/G21053.1.

Herman, A.B. & Kvaček, J. 2002: Campanian Grünbach flora of lower Austria: preliminary floristics and palaeoclimatology. *Annalen des Naturhistorischen Museums in Vien*, Vol.113A, 1-21.

Hickey, L.J. & Doyle, J.A. (1977). Early Cretaceous fossil evidence for Angiosperms evolution. *The Botanical Review*, Vol.43, No.1, 3-104, ISSN 0006-8101, doi:10.1007/BF02860849.

Hochuli, P.A.; Heimhofer, U. & Weissert, H. (2006). Timing of early angiosperm radiation: recalibrating the classical succession. *Journal of the Geological Society*, Vol.163, No.4, (July 2006), 587-594, ISSN 0016-7649, doi: 10.1144/0016-764905-135.

Kelch, D.G. (1998). Phylogeny of Podocarpaceae: comparison of evidence from morphology and 18S rDNA. *American Journal of Botany*, Vol.85, No.7, (July 1998), 986-996, ISSN 0002-9122, doi:10.2307/2446365.

Keller, G.; Han, Q.; Adatte, T. & Burns, S.J. (2001). Palaeoenvironment of the Cenomanian-Turonian transition at Eastbourne, England. *Cretaceous Research*, Vol.22, No.4, (August 2001), 391-422, ISSN 0195-6671, doi:10.1006/cres.2001.0264.

Kvaček, J. (2000). *Frenelopsis alata* and its microsporangiate and ovuliferous reproductive structures from the Cenomanian of Bohemia (Czech Republic, Central Europe). *Review of Palaeobotany and Palynology*, Vol.112, No.1-3, (October 2000), 51-78, ISSN 0034-6667, doi:10.1016/S0034-6667(00)00035-X.

Kvaček, J. & Dilcher, D.L. (2000). Comparison of Cenomanian floras from Western Interior North America and Central Europe. *Acta Universitatis Carolinae, Geologica*, Vol.44, 17-38.

Leckie, R.M.; Bralower, T.J. & Cashman, R. (2002). Oceanic anoxic events and plankton evolution: Biotic response to tectonic forcing during the mid-Cretaceous. Paleoceanography, Vol.17, No.3, (23 August 2002), 1041, ISSN 0883-8305, doi:10.1029/2001PA000623.

Martín-Closas, C. (2003). The fossil record and evolution of freshwater plants: a review. Geologica Acta, Vol.1, No.4, (December 2003), 315-338, ISSN 1695-6133, Doi:10.1344/105.000001619.

Masselot, G.; Nel, A.; Thomas, A. & Nel, J. (1997). Parcimonie de Wagner et biomonitoring de cours d'eau: application au bassin de la Risle (Normandie, France). *Annales de la Société Entomologique de France*, Vol.33, No.3, (September 1997), 237-258, ISSN 0037-9271.

Miller, K.G.; Sugarman, P.J.; Browning, J.V.; Kominz, M.A.; Hernández, J.C.; Olsson, R.K.; Wright, J.D.; Feigenson, M.D. & Van Sickel, W. (2003). Late Cretaceous chronology of large, rapid sea-level changes: Glacioeustasy during the greenhouse world. *Geology*, Vol.31, No.7, (July 2003), 585-588, ISSN 0091-7613, doi:10.1130/0091-7613(2003)031<0585:LCCOLR>2.0.CO;2.

Mutterlose, J.; Bornemann, A.; Luppold, F.W.; Owen, H.; Ruffell, A.H.; Weiss, W. & Wray, D. (2003). The Vöhrum section (northwest Germany) and the Aptian/Albian boundary. *Cretaceous Research*, Vol.24, No.3, (June 2003), 203-252, ISSN 0195-6671, doi:10.1016/S0195-6671(03)00043-0.

Nel, A.; Nel, J.; Masselot, G. & Thomas, A. (1998). An investigation into the application of the Wagner parsimony method in synecology. *Biological Journal of the Linnean Society*, Vol.65, No.2, (October 1998), 165-189, ISSN 0024-4066, doi:10.1111/j.1095-8312.1998.tb00354.x.

Perochon, E.; Masselot, G. & Nel, A. (2001). Sampling problem in synecological analysis. A concrete example in freshwater biomonitoring. *Annales de la Société Entomologique de France*, Vol.37, No.3, (September 1997), 341-346, ISSN 0037-9271.

Price, G.D. & Hart, M.B. (2002). Isotopic evidence for Early to mid-Cretaceous ocean temperature variability. *Marine Micropalaeontology*, Vol.46, No.1-2, (September 2002), 45-58, ISSN 0377-8398, doi:10.1016/S0377-8398(02)00043-9.

Price, G.D. (2003). New constraints upon isotope variation during the early Cretaceous (Barremian-Cenomanian) from the Pacific Ocean. *Geological Magazine*, Vol.140, No.5, 513-522, ISSN 0016-7568, doi: 10.1017/S0016756803008100.

Retallack, G.J. 2001. A 300-million-year record of atmospheric carbon dioxide from fossil plant cuticles. Nature (London), Vol.411, No.6835, (17 May 2001), 287-290, ISSN 0028-0836, doi:10.1038/35077041.

Retallack, G. & Dilcher, D.L. (1981a). Early angiosperm reproduction: *Prisca reynoldsii*, gen. et sp. nov. from mid-Cretaceous coastal deposits in Kansas, U.S.A. *Palaeontographica B*, Vol.179, No.5-5, (January 1981), 103-137, ISSN 0375-0299.

Retallack, G. & Dilcher, D.L. (1981b). A coastal hypothesis for the dispersal and rise to dominance of flowering plants. In: *Palaeobotany, Palaeoecology, and Evolution, Volume 2*, K.J. Niklas (Ed.), pp.27-77. Praeger Publishers, New York, ISBN 978-0030566561.

Ruffell, A.H. & Batten, D.J. (1990). The Barremian-Aptian arid phase in Western Europe. *Palaeogeography, Palaeoclimatology, Palaeoecology*, Vol.80, No.3-4, (November 1990), 197-212, ISSN 0031-0182, doi:10.1016/0031-0182(90)90132-Q.

Rydin, C.; Mohr, B. & Friis, E.M. (2003). *Cratonia cotyledon* gen. et sp. nov.: a unique Cretaceous seedling related to *Welwitschia*. *Proceedings of the Royal Society of London Biology Letters*, Vol. 270, No.Suppl.1, (7 August 2003), S29-S32, ISSN 0962-8452, doi: 10.1098/rsbl.2003.0044.

Rydin, C.; Pedersen, K.R.; Crane, P.R. & Friis, E.M. (2006). Former diversity of *Ephedra* (Gnetales): evidence from Early Cretaceous seeds from Portugal and North America. *Annals of Botany*, Vol.98, No.1, (July 2006), 123-140, ISSN 0305-7364, doi: 10.1093/aob/mcl078.

Schneider, H.; Schuettpelz, E.; Pryer, K.M.; Cranfill, R.; Magallon, S. & Lupia, R. (2004). Fern diversified in the shadow of angiosperms. *Nature*, Vol.428, No.6982, (1 April 2004), 553-557, ISSN 0028-0836, doi:10.1038/nature02361.

Sepkoski, J.J.;Jr. (1986). Biological extinction in earth history. *Science*, Vol. 231, No.4745, (28 March 1986), 1528-1533, ISSN 0036-8075, doi: 10.1126/science.11542058.

Spicer, R.A. & Herman, A.B. (2010). The late Cretaceous environment of the Arctic: a quantitative reassessment based on plant fossils. *Palaeogeography, Palaeoclimatology, Palaeoecology*, Vol.295, No.3-4, (15 September 2010), 423-442, ISSN 0031-0182, doi:10.1016/j.palaeo.2010.02.025.

Steuber, T.; Rauch, M.; Masse, J.P.; Graaf, J. & Malkoč, M. (2005). Low-latitude seasonality of Cretaceous temperatures in warm and cold episodes. *Nature*, Vol.437, No.7063, (27 October 2005), 1341-1344, ISSN 0028-0836, doi:10.1038/nature04096.

Stoll, H.M. & Schrag, D.P. (2000). High-resolution stable isotope records from the Upper Cretaceous Rocks of Italy and Spain: glacial episodes in a greenhouse planet? *Geological Society of America Bulletin*, Vol.112, No.2, (February 2000), 308-319, ISSN 0016-7606, doi:10.1130/0016-7606(2000)112<308:HSIRFT>2.0.CO;2.

Sun, G.; Dilcher, D.L.; Zheng, S. & Zhou, Z. (1998). In Search of the First Flower: A Jurassic Angiosperm, *Archaefructus*, from Northeast China. *Science*, Vol.282, No.5394, (27 November 1998), 1692-1695, ISSN 0036-8075, doi:10.1126/science.282.5394.1692.

Sun, G.; Ji, Q.; Dilcher, D.L.; Zheng, S.; Nixon, K.C.; Wang, X. (2002). Archaefructaceae, a new basal Angiosperm family. *Science*, Vol.296, No.5569, (3 May 2002), 899-904, ISSN 0036-8075, doi:10.1126/science.1069439.

Taylor, D.W. & Hickey, L.J. (1996). Evidence for and implications of an herbaceous origin for Angiosperms, In: *Flowering plant origin, evolution, and phylogeny*, D.W. Taylor & L.J. Hickey (Eds.), 232-266. Chapman and Hall, New York, NY, ISBN 978-0-412-05341-2.

Veizer, J.; Godderis, Y. & Francois, L.M. (2000). Evidence for decoupling of atmospheric CO2 and global climate during the Phanerozoic eon. *Nature*, Vol.408, No.6813, (7 December 2000), 698-701, ISSN 00280836, doi:10.1038/35047044.

Wagner,W.H.; Jr. (1961). Problems in the classification of ferns, In: *Recent advances in botany, vol. 1 From lectures and symposia presented to the IX International Botanical Congress, Montreal, 1959*, 841-844. Toronto University Press, Toronto.

Wang, H. & Dilcher, D.L. (2006). Aquatic angiosperms from the Dakota Formation (Albian, Lower Cretaceous), Hoisington III Locality, Kansas, USA. International Journal of Plant Sciences, Vol.167, No.2, (March 2006), 385-401, ISSN 1058-5893, doi:10.1086/499502.

Wang, X.Q.; Tank, D.C. & Sang, T. (2000). Phylogeny and Divergence Times in Pinaceae: Evidence from Three Genomes. *Molecular Biology and Evolution*, Vol.17, No.5, (May 2000), 773-781, ISSN 0737-4038, doi:10.1093/molbev/.

Weissert, H. & Lini, A. (1991). Ice Age interludes during the time of Cretaceous greenhouse climate? *Controversies in Modern Geology*. In: D.W. Muller, J.A. McKenzie & H. Weissert (Eds.). London, Academic Press, 173-191, ISBN 978-0-125-10340-4.

Wolfe, J.A. (1993). A method of obtaining climatic parameters from leaf assemblages. *U.S. Geological Survey Bulletin*, Vol.2040, 1-73.

Wortmann, U.G.; Herrle, J.O. & Weissert, H. (2004). Altered carbon cycling and coupled changes in Early Cretaceous weathering patterns: evidence from integrated carbon isotope and sandstone records of the western Tethys. *Earth and Planetary Science*

Letters, Vol. 220, No.1-2, (30 March 2004), 69-82, ISSN 0012-821X, doi:10.1016/S0012-821X(04)00031-7.

Ziegler, P.A. (1988). Evolution of the Arctic–North-Atlantic and the Western Tethys, Memoirs of the American Association of Petroleum Geologists, Vol.43, 198 pp.

3

Diversity and Genetic Erosion of Ancient Crops and Wild Relatives of Agricultural Cultivars for Food: Implications for Nature Conservation in Georgia (Caucasus)

Maia Akhalkatsi, Jana Ekhvaia and Zezva Asanidze
Ilia State University
Georgia

1. Introduction

The interpretation of a healthy diet is one of the dilemmas for our modern civilization. Advances in agriculture are mainly directed at increasing food production to solve problems of a growing human population. However, food security remains a problem to ensure healthy food and to prevent human disease. These two tendencies often do not coincide. At present, the selective breeding programs of crops are mainly oriented toward the production of high-yielding varieties of genetically enhanced cultivars of cereals that have increased growth rates, increasing the percentage of usable plant parts and resistance against crop diseases. This initiative is linked to what began in the 1960s and was named by William Gaud (of USAID) a "Green Revolution" (Davies, 2003). It was a product of globalization as evidenced in the creation of international agricultural research centres to introduce new crop varieties around the world. This process caused a significant increase in total cereal production and daily calorie supply in developing countries between the 1960s and 1990s (Davies, 2003). However, this process has caused the gradual replacement of traditional crop varieties, and as a result has had a dramatic effect on agrobiodiversity in many countries. Particularly impacted have been the traditional landraces used by local peoples for thousands of years and this has affected the health of these communities.

Georgia, located in the South Caucasus, owns one of the oldest agricultural traditions. The name of the country is "Sakartvelo" in the Georgian language but its common name "Georgia" is semantically linked to Greek (γεωργία, transliterated geōrgía) and Latin (georgicus) roots meaning "agriculture" (Javakhishvili, 1987). Many Georgian endemic species and local varieties of wheat, barley, legumes, grapevine and fruits are known (Ketskhoveli, 1957). The traditional use of local cultivars is considered to be a reason for human longevity in the Caucasus region (Fox, 2004). Over five percent of the male Georgian centenarians were reputed to have been over age 120 in 1959 (Garson, 1991). The percentage of males over age 70 was 0.9% in 1959 and 1.07% of women were over 70. However, these values had diminished by 1970 to 0.66% and 0.86%, respectively. At present, no exact data are available, but longevity has obviously diminished (Fox, 2004).

Archaeological data clearly show that the Caucasus region (and Georgia in particular) was settled from prehistoric time and agriculture was developed during the early Neolithic era (Javakhishvili, 1987). The information about the wide chronological intervals in the archaeological materials connected with the history of mankind in the Caucasus starts from the Early Pleistocene. The 1.7-Myr-old specimens of small-brained hominids are found in the Caucasus at Dmanisi, located in Southern Georgia (Fig.1), which is the earliest known hominid site outside of Africa (Gabunia & Vekua, 1995; Finlayson, 2005). This speciman has been classified as *Homo erectus senso lato*, which is a very early type of *H. ergaster* and/or a new taxon, *H. georgicus* (Gabunia et al., 2002). The next chronological interval in the archaeological materials is connected with the period of Late Middle Palaeolithic and Early Upper Palaeolithic periods demonstrating patterns of mobility, land-use, and hunting of Neanderthal and modern human competition within the South Caucasus (Adler & Bar-Oz, 2009). Neanderthals invaded the Caucasus region at an unknown time and modern humans may have occupied the region alongside them from ~40 Ka before the present (BP). According to the archaeological material from different caves in Georgia (Tushabramishvili, 2011) and the northern Caucasus (Ovchinnikov et al., 2000), the final replacement of the Neanderthals by modern humans might be occurred here ~28 Ka BP. The Upper Palaeolithic archaeological findings at Dzudzuana Cave (Fig.1), Imereti region, Georgia, revealed remnants of wool (*Capra caucasica*) and dyed fibers of wild flax (*Linum usitatissimum*) dated to ~36–34 Ka BP (Adler & Bar-Oz, 2009). The Dzudzuana Cave flax fibers have clearly been modified, cut, twisted and dyed black, gray, turquoise and pink, most likely with locally available natural plant-derived pigments (Kvavadze et al., 2009). E. Kvavadze and colleagues (2009) surmise that this represents the production of colourful textiles for some purpose, perhaps clothing. In general, it is supposed that the microscopic flax fibres are the remains of linen and thread, which would have been used in clothing for warmth, for shoes, to sew together pieces of leather or to tie together packs.

The archaeological findings from Neolithic and Early Bronze periods are rich with plant fossils and seeds of both wild species and local landraces. The ancient findings from Neolithic period of cereal grains in Georgia were discovered from Trialeti Range, Kvemo Kartli region (Arukhlo excavations, Bolnisi district; Fig.1) and Samegrelo region (Dikha-Gudzuba, Zugdidi district; Fig.1) from 6th up to 2nd millennium BC (Melikishvili, 1970). Seven species of cultivated wheat - *Triticum aestivum, T. spelta, T. carthlicum, T. macha, T. monococcum, T. dicoccum, T. compactum* and one wild relative *Aegilops cylindrica* have been discovered in Arukhlo, Kvemo Kartli region. Other cereals: millet - *Panicum milleaceaum*, barley - *Hordeum vulgare*, Italian millet - *Setaria italica*, oats - *Avena sativa*, wild lentil - *Lens ervoides* and pea -*Pisum sativum* have been found in the same site. The wheat fields in Arukhlo were irrigated. Very recent studies on einkorn wheat domestication using amplified fragment length polymorphism (AFLP) show that *T. boeoticum* was domesticated in southeast Turkey in the Karacadag Mountains close to Diyarbakir (Heun et al., 1997). Old Georgian kingdom Diauehi (Diaokhi) is adjacent region to this place. Therefore, it might be considered to be an area where cultivation of cereals occurred in very early historical time. The earliest archaeological finding of cultivar grapevine pips are found in Shulaveri (Fig.1), located near Dmanisi in southern Georgia and dated to ~8.000 years BP (Ramishvili, 1988). A wide range of carbonised seeds, including wild and domesticated grape (*Vitis vinifera, V. vinifera* subsp. *sylvestris*), wheat (*Triticum* sp.), pea (*Pisum sativum*), rowan (*Sorbus* sp.) and walnut (*Juglans regia*) are found in soil samples in Nokalakevi (Fig.1), Western Georgia, dated to the Hellenistic period (Grant et al., 2009).

Fig. 1. Map of Georgia. The administrative regions: 1. Abkhazia; 2. Samegrelo-Upper Svaneti; 3. Guria; 4. Adjara; 5. Racha-Lechkhumi; 6. Imereti; 7. Meskheti- Javakheti; 8. Shida Kartli; 9. Kvemo Kartli; 10. Mtskheta-Mtianeti; 11. Kakheti. The places of archaeological excavations are indicated: Dikha-Gudzuba, Nokalakevi, Dzudzuana cave, Arukhlo, Dmanisi and Shulaveri.

According to N. I. Vavilov (1992), the origin of ancient crop varieties and landraces in Georgia coincides with the period of their primary domestication. Georgia is often considered part of the Near East where many field crops were domesticated. N. I. Vavilov (1992) determined 8 centres of crop origin and diversity. Among them was the fourth centre, which included the South Caucasus, Asia Minor, Iran and Turkmenistan. The main crops domesticated in this centre (which includes Georgia) are wheat, rye, oats, seed and forage legumes, herbs, fruits, and grapes for winemaking; 83 species all tolled.

The problem is that there are no concrete data to assess either the current status of local varieties or information about the domestication process in Georgia. The fundamental work on domestication and origin of wheat and barley in this region was done by the famous Georgian botanist V. Menabde (1938, 1948). The agricultural evidence was reported by several other Georgian authors (Ketskhoveli, 1957; Khomizurashvili, 1973; Akhalkatsi et al., 2010). We have studied domestication of wild grapevine (*Vitis vinifera* subsp. *sylvestris*) and wild pear (*Pyrus caucasica*) using morphometric and systematic molecular methods (Ekhvaia & Akhalkatsi, 2010; Ekhvaia et al., 2010; Asanidze et al., 2011) confirming genetic relationships between wild populations and local cultivars of grape and pear. However, complete evaluation of diversity of Georgian local cultivars and crop wild relatives (CWR) has not yet been complete.

There are many threats to these oldest of crops in the modern period. In our opinion, the main threat to agrobiodiversity in Georgia is the loss of landraces and ancient crop varieties. Protection measures in the country are still not being implemented at an appropriate rate. National policies and comprehensive measures are urgently needed to address the problem of conserving the genetic resources of ancient crops in Georgia. Thus, we suggest that it is necessary to establish a general overview of the types of crops that are current landraces and primitive forms occurring in Georgia and to publish lists of indigenous landraces and CWRs of cereals, legumes, vegetables and fruits representing direct ancestors, and endemic, rare or endangered species, in order to evaluate the sustainability of their traditional use in terms of nature conservation.

2. Landrace assessment

Agriculture in Georgia is characterized by a great diversity of local landraces, varieties and even endemic species of crops. These varieties reveal a high level of adaptation to local climatic conditions and often have high resistance to diseases. Georgians have used these crops for a very long time and their healthy life, reflected by the longevity of individuals in the population, was considered to be connected to their good food. However, there are many threats to these oldest of crops in the modern period, particularly since the 1950s. The loss of local and ancient crop varieties should be considered to be the main threat to agrobiodiversity in Georgia.

The known diversity and distribution of local landraces is based on data obtained from archaeological reports, historical manuscripts, ethnography, and botanical field expeditions in different regions of Georgia since 1920s. The oldest known text about Georgian cultivars is a XVII century by the work of Vakhushti Batonishvili "Geographic Description of Georgian Kingdom" (Batonishvili, 1991). Active investigation of Georgian crops began in the 1920s (Ketskhoveli, 1928, 1957; Menabde, 1938, 1948; Dekaprelevich, 1947). These investigations revealed that ancient cultivars of grapevine, fruits, wheat, barley, rye, oats, common millet, Italian millet, legumes, flax, and a number of herbal and spice plants, were still being cultivated in Georgia. The rapid loss of local cultivars of cereals, legumes and flax began in the 1950s and reached an extreme in the 1990s (Akhalkatsi et al., 2010). At present, almost all of Georgia's ancient crops are maintained in conservatory collections and seed banks, but none are present in peasant house gardens in lowland areas. Only the mountain areas contain depositories of the ancient crops of Georgia, where some number of landraces still exist. The process of genetic erosion of ancient crop varieties, however, has begun even in these regions since the 1990s and this presents great concern about the loss of aboriginal crops adapted to high mountain areas (Pistrick et al., 2009).

Monitoring of crop diversity is now conducted by international nature conservation institutions and Georgian scientific and nongovernmental organizations to preserve the genetic resources of local cultivars. One of the problems is the deficit of information about the current state of ancient crops and recommendations for their conservation are inadequate. Therefore, it is necessary to assess research needs and implications for conservation and to formulate recommendations for the conservation and on-farm maintenance of Georgian landraces.

2.1 Diversity of ancient crop varieties

Reports of the diversity of local landraces in Georgia has to present been published primarily in Georgian- and Russian-language scientific publications (Ketskhoveli, 1928, 1957; Menabde, 1938, 1948; Dekaprelevich, 1947; Kobakhidze, 1974). Databases and international periodicals lack descriptions of this diversity, taxonomy and discussions of the conservation value of landraces. In our opinion it is important to spread information about diversity and conservation needs of local cultivars of Georgia worldwide to support the evaluation of their roles in healthy life of human. Some crops, such as grapevine, wheat, barley and fruit trees are characterized by the highest diversity of landraces in Georgia.

Grapevine - *Vitis vinifera* L. (Vitaceae) shows greatest genetic and morphological variability. About 500 names of autochthonous grapevine varieties known from Georgia are characterized by a wide range of colour gamma and shapes of berries and pips (Javakhishvili, 1987; Ketskhoveli et al., 1960), which points to an evolutionary centre in this region (Vavilov, 1992). These cultivars showed great ampelometric variability and broad adaptability to different climate and soil conditions (Ketskhoveli et al., 1960). Each province of Georgia possesses its own grapevine cultivars adapted to local climate. The varieties are of three forms: 1) *Babilo* is an old grapevine with stem more than 20 cm in diameter clambering on trees (Fig.2A). 2) *Maghlari* represents varieties that climb tree trunks (alder, persimmon, mulberry, cherry, beech, chestnuts, etc.) distributed mainly in peasant orchards in western and southern Georgia (Fig.2B). 2) *Dablari* is used to create typical vineyards (Fig.2C) found in commercial areas. The total area of vineyards in Georgia was 40.000 ha in the 1980s. It has diminished to ca. 25.000 ha today (Bedoshvili, 2010). Forty-four percent of this territory is located in Kakheti region, 26% in Imereti, 15% in Kartli and 15% in almost all regions of Georgia except in the high mountain regions of Khevi, Khavsureti and Tusheti. Forty-one cultivars of grapevine are used as commercial varieties in Georgia. Among them, 27 are technical varieties used for winemaking and 14 are table grapes (Bedoshvili, 2010). Ninety-seven percent of total annual yield is used for winemaking and only 3% as table grapes. Wine is made from landraces: 'Rkatsiteli' (55%); 'Tsolikauri' (10.2%); 'Chinuri' (7%); 'Saperavi' (4%); 'Kakhuri Mtsvane' (3.3%); and, several local and introduced cultivars (20.5%).

A B C

Fig. 2. Different types of vineyeards: A - *Babilo,* an old grapevine 'Meskhuri Shavi' (diameter 32 cm) belonged to the family of Gogi Natenadze in village Chachkari, Meskheti; B - *Maghlari,* clambering introduced cultivar *Vitis labrusca* 'Odessa' on beech tree in village Naghvarevi, Adjara. C - *Dablari,* typical vineyeard of landraces 'Rkatsiteli' and 'Saperavi' in village Shilda, Kakheti. Photos by Maia Akhalkatsi.

The first known threat to the grapevine in Georgia occurred during the occupation of Georgian territory by the Muslims during the medieval period. They destroyed vineyards and/or led to the destruction of human settlements, where until now are presented local wilding grape cultivars clambering on trees. The vineyards grow on the terraces of

Mediterranean type in the historic province of Tao-Klarjeti located now in southern Georgia and in the province of Artvin, Turkey. Since 15th century, the Seljuk Turks occupied this territory and vine terraces disappeared and were covered with trees or grasses. However, we have found peasants in some villages of Meskheti province searching for old cultivars in abandoned settlements. They are replanting naturalized grape cultivars to house gardens. We have found ancient grapevine varieties 'Meskhuri Shavi' and 'Meskhuri Mtsvane' to be frost resistant and growing in high mountain areas in villages Zemo Vardzia (1322 m a.s.l.) and Chachkari (1264 m a.s.l.), Aspindza district; village Zazalo, (1486 m a.s.l.), Adigeni district; and, Karzameti castle near boundary to Turkey, 1450 m a.s.l. One of the oldest Georgian grape cultivar 'Krikina', which is morphologically nearly identical to wild grapevine (*V. vinifera* subsp. *sylvestris* [C. C. Gmel.] Hegi), was found in village Rveli, Borjomi district, in house garden of Gaioz Tabatadze who replanted this cultivar from the ruins of the historic village Baniskhevi where his ancestors lived. The grape variety 'Shonuri' ('Lushnu' in Svanetian) adapted to high mountain areas grows in Upper Svaneti province from 1045 to 1400 m a.s.l. Landraces adapted to high elevations are rare and are usually replaced with the introduced grape cultivar *V. labrusca* 'Odessa'. It is widespread in mountainous villages in all regions of Georgia. Some other rare grapevine cultivars - 'Kachichi', 'Saperavi', 'Sebeli', 'Jvarisa', have been found on lower elevations in village Gvimbrala, Lentekhi district, Lower Svaneti province. Rare grape cultivars - 'Aladasturi', 'Tsulukidzis Tetra' and 'Tskhvedanis Tetra' have been found in village Tabori and 'Usakhelouri' in village Zubi, Tsageri district, Lechkhumi province. Rare landrace 'Chkhaveri' was found in village of Merisi, at 474 m a.s.l. Adjara province.

In 1860, the *V. vinifera* was virtually wiped out in the places of its origin, when an aphid, *Phylloxera vastatrix* was accidentally introduced into France, and within a few years had ravaged all vineyards in Europe and in Georgia as well. Currently, almost all Georgian grape varieties are grafted on rootstocks of American grapevines - *V. riparia, V. rupestris* and *V. berlandieri* and their hybrids are resistant to *Phylloxera*. This disaster made it necessary to undertake urgent steps for *ex situ* conservation of old, endangered and autochthonous grapevine varieties by establishing living collections in Georgia; this was begun in the 1930s. The collections of plant genetic resources were established in research institutes, which have been under reforms since 1990s and operating with diminishing funding to maintain the collections. In 2003, 929 varieties were protected in the living collections. Among them, 701 were cultivars obtained from selective breeding and only 248 of the 524 autochthonous Georgian varieties remain. These collections of the State Agricultural University were located in Dighomi (573 cultivars) and Mukhrani (155 cultivars), and, the collections in Telavi (226 cultivars) and Skra (75 cultivars) belonged to the Georgian research Institute of Horticulture, Viticulture and Winemaking (Maghradze, et al., 2010). Recently, these collections have been closed. Nevertheless, some effort has been made to establish new collections in Telavi (573 accessions), Skra (440) and Vachebi (312) in 2008. Three other new collections were set up by Saguramo "Centre for Grapevine and Fruit Tree Planting Material Propagation" (ca 400 accessions), "Kindzmarauli" and "Shumi" wineries (as a total 149 accessions). Two new collections were established in Italy by the University of Milan (Maghradze, et al., 2010). Some Georgian cultivars are in living collections abroad in Russia, Moldova and Germany as well. A small living grapevine collection exists in the G. Eliava

National Museum in Martvili district, Samegrelo province, founded in 1972 and containing 24 old Colchic grapevine varieties (Eliava, 1992). Seven cultivars of Meskheti region were collected in the research station of Biological Farming Association Elkana in village Tsnisi, Akhaltsikhe district. Many grape landraces are extinct and do not exist even in living collections.

Wheat - *Triticum* L. (Poaceae) also shows high diversity in Georgia. Nineteen species of wheat from the 26 known species of the genus *Triticum* have been historically distributed in Georgia (Tab.1). Some of them are endemic species: *T. timopheevii, T. zhukovskyi, T. macha, T. carthlicum* and *T. palaeo-colchicum*. Sixteen species, 144 varieties, and 150 forms of wheat were registered in Georgia in the 1940s (Menabde, 1948).

N	*Taxon names by Menabde, 1948, 1961*	*Taxon accepted name by ARS-GRIN, USDA 2011*	*Status*	*Ploidy levels*
	T. aegilopoides Balansa ex Körn. (=*T. boeticum* Boiss.)	*T. monococcum* subsp. *aegilopoides* (Link) Thell.	W	2n
	T. monococcum L.	*T. monococcum* L.	PS	2n
	T. timopheevii Zhuk.	*T. timopheevii* (Zhuk.) Zhuk. subsp. *timopheevii*	EG, PS	4n
	T. chaldicum Menabde	*T. timopheevii* (Zhuk.) Zhuk. subsp. *armeniacum* (Jakubz.) Slageren	W	4n
	T. dicoccoides Körn.	*T. turgidum* L. subsp. *dicoccoides* (Körn. ex Asch. & Graebn.) Thell.	W	4n
	T. palaeo-colchicum Menabde	*T. turgidum* L. subsp. *palaeocolchicum* Á. Löve & D. Löve	EG, SP	4n
	T. dicoccum Schuebl.	*T. turgidum* L. subsp. *dicoccon* (Schrank) Thell.	SP	4n
	T. durum Desf.	*T. turgidum* L. subsp. *durum* (Desf.) Husn.	SP	4n
	T. turgidum L.	*T. turgidum* L. subsp. *turgidum*	SP	4n
	T. carthlicum Nevski (=*T. ibericum* Menabde; *T. persicum* Vavilov ex Zhuk.)	*T. turgidum* L. subsp. *carthlicum* (Nevski) Á. Löve & D. Löve	EG, SP	4n
	T. polonicum L.	*T. turgidum* L. subsp. *polonicum* (L.) Thell.	SP	4n
	T. turanicum Jakubz.	*T. turgidum* L. subsp. *turanicum* (Jakubz.) Á. Löve & D. Löve	IS	4n
	T. abyssinicum Vavilov	*T. dicoccon* subsp. *abyssinicum* Vavilov	IS	4n
	T. vulgare Villars	*T. aestivum* L.	SP	6n
	T. macha Dekapr. & Menabde	*T. aestivum* L. subsp. *macha* (Dekapr. & V.L. Menabde) Mackey	EG, PS	6n
	T. spelta L.	*T. aestivum* subsp. *spelta* (L.) Thell.	IS	6n
	T. sphaerococcum Percival	*T. aestivum* subsp. *sphaerococcum* (Percival) Mackey	IS	6n
	T. compactum Host	*T. aestivum* subsp. *compactum* (Host) Mackey	SP	6n
	T. zhukovskyi Menabde & Ericzjan	*T. zhukovskyi* V.L. Menabde & Eritzjan	EG, SP	6n

Table 1. List of wheat species distributed in Georgia by V. Menabde (1948, 1961). The accepted names are added from web-page: http://www.ars-grin.gov/cgi-bin/npgs/html/splist.pl?28515. The status of species is based on phylogenetic studies of V. Menabde (1948, 1961): EG - endemic of Georgia; W- wild; PS - primary species; SP - secondary species; IS - introduced species. Ploidy levels are indicated.

According to V. Menabde (1948), three species from the list are wild - *T. boeticum* (2n=14), *T. dicoccoides* (2n=28), *T. timopheevii* subsp. *armeniacum* (2n=28); they were mixed with cultivars in the wheat fields and did not exist in natural habitats in Georgia. Sites of *T. boeoticum* are concentrated in south-eastern Turkey, where this species was probably domesticated (Heun et al., 1997). The current distribution indicates that its weedy races have spread with cultivated cereals far to the west and east. There is evidence that it was found in fields with *T. monococcum* in Georgia (Menabde, 1948). Since the 1930s their number has diminished and all of these species had disappeared after the 1960s, when non-aboriginal cultivars were introduced in *kolkhozis*—agricultural farming corporations in Soviet times, changing the species composition in wheat fields. At present, none of these species occur in agricultural fields of Georgia.

Three species from the list (Tab.1) are considered by V. Menabde (1948) as primary species (close to the first domesticated species): *T. monococcum* (2n=14), *T. timopheevii* (2n=28) and *T. macha* (2n=42). First two species, *T. monococcum* - 'Gvatsa Zanduri' and *T. timopheevii* - 'Cheltha Zanduri', in Georgian, are old autochthonous wheat species distributed mainly in western Georgia - Racha-Lechkhumi, Imereti and Samegrelo. *T. timopheevii* was growing in a small area in western Georgia together with its hexaploid derivative - *T. zhukovskyi*, and cultivated einkorn - *T. monococcum* (Menabde, 1961). These three species represent polyploid series of wheat *Zanduri*, which was possible to find in peasant farms till 1990s. *T. macha* is archaeological findings in Dikha-Gudzuba and Shulaveri excavations dated by Neolithic period and was cultivated in Racha-Lechkhumi, Imereti and Samegrelo up to 1950s (Dekaprelevich, 1947).

Nine native species of wheat - *T. palaeo-colchicum, T. dicoccum, T. durum, T. turgidum, T. carthlicum, T. polonicum, T. aestivum, T. zhukovskyi* and *T. compactum*, are considered by V. Menabde (1948, 1961) as secondary species originated by hybridization with wild and primary species of *Triticum, Aegilops* spp., *Thinopyrum intermedium* (Host) Barkworth & D. R. Dewey subsp. *intermedium* (=*Agropyron glaucum* [Desf. ex DC.] Roem. & Schult.), and *Thinopyrum elongatum* (Host) D. R. Dewey (=*Agropyron elongatum* [Host] P. Beauv.). *T. aestivum, T. carthlicum* and *T. durum* have many varieties and cultivars. The four species in the list (Tab.1) - *T. abyssinicum, T. spelta, T. sphaerococcum* and *T. turanicum* represent geographical races introduced from different regions in the historically different times.

The traditional wheat fields in all regions of Georgia usually contain several species and varieties (Eritzjan, 1956; Zhizhilashvili & Berishvili, 1980). Bread wheat fields contain: *T. aestivum* var. *erythrospermum* 'Tetri dolis puri', *T. aestivum* var. *ferrugineum* 'Tsiteli dolis puri', *T. aestivum* var. *lutescens* 'Upkho tetri dolis puri', *T. aestivum* var. *milturum* 'Upkho tsiteli dolis puri', *T. compactum* 'Kondara khorbali'. Usually, this combination of wheat taxa is associated with wild weed *Makhobeli* - *Cephalaria syriaca* (L.) Schrad. ex Roem. & Schult. (Dipsacaceae) occurring most often in such wheat fields. The seeds of this species are of the same size as wheat and after threshing remain in the harvest. Seeds are ground into a powder and used with wheat to make bread, cakes, etc. It adds a nice flavour but quickly goes rancid. Another combination of varieties was dominated by *T. durum* 'Shavpkha' composed by *T. durum* var. *apulicum, T. durum* var. *leucurum, T. durum* var. *murciense, T. aestivum* var. *erythrospermum, T. aestivum* var. *pseudo-barbarossa, T. aestivum* var. *lutescens, T. compactum* var. *erinaceum*. This population is adapted to dry climate in the lowland areas and in the high elevations up to 1800 m a.s.l. in Javakheti Plateau, where it is sown in early spring. The same character of adaptation to high elevation is typical for the wheat species, *T. carthlicum* 'Dika', sown on high mountain areas in spring. The combination of varieties dominated by 'Dika' is as follows: *T. carthlicum* var. *rubiginosum*, *T. carthlicum* var. *stramineum*, *T. aestivum* var. *erythrospermum*, *T. aestivum* var. *ferrugineum*, *T. compactum* var. *erinaceum*.

A B C

Fig. 3. A- Wheat field of the Georgian endemic *Triticum carthlicum* 'Dika' in research station of the Biological Farming Association Elkana, village Tsnisi, Meskheti; B - Traditional wheat bread in Meskheti; C- Six row barley field with mixture of wild rye *Svila* (*Secale segetale*) in village Shilda, Kakheti. Photos by Maia Akhalkatsi.

Wheat fields were planted throughout Georgia at elevations from 300 m to 2160 m a.s.l. We have found this highest location of soft wheat field in the Eastern Greater Caucasus, village Chero in Tusheti (Akhalkatsi et al., 2010). At present, almost none of these traditional wheat varieties and species occur in the territory of Georgia. Only aboriginal varieties of bread wheat still exist in several high mountain regions like Tusheti, Meskheti, Javakheti and Svaneti (Pistrick et al., 2009). Living collections and gene banks preserve the local varieties. The living collection of the Biological Farming Association Elkana has many landraces in village Tsnisi, Akhaltsikhe district (Fig.3A). In 2010, they sowed a 10 ha wheat field. The harvest from this field contained local cultivar *T. aestivum* var. *ferrugineum* 'Akhaltsikhis tsiteli dolis puri' and weed *Makhobeli*. The flour was baked as bread in Tbilisi and as traditional bread in Meskheti (Fig.3B).

Barley - *Hordeum vulgare* L. (Poaceae) is an ancient agricultural crop in Georgia. It was the second most important cereal in Georgia after wheat and main crop in high mountain regions used for bread, forage and production of beer, as well as an attribute of religious rituals and in the folk medicine (Javakhishvili, 1987). Two different names were used for barley in Georgian language - *Krtili* and *Keri*. *Krtili* denotes six-row winter barley (*H. vulgare* subsp. *hexastichon* [L.] Čelak.) that was sowed in autumn; *Keri* refers to two-row summer barley (*H. vulgare* subsp. *distichon* [L.] Körn.) sowed in spring (Menabde, 1938). Six-row barley was sown in lowland areas but was cultivated up to 2130 m a.s.l. in Svaneti. Two-row barley was cultivated mainly in high mountain regions. The cultivars of two-row barley *H. vulgare* var. *nutans* 'Akhaltesli' and *H. vulgare* var. *nigrum* Willd. 'Dzveltesli shavpkha' are distributed up to 2100 m a.s.l. in all high mountain areas. *H. vulgare* var. *nutans* is mixed in the field with wheat - *T. carthlicum* 'Dika', and the flour is produced from mixed wheat and barley seeds. *H. vulgare* var. *nudum* Spenn. 'Kershveli' was cultivated in Meskheti and Svaneti. Four-row barley (*H. vulgare* subsp. *tetrastichon* [Stokes] Čelak.) is rare and the cultivar - *H. vulgare* var. *pallidum* Ser. 'Tetri Keri' occurs only in the high mountain region of

Meskheti, Tusheti and Svaneti up to 2100 m a.s.l. These cultivars persist today only in high mountain regions. However, their distribution has been seriously diminished. At present, introduced varieties of barley are widely cultivated in the lowlands and their names are unknown to the local population.

Rye – *Secale cereale* L. (Poaceae) is only a local cultivar of high mountain regions of Georgia (1800-2200 m a.s.l.). Fields of *S. cereale* (2n=14) are now found only in Upper and Lower Svaneti and Meskheti. Rye was used for making alcohol and as forage. The wild species, *S. segetale* (Zhuk.) Roshev. (2n=42), called *Svila* is widespread in wheat and barley fields and is harvested together with them (Fig.3C). The bread of wheat with *Svila* is considered to be very nutricious and has good taste. An endemic species of rye is *S. vavilovii* Grossh. (2n=14). It is also called Caucasian rye. This species was found in wheat field in Georgia (Bockelman et al., 2002). We have monitored the place in village Beghleti, Khashuri district in 2008, where Georgian botanists had noted the presence of this species in the wheat fields, but cultivated plots no longer exist in that area. The village has lost of most of its residents and no agriculture is undertaken there. Introduced cultivars and commercial varieties of rye are not used in Georgia.

Oats – *Avena sativa* L. (Poaceae) is a traditionally cultivated plant distributed from 400 to 1400 m a.s.l. It is used only as forage for horses and poultry. Two varieties of oats have been described for Upper Svaneti (Ketskhoveli, 1957) - *A. sativa* var. *aurea* Körn. and *A. sativa* var. *krausei* Körn. In lowlands, usually, the origin of the seeds is unknown to local farmers. It is purchased in the market and farmers receive no information about their origin. Millet – *Panicum miliaceum* L. (Poaceae) is very old agricultural plant cultivated in all regions of Georgia. It was used as a supplementary feed (for animals and poultry) and for making alcoholic drinks. At present, it is cultivated only in high mountain regions (1000-1800 m a.s.l.). Several varieties are described in Upper and Lower Svaneti: *P. miliaceum* var. *aureum* V.M. Arnold & Shibaiev. - grain yellow or cream; *P. miliaceum* var. *subaereum* Körn. - grain grey; *P. miliaceum* var. *griseum* Körn. - grain brown; *P. miliaceum* var. *atrocastaneum* Batalin ex V.M. Arnold & Shibaiev. - grain black; *P. miliaceum* var. *badium* Körn. - grain white (Zhizhilashvili & Berishvili, 1980). The acreage of millet fields declined after introduction of maize in Georgia in 17th century. Italian millet - *Setaria italica* (L.) P. Beauv. (Poaceae) was cultivated in Colchis, Samegrelo since ancient times. The cultivar - *S. italica* subsp. *colchica* (Dekapr. & Kaspar.) Maisaya & Gorgidze was represented with 32 landraces (Maisaia et al., 2005). It can currently be found in the Samegrelo region of western Georgia. Another subspecies - *S. italica* subsp. *moharia* (Alef.) H. Scholz., is called *Kvrima* in Georgian. It was cultivated for a long time but was replaced by maize.

Legumes - peas, lentils, chickpeas, faba beans, common vetch, bitter vetch, chickling vetch, alfalfa, sainfoin and blue fenugreek are traditional crops in Georgia (Tab.2). Green Pea (*Pisum sativum*) is originated in the South Caucasus. It is grown in house gardens in small amounts for food today. Two species of pea are cultivated in Georgia - *P. sativum* with white flowers, round white or yellow seeds, and *P. arvense* with purple flowers, ridged dark coloured seeds. The third wild species *P. elatius* Steven ex M. Bieb. with dark purple flowers is often found in locations of old settlements, ruins of monasteries and churches and inside castle walls. The local cultivar of green pea, *P. sativum* subsp. *transcaucasicum* Govorov, has

14 varieties (Kobakhidze, 1974). Local varieties of Chickpea (*Cicer arietinum*) are rarely cultivated today. Three subspecies and 24 varieties were available in western Georgia - Racha-Lechkhumi, Svaneti and Imereti up to 1920s (Dekaprelevich & Menabde, 1929). In the 1970s, the same three subspecies - *C. arietinum* subsp. *mediterraneum* G. Pop., *C. arietinum* subsp. *eurasiaticum* G. Pop. and *C. arietinum* subsp. *orientalis* G. Pop., remained, but included only 6 of 24 varieties - *C. arietinum* subsp. *mediterraneum* var. *ochroleucum* A. Kob., *C. arietinum* subsp. *mediterraneum* var. *rozeum* G. Pop., *C. arietinum* subsp. *eurasiaticum* var. *aurantiacum* G. Pop., *C. arietinum* subsp. *orientalis* var. *fulvum* G. Pop., *C. arietinum* subsp. *orientalis* var. *rufescens* G. Pop., and *C. arietinum* subsp. *orientalis* var. *rufescens brunneo-punctatus* A. Kob. (Kobakhidze, 1974). Chickpeas were traditionally available in Svaneti, but by the 1970s only one farmer was sowing it in Kala community village Khe (Zhizhizlashvili & Berishvili, 1983). The Biological Farming Association Elkana is producing local cultivars of chickpea and selling them in market. Lentil (*Lens culinaris*) was represented in Georgia by two subspecies - *L. culinaris* subsp. *macrosperma* N.F. Mattos and *L. culinaris* subsp. *microsperma* N.F. Mattos; and 15 varieties (Kobakhidze, 1974). The last subspecies with small seeds was sown in high mountain areas in Javakheti. It was available in Meskheti till 1970s. Lentils were cultivated in Upper Svaneti from prehistoric times, but, at present, it is nearly extinct. In 1980s, three cultivars were described in Svaneti - 1. *L. culinaris* var. *persica* Bar. - reddish-brown seeds; 2. *L. culinaris* var. *ochroleucus nigro-punctulata* A. Kob. - light brown seeds with black dots; 3. *L. culinaris* var. *nigro-marmorata* A. Kob. - seeds have reddish-yellow background with black marbling (Zhizhizlashvili & Berishvili, 1983). The Biological Farming Association Elkana is producing local cultivars of lentil for the market. Faba bean (*Vicia faba)* is one of the oldest cultivated plants. Faba bean is ancient agricultural plant in western and southern Georgia. Three varieties and 31 subvarieties were described in Georgia with small (*V. faba* var. *minor* Beck.), medium (*V. faba* var. *equina* Pers.) and large (*V. faba* var. *major* Harz.) seeds (Kobakhidze, 1974). At present, the large seed Faba bean is widely distributed in Upper and Lower Svaneti. Two varieties are found in the Lower Svaneti: 1. *V. faba* var. *minor* subvar. *straminea* A. Kob. - compressed on sides, tip obtuse, colour light cream. 2. *V. faba* var. *equina* subvar. *ochroleucus* A. Kob. - slightly compressed on sides, tip rounded, colour yellowish (Zhizhizlashvili & Berishvili, 1983). Chickling vetch (*Lathyrus sativus*) is used as human food in a soup to called *shechamandi.* It is also a green forage, used as silage and fed as seed flour to pigs and poultry. It is now available only at the research station of the Biological Farming Association Elkana. Bitter vetch – *Vicia ervilia* is distributed in Meskheti and Javakheti. There are cultivated and wild forms of this species. It is used as a forage and for soil enrichment with nitrogen. Common vetch (*Vicia sativa*) is used as forage and for hay, especially in Upper and Lower Svaneti and Javakheti. It is a valuable forage crop, rich in proteins. More often it appears as a weed in the fields of high mountain areas among grain crops – millet, barley, rye. Sainfoin (*Onobrychis* spp.), alfalfa (*Medicago sativa*) and clover (*Trifolium* spp.) are forage legumes. A local variety of *Onobrychis transcaucasica* Grossh. 'Akhalkalakuri', is widely used. Blue fenugreek (*Trigonella caerulea*) is traditional spice plant used in almost all of the foods of Georgian cuisine. It is available in all regions of Georgia.

Flax – *Linum usitatissimum* L. (Linaceae), was one of the oldest and important field crops in Georgia. Since prehistoric times, it was used to produce excellent linens (Kvavadze et al., 2009) and to make oil from its seeds. Big millstones were used to extract the oil from the flax seeds and they remain in many historical ruins. Until recently, flax was cultivated only in Javakheti, where flax seeds were used to produce pharmacologically pure oil for medicines. According to the eighteenth century Georgian scientist and geographer Vakhushti Batonoshvili (1991), several volatile oil-bearing plants were cultivated in Georgia - roses, camphor, lavender and basil. Kenaf - *Cannabis sativa* L. (Cannabaceae), was used to produce fiber for cord and thread for sacks. The seeds were used to produce oil. A traditional use of kenaf seeds was to mix them with wheat flour and making breads that had antidepressant effects.

Traditional vegetables (Tab.2) are represented by sugar beets, spinach, carrots, radishes, turnips, onions, Welsh onion, leeks and garlic. Beet - *Beta vulgaris,* is an ancient cultivated plant whose tubers and young leaves were used in Georgian cooking. Leaves primarily came from the variety *B. vulgaris* subsp. *cicla* (L.) W.D.J. Koch 'Tsiteli Mkhali' that was grown in lower elevations up to 1400 m a.s.l. Another beet variety - *B. vulgaris* L. subsp. *vulgaris* 'Sasufre Charkhali' is rare. Carrot - *Daucus carota,* was edible as a wild species in Georgia since prehistoric times. The cultivated carrot is widespread in peasant's house gardens in lowland areas. Onion - *Allium cepa* and garlic - *A. sativum,* are ancient cultivated plants available in all regions of Georgia. Red onions are very popular in Georgian people. *A. sativum* is called 'Georgian garlic'. Another variety is 'Russian garlic' representing *A. ampeloprasum* L. Leek - *A. porrum,* is typical in western Georgia. Welsh onion - *A. fistulosum* is currently grown in several high mountain areas. Until the 1970s, it was widespread in Imereti, but at present, Chinese shallot - *A. cepa* var. *aggregatum* G. Don has completely supplanted Welsh onion. Radish - *Raphanus sativus,* is grown in lower elevations in gardens and is cultivated by farmers for the market.

N	Latin name	Family	English common name	Georgian common name
	Allium cepa L.	Liliaceae	*Onion*	*Khakhvi*
	Allium fistulosum L.	Liliaceae	*Welsh Onion*	*Chlakvi*
	Allium porrum L.	Liliaceae	*Leek*	*Prasi*
	Allium sativum L.	Liliaceae	*Garlic*	*Niori*
	Beta vulgaris L. subsp. *vulgaris*	Chenopodiaceae	*Beet*	*Charkhali*
	Brassica rapa L. subsp. *rapifera* Metzger	Brassicaceae	*Turnip*	*Talgami*
	Cannabis sativa L.	Cannabaceae	*Kenaf*	*Kanafi*
	Cicer arietinum L.	Fabaceae	*Chickpea*	*Mukhudo*
	Daucus carota L.	Apiaceae	*Carrot*	*Stafilo*
	Lathyrus sativus L.	Fabaceae	*Chickling vetch*	*Tsulispira*
	Lens culinaris Medik.	Fabaceae	*Lentil*	*Ospi*
	Linum usitatissimum L.	Linaceae	*Flax*	*Seli*
	Medicago sativa L.	Fabaceae	*Alfalfa*	*Ionja*
	Onobrychis transcaucasica Grossh.	Fabaceae	*Sainfoin*	*Espartseti*
	Pisum arvense L.	Fabaceae	*Pea*	*Barda*
	Pisum sativum L.	Fabaceae	*Pea*	*Barda*
	Raphanus sativus L.	Brassicaceae	*Radish*	*Boloki*
	Spinacia oleracea L.	Chenopodiaceae	*Spinach*	*Ispanakhi*
	Vicia ervilia (L.) Willd.	Fabaceae	*Bitter vetch*	*Ugrekheli*
	Vicia faba L.	Fabaceae	*Faba bean*	*Tsertsvi*
	Vicia sativa L.	Fabaceae	*Common vetch*	*Tsertsvela*

Table 2. Seed and forage legumes and traditional vegetables of Georgia.

Herbs are represented by numerous species (Tab.3) - parsley, coriander, tarragon, sweet basil, savory, gardencress pepperweed, dill, fennel, celery, garden lettuce, peppermint. Herbs are cultivated in small sections of house gardens even in urban settlements. Sometimes, people have herbs indoors in pots.

N	Latin name	Family	English common name	Georgian common name
	Anethum graveolens L.	Apiaceae	*Dill*	*Kama*
	Apium graveolens L.	Apiaceae	*Celery*	*Niakhuri*
	Artemisia dracunculus L.	Asteraceae	*Tarragon*	*Tarkhuna*
	Coriandrum sativum L.	Apiaceae	*Coriander*	*Kindzi*
	Foeniculum vulgare Mill.	Apiaceae	*Fennel*	*Didi Kama*
	Lactuca sativa L.	Asteraceae	*Garden lettuce*	*Salati*
	Lepidium sativum L.	Brassicaceae	*Gardencress pepperweed*	*Tsitsmati*
	Mentha piperata L.	Lamiaceae	*Peppermint*	*Pitna*
	Ocimum basilicum L.	Lamiaceae	*Sweet basil*	*Rehani*
	Petroselinum crispum (Mill.) A.W. Hill	Apiaceae	*Parsley*	*Okhrakhushi*
	Satureja hortensis L.	Lamiaceae	*Savory*	*Kondari*
	Trigonella caerulea (L.) Ser.	Fabaceae	*Blue fenugreek*	*Utskho Suneli*

Table 3. List of traditionally cultivated herbs in Georgia.

Fruits are valuable cultivars in Georgia. Wild and cultivated fruit crops reveal high species and genetic diversity in Georgia and represent rich material for future breeding activities. Many fruits have wild relatives representing the same species and direct ancestors of local cultivars (Tab.4,5).

2.2 Introduced cultivated plants

Georgia is located at the crossroads of Europe and Asia. Many cultivated plants have been introduced since ancient times to Georgia from other regions of the world (Javakhishvili 1987). Some introduced crops have become very popular and widespread. They are introduced from different countries. Such crops as cucumber (*Cucumis sativus*), found in Georgia since medieval times, eggplant (*Solanum melongena*), marigold (*Tagetes patula*), used in almost all traditional meals; and black pepper (*Piper nigrum*) were introduced from India. Watermelon (*Citrullus lanatus*) from South Africa was cultivated in the Caucasus since medieval times. Maize (*Zea mays*), sunflower (*Helianthus annuus*), tomato (*Solanum lycopersicum*), bean (*Phaseolus vulgaris*), pepper (*Capsicum annuum*), and potato (*Solanum tuberosum*) were introduced to Georgia from the Americas at about the same time as in Europe (Javakhishvili, 1987). Tea (*Camellia sinensis*) and citrus fruits (*Citrus limon, Citrus reticulata, Citrus sinensis*) came from China in the 1830s (Bakhtadze, 1947). *Nicotiana rustica*, (*tutuni* in Georgian) has been cultivated for a long time and is found in the most regions, including high mountain areas, of Georgia. *N. tabacum*, was introduced during the Soviet period and was cultivated in *kolkhozis* for commercial use.

Georgia has become a secondary centre of diversity for most of these crops. Landraces of bean, maize, potato, tomato, and cucumber that do not exist in their countries of origin can be found in Georgia. Bean (*Phaseolus* spp.; *Vigna* spp.) was introduced via Turkey to Guria and Samegrelo during the second half of the XVI century (Javakhishvili, 1987). At present,

61 varieties and 406 forms of common bean had originated in Georgia due to widespread distribution and hybridization of different species of bean: *Phaseolus vulgaris* L., *P. lunatus* L., *P. coccineus* L. (=*P. multiflorus* Lam.), *P. acutifolius* A. Gray, *Vigna radiata* (L.) R. Wilczek var. *radiata* (=*P. aureus* Roxb.), *V. angularis* (Willd.) Ohwi & H. Ohashi var. *angularis* (=*P. angularis* [Willd.] W. Wight) and *V. umbellata* (Thunb.) Ohwi & H. Ohashi (=*P. calcaratus* Roxb.) (Kobakhidze, 1965). Beans are cultivated in gardens in large amounts providing sufficient food for families and representing a cash crop for additional income. Diversity of beans remains high. Maize (*Zea mays* L.) was introduced to Georgia in 1633-1650 (Javakhishvili, 1987). The Georgian name *Simindi* originated from the old name for flour *Samindo* as flour was introduced earlier to Georgia than the initial cultivation of maize. Besides landraces such as 'Kazha simindi' from Svaneti there are some cultivars that originated in Georgia: 'Ajametis tetri', 'Abashis kviteli', 'Kartuli kruki', 'Gegutis kviteli', 'Imeruli hibridi' and 'Lomtagora'. Many cultivars were introduced from Russia, Hungary, Yugoslavia, etc., during Soviet time. The last introduction occurred in 2011, when the high-yield US corn hybrid 'Pioneer' was sown in Georgia. Corn had replaced common and Italian millet and is used as an everyday food, especially in western Georgia. **Potato (*Solanum tuberosum* L.)** was introduced to Georgia during the second half of the XIX century. Several landraces of high quality are grown in high-mountain regions: Svaneti, Racha-Lechkhumi, Khevsureti, Khevi and Adjara. Breeder's cultivars were introduced into lowland areas during Soviet time. Recently, genetically modified potato cultivars have been introduced in Georgia by international seed-distribution organizations. These modern cultivars have almost completely supplanted local landraces even in high mountain regions. Tea and citrus had high commercial value in Georgia, but in the 1990s these crops were abandoned and tea was not produced in Georgia until recently. Citrus (lemons, oranges and mandarins) were sold only in the local Georgian market. At present, this business is restored.

Information about introduced varieties has been published annually during the XX century. The latest official edition of the Catalogue of the Georgian Released Varieties of 1997 (published in 1996) listed 195 varieties of field and vegetable crops and 195 varieties of fruits. These varieties were part of the collections that existed at the end of the 1980s and beginning of the 1990s. At present, only a few of these varieties exist. The data about modern breeder's varieties introduced into Georgia during last decade are usually absent and a number of varieties have been cultivated in Georgia without going through the official procedures for release. Therefore, it is difficult to evaluate the diversity of recently introduced cultivars.

3. Crop wild relative assessment

The CWR are taxa related to species of direct socio-economic importance, which includes the progenitors of crops. According to modern concept of wild relatives, under CWR we should understand all species related to any cultivated plants, as well as to wild species of ornamental, food, fodder and forage, medicinal plants, condiments, forestry species and plants used for industrial purposes, such as oils and fibers i.e. to all plants of economic importance (Laguna, 2004). Although, "classical" definition of CWR is restricted only to species related to cultivated crops, including such important field crops as wheat, barley, rye, oats, sorghum, common and Italian millet, grain and forage legumes (such as *Phaseolus, Vicia, Vigna, Lens, Lathyrus, Cicer*) and some vegetables and industrial crops.

The flora of the Caucasus harbours a remarkable concentration of economically important plants, particularly CWRs such as cereals, legumes, fruits, vegetables, herbs and technical plants like flax. The list of CWRs in Georgia was published in Plant Genetic Resources (PGR) Forum - CWR Catalogue of Europe and the Mediterranean (Maxted et al., 2008). This catalogue listed 1784 species of vascular plants, representing 43% of the 4130 vascular plant species found in Georgia. These are mainly wild species that also have considerable economic importance providing food, fuel, timber, forage, hay and habitats for animal life. A large number of taxa used in folk and scientific medicine are also included among economically valuable plants. However, this list is not detailed enough to assess the economic value of CWRs representing the same species or direct ancestors of crop plants. There is no information on the status of endangered and endemic species. Thus, we developed a general description of vegetation types and separated CWR endemic species and species genetically closely related to crops.

3.1 Flora and vegetation

Georgia is located between 41°02′ and 43°34′ latitude north and between 40° and 46°43′ longitude east. It borders the Russian Federation to the north, Turkey and Armenia to the south, Azerbaijan to the east, and has approximately 310 km of coastline along the Black Sea to the west. Georgian territory (69.700 km^2) covers two separate mountain systems: the Greater Caucasus Range which trends north-west to east-southeast between the Black Sea and Caspian Sea; and the Lesser Caucasus Mountains, which run parallel to the greater range at a distance to the south that averages about 100 kilometres. Two thirds of the country is mountainous with an average height of 1.200 m.a.s.l., with the highest peaks of Mount Shkhara (5.184 m.a.s.l.) in the western Greater Caucasus and Mount Didi Abuli (3.301 m.a.s.l.) in the Lesser Caucasus. Colchis, Kartli and Alazeni valleys and Iori plateau represent intermontane lowlands located between these two mountain systems. Geologically, the Caucasus consists of Meso- to Cenozoic deposits. Ancient Precambrian and Paleozoic formations are rare (Neidze et al., 2008).

The Likhi Range divides the country into eastern and western halves that differ in climate and landscapes. Western Georgia has a humid subtropical climate with annual precipitation ranging from 1000–4000 mm. Temperatures fluctuate between the winter averages of 2.8° to 6.7° C and the summer averages of 22.7° to 23.8° C. Eastern Georgia has a more continental climate, due to the barrier of the Likhi Range, which bars the warm Black Sea winds from this area. The temperatures vary from the January averages of 0-2.2° C to the July mean of 27.8° C. Annual precipitation is considerably less in eastern Georgia and ranges from 400–1600 mm. Southern Georgia has a continental climate. The local winters are cold. The frosts are - 25° C and in July temperatures rise to 40° C. Annual rainfalls are usually less than 600-1000 mm (Neidze et al., 2008).

Soil types vary in Georgia. The most widespread types in the lowlands of western Georgia are bog, podzolic, red, yellow and hilly piedmonts, which are mainly acidic. Mountain-forest and mountain-meadow soils occur in higher elevations. Chestnut and chernozem soils are widespread in the lowlands of eastern Georgia and are characterized by neutral pH. Brown humid-sulphates, saline soils of steppes and semi-deserts, as well as intermediate forest-steppe and mountain-meadow soils occur in semi-desert areas. Alluvial soils are found along the rivers throughout Georgia. Brown soils are typical for the Georgian forest zone in the range of 800-2000 m a.s.l. (Sabashvili, 1970).

Western Georgia's landscape ranges from see-level swamps and lowland temperate rainforests to eternal subnival zone and glaciers, while the eastern part of the country contains temperate forests and semi-arid plains in lower elevations and alpine and subnival zones. Main rivers are R. Mtkvari, R. Rioni, R. Enguri and R. Alazani. There are 70 natural lakes and 11 artificial reservoirs. The lower section of the Rioni River is located in the Colchis valley and was naturally occupied by marshes and lagoons, but in 1960s this area was the site of a large reclamation-drainage project and it was converted to agricultural land. The majority of the forests that covered the Colchis plain are now virtually gone; the exceptions are those included in national parks and reserves. The Mtkvari River basin, which includes the major parts of southern and eastern Georgia, is drier. It is covered with semi-arid vegetation and temperate forests. Forests, in total, amount to 40% of Georgia's territory while the alpine zone accounts for roughly 10% of the land. Much of the natural habitat in the lowland areas of western Georgia has disappeared over the last 100 years because of agricultural development and urbanization.

The vegetation of Georgia belongs to three floristic provinces - Euxine, Caucasian and Armeno-Iranian (Takhtajian, 1986). The Euxine and Caucasian provinces belong to the Circumboreal Region, the Boreal Subkingdom and the Holarctic Kingdom and the Armeno-Iranian Province belongs to the Irano-Turanian Region, the Tethyan (Mediterranean) Subkingdom and the Holarctic Kingdom. There are following vegetation zones: 1. Colline zone (0-400 m a.s.l.), which includes coastal and halophytic habitats in western Georgia and dry open woodlands and semi-deserts in eastern Georgia; 2. Lower montane zone (400-800 m a.s.l.) is used as arable land. The natural vegetation in western Georgia is represented by small remnant areas of Colchic broad-leaved mixed forest. Oak-hornbeam forests and dry scrublands occur in eastern Georgia; 3. Middle montane zone (800-1500 m a.s.l.) is primarily used for agriculture. Broad-leaved mixed forests, mountain xerophytes scrublands, and mountain steppes are represented; 4. Upper montane zone (1200-2050 m a.s.l.) is covered by beech and broadleaf-coniferous mixed forests; 5. Subalpine zone (1900-2400[2500] m a.s.l.) is a treeline ecotone, with tall herbaceous vegetation, shrublands and polydominant subalpine grass and herb meadows used as pastures or arable land; 6. Alpine zone (2500–2900 m a.s.l.) has alpine meadows and snowbed communities. Vegetation is mostly used for grazing and is of considerably lower quality than the subalpine vegetation, both by biomass volume and typological diversity; 7. Subnival zone (2900-3300 m a.s.l.) is patchy highest limits of vegetation. 8. Nival zone (3300-5184 m a.s.l.) covered by glaciers. 9. Azonal vegetation type is represented by fragments of wetlands rich in boreal type flora, halophytic desert vegetation and rocky areas (Nakhutsrishvili, 1999).

Flora of Georgia is represented by 4,130 species of vascular plants. Among them are 79 pteridophytes, 17 gymnosperms and 4034 angiosperms (Nakhutsrishvili, 1999). The 10 leading families are Asteraceae (538 species), Poaceae (332 species), Fabaceae (317 species), Rosaceae (238), Brassicaceae (183), Scrophulariaceae (179), Apiaceae (177), Lamiaceae (149), Caryophyllaceae (135) and Liliaceae (129). High endemism is characteristic of the Caucasus and represents one of the world's hot spots of biodiversity. Out of all, 1304 (32.3%) species are endemics of the Caucasus ecoregion and 261 (6.6%) are endemics of Georgia (Schatz et al., 2009). There are 17 endemic genera in the flora of the Caucasus. Most of them are represented by one species: *Agasyllis latifolia* (M. Bieb.) Boiss., *Alboviodoxa elegans* (Albov) Woronow, *Charesia akinfievii* (Schmalh.) E. Busch, *Cladochaeta candissima* (M. Bieb.) DC., *Gadellia lactiflora* (M. Bieb.) Schulkina, *Mandenovia komarovii* (Manden.) Alava, *Paederotella pontica* (Rupr. ex Boiss.) Kem.-Nath., *Petrocoma hoefftiana* (Fisch.) Rupr., *Pseudobetckea caucasica* (Boiss.) Lincz., *Pseudovesicaria digitata* (C. A. Mey.) Rupr., *Sredinskya grandis* (Trautv.) Fed., *Symphyoloma graveolens* C. A. Mey., *Trigonocaryum involucratum* (Steven) Kusn., *Woronowia speciosa* (Albov) Juz. Two genera contain two species of each: *Chymsydia agasylloides* (Albov) Albov, *C. colchica* (Albov) Woronow, *Grossheimia macrocephala* (Muss.-Puschk. ex Willd.) Sosn. & Takht., *G. polyphylla* (Ledeb.) Holub. One endemic genus is represented by 5 species: *Kemulariella caucasica* (Willd.) Tamamsch., *K. rosea* (Steven ex M. Bieb.) Tamamsch., *K. abchasica* (Kem.-Nath.) Tamamsch., *K. tugana* (Albov) Tamamsch., and *K. colchica* (Albov) Tamamsch.

For conservation action to be effective, it is important to understand not just the needs of individual species, but also the context in which conservation efforts will need to take place. Therefore, it is important to evaluate the conservation values of the species that contribute most to human health and to develop conservation measures to avoid their extinction.

3.2 Diversity of crop wild relatives in Georgia

Flora of the Caucasus region is rich as there are high concentrations of economically important and edible plants, particularly wild crop relatives such as grapevine, wheat, barley, rye, oats, seed and forage legumes, fruits and vegetables. The Caucasus is considered to be the centre of evolution for many unique life forms and is a natural museum for rich genetic resources (Vavilov, 1992).

We identified the number of species of the genera that are traditional crops in Georgia (Tab.4). A total of 20 plant families, 76 genera and 479 species were identified as wild relatives of ancient crops in Georgia. Most of these plant species are closely related genetically to landraces and might be their progenitor species.

CWR are commonly defined in terms of wild species related to agricultural and horticultural crops. As such a broad definition of a CWR would be any wild taxon belonging to the same genus as a crop. A working definition of a CWR was provided by N. Maxted and colleagues (Maxted et al., 2006): "*A crop wild relative is a wild plant taxon that has an indirect use derived from its relatively close genetic relationship to a crop; this relationship is defined in terms of the CWR belonging to gene pools 1 or 2, or taxon groups 1 to 4 of the crop*".

According to gene pool concept three gene pools are distinguished as follows: (1) Primary Gene Pool (GP-1) within which GP-1A are the cultivated forms and GP-1B are the wild or weedy forms of the crop; (2) Secondary Gene Pool (GP-2) which includes the coenospecies (less closely related species) from which gene transfer to the crop is possible but difficult using conventional breeding techniques; (3) Tertiary Gene Pool (GP-3) which includes the species from which gene transfer to the crop is impossible, or if possible requires sophisticated techniques, such as embryo rescue, somatic fusion or genetic engineering. The taxon group concept is used to establish the degree of CWR relatedness of a taxon. Application of the taxon group concept assumes that taxonomic distance is positively related to genetic distance. CWR rank of taxon groups is defined as follows: (1) Taxon Group 1A - crop; (2) Taxon Group 1B - same species as crop; (3) Taxon Group 2 - same series or section as crop; (4) Taxon Group 3 - same subgenus as crop; (5) Taxon Group 4 - same genus; (6) Taxon Group 5 - same tribe but different genus to crop (Maxted et al., 2006).

Families	*Number of genera*	*Number of species*	*Genera with number of species*
Apiaceae	8	17	*Anethum (1), Apium (2), Carum (5), Coriandrum (1), Daucus (1), Foeniculum (1), Pastinaca (5), Petroselinum (1)*
Asparagaceae	1	3	*Asparagus (3)*
Asteraceae	3	16	*Cichorium (1), Lactuca (7), Scorzonera (8)*
Betulaceae	1	6	*Corylus (6)*
Brassicaceae	5	20	*Brassica (4), Lepidium (8), Raphanus (2), Rorippa (4), Sinapis (2),*
Cannabaceae	2	3	*Cannabis (2), Humulus (1)*
Chenopodiaceae	2	3	*Beta (2), Spinacia (1)*
Cornaceae	1	1	*Cornus (1)*
Fabaceae	10	154	*Cicer (1), Lathyrus (20), Lens (3), Lotus (5), Medicago (21), Onobrychis (19), Pisum (1), Trifolium (40), Trigonella (10), Vicia (34)*
Grossulariaceae	2	4	*Grossularia (1), Ribes (3)*
Juglandaceae	1	1	*Juglans (1)*
Lamiaceae	4	19	*Mentha (4), Origanum (1), Satureja (3), Thymus (11)*
Liliaceae	2	39	*Allium (36)*
Linaceae	1	12	*Linum (12)*
Moraceae	2	3	*Ficus (1), Morus (2)*
Poaceae	16	64	*Aegilops (7), Agropyron (2), Avena (8), Brachypodium (3), Cynosorus (2), Elymus (4), Elytrigia (9), Echinochloa (3), Hordeum (5), Hordelymus (1), Panicum (5), Psathyrostachis (1), Secale (5), Setaria (6), Sorghum (1), Taeniatherum (2)*
Punicaceae	1	1	*Punica (1)*
Rosaceae	12	110	*Amygdalus (1), Cerasus (4), Crataegus (8), Cydonia (1), Fragaria (3), Malus (1), Mespilus (1), Prunus (2), Pyrus (11), Rosa (30), Rubus (36), Sorbus (12)*
Staphyleaceae	1	2	*Staphylea (2)*
Vitaceae	1	1	*Vitis (1)*
Total: 20	76	479	

Table 4. List of wild relatives of ancient crops in Georgia.

Family	*Crop*	*Taxon*	*GP*	*TG*
Apiaceae	*Daucus carota*	*Daucus carota* L.	GP1B	TG1B
Apiaceae	*Coriandrum sativum*	*Coriandrum sativum* L.	GP1B	TG1B
Asparagaceae	*Asparagus officinalis*	*Asparagus caspius* Schult. & Schult. fil.	GP1B	TG1B
Asparagaceae	*Asparagus officinalis*	*Asparagus officinalis* L.	GP1B	TG1B
Asparagaceae	*Asparagus officinalis*	*Asparagus verticillatus* L.	GP1B	TG1B
Betulaceae	*Corylus avellana*	*Corylus avellana* L.	GP1B	TG1B
Betulaceae	*Corylus avellana*	*Corylus iberica* Wittm. ex Kem.-Nath.	GP2	TG2
Betulaceae	*Corylus avellana*	*Corylus colchica* Albov	GP2	TG2
Brassicaceae	*Brassica oleracea*	*Brassica juncea* (L.) Czern.	GP2	TG2
Brassicaceae	*Brassica oleracea*	*Brassica napus* L.	GP2	TG2
Brassicaceae	*Brassica oleracea*	*Sinapis arvensis* L.	GP2	TG2
Cannabaceae	*Cannabis sativa*	*Cannabis sativa* L.	GP1A	TG1A
Cannabaceae	*Humulus lupulus*	*Humulus lupulus* L.	GP1A	TG1A
Chenopodiaceae	*Beta vulgaris*	*Beta maritima* L.	GP2	TG2
Fabaceae	*Pisum sativum*	*Pisum elatius* M. Bieb.	GP1B	TG1B
Fabaceae	*Cicer arietinum*	*Cicer caucasica* Bornm.	GP2	TG2
Fabaceae	*Lathyrus sativus*	*Lathyrus tuberosus* L.	GP2	TG2
Fabaceae	*Lens culinaris*	*Lens nigricans* (M. Bieb.) Webb & Berth.	GP2	TG2
Fabaceae	*Lens culinaris*	*Lens ervoides* (Brign.) Grande	GP2	TG2
Fabaceae	*Lens culinaris*	*Lens culinaris* Medik. subsp. *orientalis* (Boiss.) Ponert	GP1B	TG1B
Fabaceae	*Vicia faba*	*Vicia johannis* Tamamsh.	GP2	TG2
Fabaceae	*Vicia faba*	*Vicia narbonensis* L.	GP2	TG2
Fabaceae	*Vicia sativa*	*Vicia sativa* L.	GP1A	TG1A
Grossulariaceae	*Ribes rubrum*	*Ribes alpinum* L.	GP2	TG2
Grossulariaceae	*Ribes rubrum*	*Ribes caucasicum* M. Bieb.	GP2	TG2
Juglandaceae	*Juglans regia*	*Juglans regia* L.	GP1A	TG1A
Lamiaceae	*Satureja hortensis*	*Satureja laxiflora* K. Koch	GP2	TG2
Lamiaceae	*Satureja hortensis*	*Satureja spicigera* (K. Koch) Boiss.	GP2	TG2
Linaceae	*Linum usitatissimum*	*Linum bienne* Mill.	GP1B	TG1B
Linaceae	*Linum usitatissimum*	*Linum usitatissimum* L.	GP1A	TG1A
Moraceae	*Morus alba*	*Morus alba* L.	GP1A	TG1A
Moraceae	*Morus nigra*	*Morus nigra* L.	GP1A	TG1A
Moraceae	*Ficus carica*	*Ficus carica* L.	GP1A	TG1A
Poaceae	*Triticum aestivum*	*Aegilops cylindrica* Host	GP1B	TG5
Poaceae	*Triticum aestivum*	*Aegilops triuncialis* L.	GP2	TG5
Poaceae	*Triticum aestivum*	*Aegilops tauschii* Coss.	GP1B	TG5
Poaceae	*Hordeum hexastichon*	*Hordeum bulbosum* L.	GP1B	TG2
Poaceae	*Hordeum distichon*	*Hordeum spontaneum* K. Koch	GP1B	TG1B
Poaceae	*Avena sativa*	*Avena barbata* Pott ex Link	GP2	TG2
Poaceae	*Avena sativa*	*Avena sterilis* L.	GP2	TG2
Poaceae	*Secale cereale*	*Secale strictum* subsp. *anatolicum* (Boiss.) K. Hammer	GP2	TG2
Poaceae	*Secale cereale*	*Secale strictum* subsp. *kuprijanovii* (Grossh.) K. Hammer	GP2	TG2
Poaceae	*Secale cereale*	*Secacle cereale* L. subsp. *segetale* Zhuk.	GP1A	TG1A
Poaceae	*Panicum miliaceum*	*Panicum capillare* L.	GP2	TG2
Poaceae	*Panicum miliaceum*	*Panicum sumatrense* Roth	GP2	TG2

Family	*Crop*	*Taxon*	*GP*	*TG*
Poaceae	*Panicum miliaceum*	*Panicum dichotomiflorum* Michx.	GP2	TG2
Poaceae	*Setaria italica*	*Setaria viridis* (L.) P. Beauv.	GP2	TG2
Poaceae	*Setaria italica*	*Setaria verticillata* (L.) P. Beauv.	GP2	TG2
Poaceae	*Setaria italica*	*Setaria glauca* (L.) P. Beauv.	GP2	TG2
Poaceae	*Setaria italica*	*Setaria intermedia* Roem. & Schult.	GP2	TG2
Punicaceae	*Punica granatum*	*Punica granatum* L.	GP1A	TG1A
Rosaceae	*Pyrus communis*	*Pyrus caucasica* Fed.	GP1B	TG1B
Rosaceae	*Pyrus communis*	*Pyrus balansae* Decne.	GP1B	TG1B
Rosaceae	*Malus domestica*	*Malus orientalis* Uglitzk.	GP2	TG2
Rosaceae	*Cydonia oblonga*	*Cydonia oblonga* Mill.	GP1B	TG1B
Rosaceae	*Prunus domestica*	*Prunus domestica* L. subsp. *insititia* (L.) C. K. Schneid.	GP1A	TG1A
Rosaceae	*Prunus domestica*	*Prunus spinosa* L.	GP1B	TG1B
Rosaceae	*Prunus cerasifera*	*Prunus cerasifera* Ehrh. var. *divaricata* (Ledeb.)L.H.Bailey	GP1A	TG1A
Rosaceae	*Cerasus avium*	*Cerasus avium* (L.) Moench	GP1B	TG1B
Rosaceae	*Cornus mas*	*Cornus mas* L.	GP1A	TG1A
Rosaceae	*Mespilus germanica*	*Mespilus germanica* L.	GP1A	TG1A
Rosaceae	*Rubus idaeus*	*Rubus idaeus* L.	GP1A	TG1A
Rosaceae	*Amygdalus communis*	*Amygdalus georgica* Desf.	GP2	TG2
Staphyleaceae	*Staphylea pinnata*	*Staphylea pinnata* L.	GP1A	TG1A
Staphyleaceae	*Staphylea colchica*	*Staphylea colchica* Steven	GP1A	TG1A
Vitaceae	*Vitis vinifera*	*Vitis vinifera* subsp. *sylvestris* (C.C.Gmel.) Hegi	GP1B	TG1B

Table 5. Gene pool and taxon group of wild relatives to Georgian ancient crops. GP- Gene Pool; TG-Taxon Group.

Thus, the combined use of the gene pool and taxon group concept proposed above provide the most pragmatic means available to determine whether a species is a CWR and how closely related a CWR is to its crop. We have determined 66 species of CWR belonging to 43 genera and 17 families, which can be assigned as Primary (GP-1) and Secondary Gene Pool (GP-2) and Taxon Group 1 and 2 (Tab.5). Seventeen (25.75%) are wild species but used as crops by collecting in the natural habitats and they were identified as GP1A. The same CWR species as crop were 19 (28.8%). Different species were 30 (45.45%), but representing direct progenitors whose genome is involved in the evolution of cultivars. Almost the same numbers were obtained during taxon group classification: TG1A - 17 species (25.75%), TG1B - 16 (24.25%), TG2 - 30 (45.45%), TG 5 - 3 (4.55%). The last 3 species belonging to the Taxon Group 5 are *Aegilops*, a wild relative of wheat. Goatgrass (*Aegilops tauschii*) is considered to be a donor of D genome of bread wheat genomic constitution = AABBDD (Petersen et al., 2006). The distribution area of this species is wide, however, D genomes of all forms of *T. aestivum* were found to be most closely related to accessions collected in Georgia, Armenia, Nakhitshevan and Azerbaijan (Dvorák et al., 1998). Thus, the germplasm of the populations of goatgrass in the South Caucasus needs conservation and should be preserved both *in situ* in protected areas and *ex situ* in seed collections.

Barley is one of the oldest crops to be domesticated from its wild progenitor *Hordeum spontaneum* (Badr et al., 2000, Kilian et al., 2006). We have found *H. spontaneum* in Georgia in three different places. This species was not included in the list of "Flora of Georgia' and it is a new species for Georgia. It is assumed that *H. spontaneum* might have evolved from *H. bulbosum* by fixation of the genes controlling self-compatibility and annual habit (Cass et al., 2005). This last species is widespread in Georgian regions.

Most fruit trees in Georgia are wild in forests and have cultivars domesticated from these wild ancestors. An economically important Georgian fruit crop is grape, which has a wild relative species *Vitis vinifera* subsp. *sylvestris.* We have found 9 populations of wild grapevine and conducted studies to reveal genetic and morphological relations between wild grapevine and landraces in Georgia (Ekhvaia et al., 2010).

Many fruits are domesticated in the Caucasus from wild ancestors representing Primary Gene Pool (GP-1B) to be the wild or weedy forms of the crops (Tab.5). The fruit crops (GP1A) and ancestor species (GP-1B) are the following: Pome fruits - pear (*Pyrus communis, P. caucasica*), apple (*Malus domestica, M. orietalis*), quince (*Cydonia oblonga*); stone fruits - plum (*Prunus domestica, P. domestica* var. *insititia, P. spinosa*), Myrobalan (*Prunus vachushti*), sour-plum (*Prunus cerasifera* var. *divaricata*), cherries (*Cerasus avium, C. vulgaris*), Cornel cherry (*Cornus mas*), medlar (*Mespilus germanica*), mulberry (*Morus alba, M. nigra*), pomegranate (*Punica granatum*); berries - red raspberry (*Rubus idaeus*), currant (*Ribes rubrum, R. nigra, R. alpinum, R. biebersteinii*), fig (*Ficus carica*), bladdernut (*Staphylea pinnata,* used flowers for marinade), and nuts - such as hazelnut (*Corylus avellana*), almond (*Amygdalus communis*), and walnut (*Juglans regia*), etc.

We evaluated CWRs endemic for the Caucasus (Tab.6). The endemic species of the same genus as crop were calculated. The number of endemic species from the total 479 CWRs of agricultural cultivars for food is 114 (23.8%).

4. Domestication events

4.1 Domestication of grapevine in Georgia

The grapevine was among the first fruits to be cultivated in Georgia (Javakhishvili, 1987). There are many arguments to confirm the fact that domestication events of grape took place in Georgia. One of the indicators is archaeological evidence. The 1,5-Myr-old petrified specimen of wild grapevine leaf was found in Georgia in the Meskheti province (Fig.4A). A confirmation for long lasting cultivation of grapevine in Georgia is archaeological remains of berries and seeds of domesticated grapes dated ~6.000 BC (in the vicinity of village Shulaveri [Fig.1], southeast Georgia; Ramishvili 1988). Other archaeological evidence of prehistoric winemaking was found in proximity of the Caucasus in northern Iran at the Hajji Firuz Tepe site in the northern Zagros Mountains dated to about 5.400–5.000 BC (McGovern, 2003) and in the Levant where archaeological findings are dated from ca. 4.000-3.200 BC (Zohary & Hopf 2000). Georgian traditions based on winemaking and grape culture to a high degree might be considered to be a second indicator of Caucasian origin of the grapevine. However, the primary scientific argument should be premised on N. I. Vavilov's (1992) idea that the centres of origin of cultivars should be characterized by high genetic and morphological variability of both wild and cultivated taxa. Five hundred is a very high number of known autochthonous grapevine varieties found in such a small territory (Javakhishvili, 1987; Ketskhoveli et al., 1960). These cultivars showed high morphological variability of leaf form, colour and shape of berries and shape and structure of pips. They are adapted to wide array of climatic conditions (Ketskhoveli et al., 1960). Each province of Georgia possesses unique grapevine cultivars adapted to local climate. The landraces of western Georgia grow in humid subtropical climate and other cultivars are adapted to moderate climates in eastern Georgia. Several local cultivars are growing in high mountain regions in Meskheti and Svaneti up to 1500 m a.s.l.

Besides the cultivars, there is high morphological and genetic diversity of wild grapevine populations in the Caucasus. All five haplotypes detected by using cpDNA microsatellite markers have been found in the Caucasian ecoregion suggesting that this area is possibly the centre of origin of both wild and cultivated grapevines (Grassi et al., 2006). However, only one provenance from Georgia has been analyzed in this study despite the number of populations of wild grapevine found in Georgia today that display morphological diversity (Ramishvili, 1988). We carried out a detailed comparative morphometric study of nine populations of wild grapevine, *V. vinifera* subsp. *sylvestris*, growing in the four river basins of the Ajaristskali, Mtkvari, Alazani and Iori located in western, southern and eastern Georgia.

Family	*Taxon*	*Family*	*Taxon*
Apiaceae	*Carum alpinum* (M. Bieb.) Benth. & Hook. ex B. D. Jacks.	Lamiaceae	*Thymus nummularius* M. Bieb.
Apiaceae	*Carum grossheimii* Schischk.	Lamiaceae	*Thymus tiflisiensis* Klokov & Des. - Shost.
Apiaceae	*Carum porphyrocoleon* (Freyn & Sint.) Woronow	Liliaceae	*Allium albovianum* Vved.
Apiaceae	*Pastinaca armena* Fisch. & C. A. Mey.	Liliaceae	*Allium candolleanum* Albov
Apiaceae	*Pastinaca aurantiaca* (Albov) Kolak.	Liliaceae	*Allium gramineum* K. Koch
Apiaceae	*Pastinaca pimpinellifolia* M. Bieb.	Liliaceae	*Allium gunibicum* Miscz. ex Grossh.
Asparagaceae	*Asparagus caspius* Schult. & Schult. fil.	Liliaceae	*Allium kunthianum* Vved.
Asteraceae	*Scorzonera charadzeae* Papava	Liliaceae	*Allium leucanthum* K. Koch
Asteraceae	*Scorzonera czerepanovii* R. Kam.	Liliaceae	*Allium otschiauriae* Tscholokaschvili
Asteraceae	*Scorzonera dzhawakhetica* Sosn. ex Grossh.	Liliaceae	*Allium ponticum* Miscz. ex Grossh.
Asteraceae	*Scorzonera ketzkhowelii* Sosn. ex Grossh.	Liliaceae	*Allium szovitsii* Regel
Asteraceae	*Scorzonera kozlowskyi* Sosn. ex Grossh.	Linaceae	*Linum hypericifolium* Salisb.
Asteraceae	*Scorzonera seidlitzii* Boiss.	Poaceae	*Elymus buschianus* (Roshev.) Tzvelev
Betulaceae	*Corylus abchasica* (Kem.-Nath.) Kem.-Nath.	Poaceae	*Elymus troctolepis* (Nevski) Tzvelev
Betulaceae	*Corylus colchica* Albov	Poaceae	*Elytrigia gracillima* (Nevski) Nevski
Betulaceae	*Corylus egrissiensis* Kem.-Nath.	Poaceae	*Elytrigia sinuata* (Nevski) Nevski
Betulaceae	*Corylus imeretica* Kem.-Nath.		
Betulaceae	*Corylus kachetica* Kem.-Nath.	Poaceae	*Secale strictum* subsp. *anatolicum* (Boiss.) K. Hammer
Betulaceae	*Corylus x fominii* Kem.-Nath.	Poaceae	*Secale strictum* subsp. *kuprijanovii* (Grossh.) K. Hammer
Fabaceae	*Cicer caucasicum* Bornm.	Rosaceae	*Amygdalus georgica* Desf.
Fabaceae	*Lathyrus colchicus* Lipsky	Rosaceae	*Crataegus caucasica* K. Koch

Family	*Taxon*	*Family*	*Taxon*
Fabaceae	*Lathyrus cyaneus* (Steven) K. Koch	Rosaceae	*Crataegus eriantha* Pojark.
Fabaceae	*Lotus caucasicus* Kuprian. ex Juz.	Rosaceae	*Pyrus demetrii* Kuthatheladze
Fabaceae	*Medicago dzawakhetica* Bordz.	Rosaceae	*Pyrus eldarica* Grossh.
Fabaceae	*Medicago hemicycla* subsp. *medidaghestanica* Sinskaya	Rosaceae	*Pyrus fedorovii* Kuthatheladze
Fabaceae	*Onobrychis angustifolia* Chinth.	Rosaceae	*Pyrus georgica* Kuthatheladze
Fabaceae	*Onobrychis biebersteinii* Sirj.	Rosaceae	*Pyrus ketzkhovelii* Kuthatheladze
Fabaceae	*Onobrychis cyri* Grossh.	Rosaceae	*Pyrus oxyprion* Woronow
Fabaceae	*Onobrychis grossheimii* Kolak. ex Fed.	Rosaceae	*Pyrus sachokiana* Kuthatheladze
Fabaceae	*Onobrychis iberica* Grossh.	Rosaceae	*Pyrus salicifolia* Pall.
Fabaceae	*Onobrychis kachetica* Boiss. ex Huet.	Rosaceae	*Pyrus takhtadzhianii* Fed.
Fabaceae	*Onobrychis kemulariae* Chinth.	Rosaceae	*Rosa buschiana* Chrshan.
Fabaceae	*Onobrychis komarovii* Grossh.	Rosaceae	*Rosa didoensis* Boiss.
Fabaceae	*Onobrychis meschetica* Grossh.	Rosaceae	*Rosa doluchanovii* Manden.
Fabaceae	*Onobrychis oxytropoides* Bunge	Rosaceae	*Rosa ermanica* Manden.
Fabaceae	*Onobrychis petraea* (M. Bieb. ex Willd.) Fisch.	Rosaceae	*Rosa galushkoi* Demurova
Fabaceae	*Onobrychis sosnowskyi* Grossh.	Rosaceae	*Rosa hirtissima* Lonacz.
Fabaceae	*Onobrychis transcaucasica* Grossh.	Rosaceae	*Rosa irysthonica* Manden.
Fabaceae	*Trifolium fontanum* Bobrov	Rosaceae	*Rosa kozlowskii* Chrashan.
Fabaceae	*Trifolium ruprechtii* Tamamsch. & Fed.	Rosaceae	*Rosa marschalliana* Sosn.
Fabaceae	*Trifolium sintenisii* Freyn	Rosaceae	*Rosa oplisthes* Boiss.
Fabaceae	*Vicia abbreviata* Fisch. ex Spreng.	Rosaceae	*Rosa ossethica* Manden.
Fabaceae	*Vicia alpestris* Steven	Rosaceae	*Rosa oxyodon* Boiss.
Fabaceae	*Vicia antiqua* Grossh.	Rosaceae	*Rosa prilipkoana* Sosn.
Fabaceae	*Vicia caucasica* Ekutim.	Rosaceae	*Rosa pulverulenta* M. Bieb.
Fabaceae	*Vicia ciliatula* Lipsky	Rosaceae	*Rosa teberdensis* Chrshan.
Fabaceae	*Vicia grossheimii* Ekutim.	Rosaceae	*Rosa transcaucasica* Manden.
Fabaceae	*Vicia iberica* Grossh.	Rosaceae	*Rosa tuschetica* Boiss.
Fabaceae	*Vicia sosnowskyi* Ekutim.	Rosaceae	*Sorbus buschiana* Zinserl.
Grossulariace-ae	*Ribes biebersteinii* Berland. ex DC	Rosaceae	*Sorbus caucasica* Zinserl.
Lamiaceae	*Satureja bzybica* Woronow	Rosaceae	*Sorbus caucasigena* Kom.
Lamiaceae	*Thymus caucasicus* Willd. ex Ronniger	Rosaceae	*Sorbus colchica* Zinserl.
Lamiaceae	*Thymus collinus* M. Bieb.	Rosaceae	*Sorbus migarica* Zinserl.
Lamiaceae	*Thymus coriifolius* Ronniger	Rosaceae	*Sorbus subfusca* (Ledeb.) Boiss.
Lamiaceae	*Thymus grossheimii* Ronniger	Rosaceae	*Sorbus fedorovii* Zaikonn.
Lamiaceae	*Thymus karjaginii* Grossh.	Rosaceae	*Sorbus velutina* (Albov) C.K. Schneid.
Lamiaceae	*Thymus ladjanuricus* Kem.-Nath.	Staphyleaceae	*Staphylea colchica* Steven

Table 6. One hundred fourteen endemic species of the Caucasus ecoregion related to the ancient crops in Georgia (Schatz et al., 2009).

The results reveal high morphological diversity of wild grapevine growing in Georgia. Morphological characters such as shape of leaf blade, number of lobes, pubescence type, coloration of internodes, leaves and berry skin, leaf vein lengths and angles between them and form of petiole sinus show high variability both within and among populations. The variability was related to the skin colour of berries. Some wild grapes had white berries, most had blue-black coloration. White-fruited phenotype is considered to be determined by the variation present in the gene VvmybA1, a transcriptional regulator of anthocyanin biosynthesis (This et al., 2006). The wild ancestor, however, should be considered to be black colour grapevine, which is most common in Georgian wild populations, eventhough the mutation leading to the white-fruited wild grapevine has been found in other Georgian wild populations (Ramishvili, 1988). On a phenotypical basis of our investigation (Ekhvaia & Akhalkatsi, 2010) it can be confirmed that the infraspecific evolution of *Vitis vinifera* subsp. *sylvestris* has produced three population groups south of the Great Caucasus Range. Overall, there are three phenetically distinct morphometric groups of western, southern and eastern Georgian wild grapevines. These three groups can be readily distinguished by the length of main leaf veins and lengths of nectaries in male flowers. This conclusion differs from the classical classification of Georgian wild and cultivated grapevine (Ramishvili, 1988) that considers two centres of origin of grapes in the South Caucasus region: (1) an Alazani origin with whole eastern and southern Georgia and adjacent territories of Azerbaijan and Armenia, and (2) a Colchic centre of origin which includes the entire western Georgian region with the Black Sea coastal zone. Our data clearly show a separated group in the southern Georgian population located in the territory of historical Tao-Klarjeti, a region of Georgia with many aboriginal grapevine cultivars. Therefore, it is of high importance to study aboriginal grape varieties in the place of its supposed domestication and to determine genetic relations among native grapevine cultivars and local wild populations.

A B

Fig. 4. A- The 1,5-Myr-old petrified specimen of wild grapevine leaf. National Museum of Akhaltsikhe, Georgia; B-Stone carving on the medieval church Ananuri, Mtskheta-Mtianeti region, Georgia. Photos by Maia Akhalkatsi.

Molecular study based on nuclear microsatellite (SSR) markers revealed close genetic relationships between wild grape and local cultivars in Georgia (Ekhvaia et al., 2010, 2011). Twenty-four Georgian autochthonous and 45 accessions of wild *V. vinifera* subsp. *sylvestris* were analyzed at 17 microsatellite loci (VrZAG21, VrZAG47, VrZAG62, VrZAG64, VrZAG79, VrZAG83, VVMD7, VVMD24, VVMD25, VVMD27, VVMD28, VVMD32, VVMD34, VVS2, VVS4, scu04vv and scu14vv). Six accessions of the American rootstocks -

Fercal ('Cabernet -Sauvignon' x *Vitis berlandieri*), Telecki 5C (*V. berlandieri* x *V. riparia*), Malegue 44-53 (*V. riparia* X [*V. cordifolia* X *V. rupestris*]), Couderc 3309 (*Riparia tomenteuse x Rupestris* Martin), cultivar *V. labrusca* 'Odessa' and *V. riparia* naturalized in Georgia were used as outgroup. Thirty-seven accessions of wild grapevine were collected from different regions of Georgia and 8 wild accessions were sampled in Turkey's Artvin province adjacent to Georgia. All individuals within the studied populations of wild grapevine were identified as dioecious plants with male or female flowers. Genotype analysis at the most studied loci showed that Georgian cultivated and wild grapevine was characterized by high level of genetic variability. Genetic structure also was analyzed using F statistics. The low level of genetic differentiation (F_{st}=0.03) between Georgian cultivated and wild grapevines demonstrates that *in situ* domestication of wild germplasm took place within local populations. This means that autochthonous Georgian cultivars should be originated from local wild grapevine (Ekhvaia et al., 2010).

The dendrogram generated using Dice coefficient identified eight major clusters within the 75 different genotypes defined at 0.22 similarity level (Fig.5). Clusters A consist of 13 Georgian local cultivars and 8 wild grapevine accessions from different regions of Georgia. One example confirming the genetic linkage between cultivated and local wild grapevine is placement of the famous Georgian red cultivar 'Tavkveri' in cluster A, where it is closely linked to the wild accession WT46Tbilisi5 (GS value 0.96; Fig. 5) due to identical alleles at 33 out of 34 alleles. Thus, the hypothesis that this cultivar could have been selected from local wild grapevine can be considered, especially because like wild grapevine 'Tavkveri' is characterized by presence of functionally female flowers. The fact that the 5 ancient Georgian cultivars 'Chvitiluri', 'Kachichi', 'Shonuri', 'Saperavi' and 'Uchakhardani' fall within cluster B (Fig.5), which mainly contains wild accessions allowed us to suppose that these cultivars were derived from the earliest local domestication events. Cluster G shows the genetic similarity of most ancient Georgian cultivars 'Meskhuri Shavi' and 'Krikina' adapted to high mountain climate conditions in Meskheti.

In conclusion, it should be mentioned that the Georgian cultivated and wild grapevines represent a unique and interesting genetic resources that is characterized by a high similarity level between wild and cultivated grapevines. The admixture found among local Georgian cultivars and wild grapevine indicates the possibility that these cultivars are derived from ancestral domestication of local wild types. Thus, the obtained data are supporting that Georgia is one of the oldest centres of domestication of grapevine and harbour of valuable genetic resources for grape breeding.

It should be mentioned that wild grapevine populations occurring nowadays on the territory of Georgia are threatened by different impacts in their natural habitats and need to be protected. The confirmation of threatened status of the Georgian wild grapevine might be detected low level of heterozygous individuals found for the most of the studied loci, which reflects the isolated status and the reduced number of individuals in the wild populations. Therefore, it is necessary to conserve wild forms and aboriginal cultivars of grape for the maintenance of genetic variability and to avoid genetic erosion of valuable genetic resources for grape breeding in Georgia.

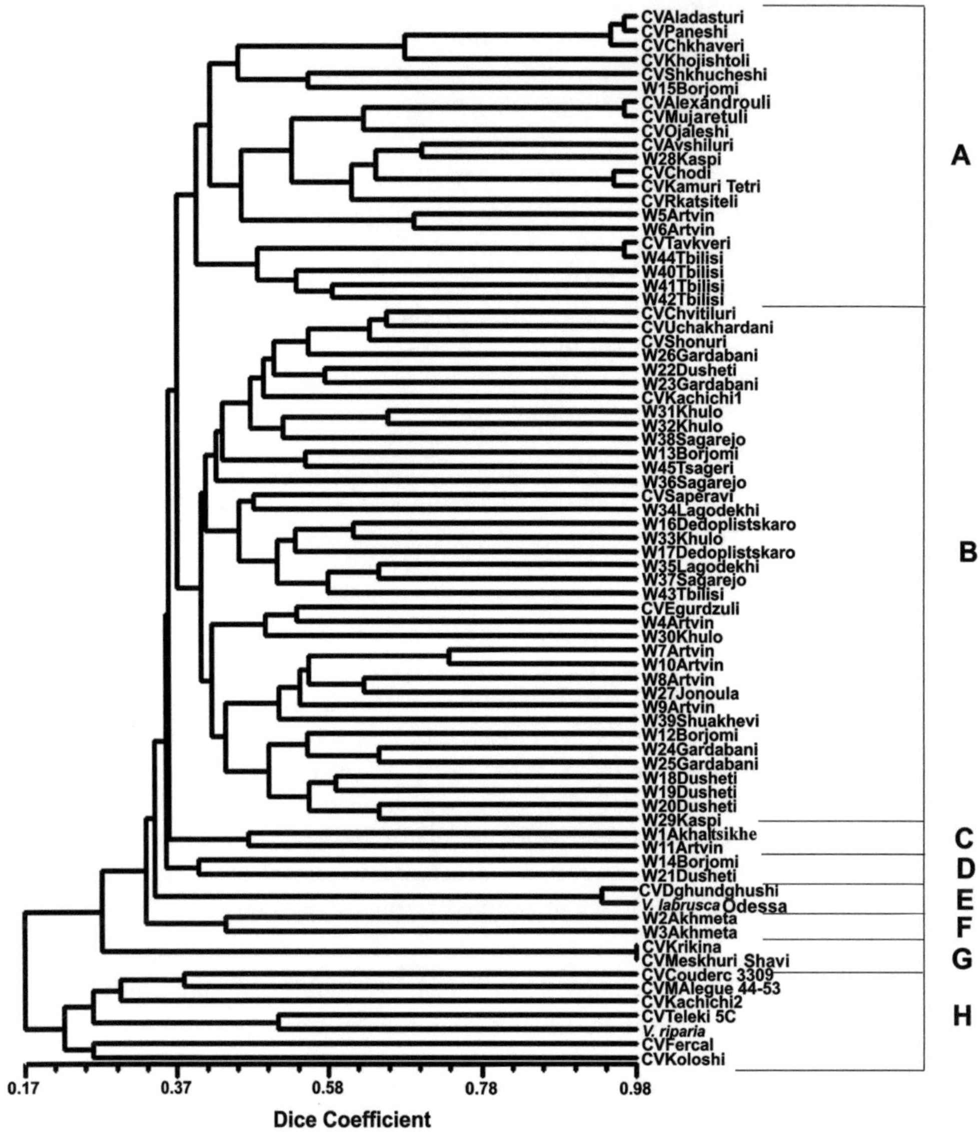

Fig. 5. Dendrogram of 75 accessions: Twenty-four Georgian local cultivars (*Vitis vinifera* subsp. *vinifera)*, 45 Georgian wild grapevine (*V. vinifera* subsp. *sylvestris*), 4 rootstock cultivars (Couderc 3309, Fercal, MAlegua 44-53 and Teleki 5C), introduced cultivar *V. labrusca* 'Odessa' and naturalized *V. riparia* constructed by unweighted arithmetic average clustering (UPGMA) method based on Dice's coefficient of shared SSR polymorphisms.

4.2 Domestication of pear in Georgia

Many pear cultivars occur in Georgia from pre-historic period indicating the early domestication event of this cultivated fruit tree (Javakhishvili, 1987). In total, 11 species of wild pear occur in Georgia (Kuthatheladze, 1980). They are distributed in different regions of Georgia, what is caused by the variable geographical relief and habitat diversity of the country. *Pyrus caucasica* Fed., the endemic species of the Caucasus is most widespread among the wild pears of Georgia and is considered as main progenitor species of local pear cultivars (Khomizurashvili, 1973). Moreover, *P. caucasica* and *P. pyraster* (L.) Burgsd. are regarded as the main wild progenitors, from which the cultivated European pear (*P. communis* L.) has probably evolved (Zohary & Hopf, 2000; Volk et al., 2006; Yamamoto & Chevreau, 2009).

According to the literature data (Khomizurashvili, 1973) introduced cultivars of pear from Europe and Russia appeared in Georgia at the end of 19th century and before there were existed only the local cultivars. However, we assume that the process of cultivar introduction might have started much earlier, as Georgia had cultural contacts to Asian and European countries since antique period. It is also remarkable that Greece is considered to be a first provider of selective cultivars of pear to the ancient world (Jackson, 2003) and earliest relations between Georgia and Greece should be considered as possible way of introduction of European pear cultivars in the Caucasus.

The local Georgian names of the cultivated pear *Mskhali* and wild Caucasian pear *Panta* exists in all Georgian dialects; they do not have analogues in any other languages (Javakhishvili, 1987). The Georgian names of cultivated and wild pears are linked with geographic objects such as mountains (Skhaltbis Range in Kartli, Mt. Mskhal-Gori in Kakheti's Kavkasioni), rivers (R. Skhaltba), or villages (Pantiani, Skhalta, Skhlobani, etc.; Javakhishvili, 1987). The name of wild Caucasian pear *Panta* is used among cultivars called 'Panta Mskhali', i.e. cultivar with name of wild pear. Moreover, the classification of Georgian pear cultivars (Khomizurashvili, 1973) contains a group of landraces with the same name. This classification system divides Georgian cultivars into four groups: 'Gulabi', 'Panta Mskhali', 'Kalos Mskhali', and 'Khechechuri'. The name of each group represents the name of a cultivar, which is considered to be a typical representative of a group. In the 'Gulabi' group, there are included both local and introduced cultivars to have most high economic values, big juicy fruits with sweet taste. The 'Panta Mskhali' group contains local varieties with small fruits becoming black after maturation. This is a character feature of wild Caucasian pear. The 'Kalos Mskhali' group includes local cultivars having bigger fruits than the second group. The 'Khechechuri' group matures in late autumn with juicy fruits containing a big amount of stone cells. According to N. Khomizurashvili (1973), the last three groups are originated by direct domestication of wild pear in Georgia. Although, some signs of selective breedings are remarkable as well. Relationships between wild *P. caucasica* and local cultivars are mirrored by a high morphological variability of leaf and fruit forms. This idea was for the first time confirmed by statistical methods of taxonomic identification and relationships among taxa in our study (Asanidze et al., 2011). We decided to conduct comparative morphometric study of cultivars recently occurred in Georgia and reveal their relationship to local wild pear species. The results have to determine local cultivars originated by direct domestication events in ancient time and discriminate from cultivars, which will have relationships with other wild species - *P. pyraster* or *P. pyrifolia,* considered as wild ancestors of European and Far East pear cultivars respectively.

We carried out the investigation to determine morphological characteristics of leaves, young shoots and fruits differentiating local and introduced pear cultivars of Georgia to reveal the relationships between cultivars and wild ancestor species of pear by statistical methods used in plant morphology (Asanidze et al., 2011). A total of 214 wild and cultivated pear trees have been sampled in natural habitats, living collections and peasant grounds in different regions of Georgia. Wild pear species were determined according to Sh. Kuthatheladze (1980). The pear accessions evaluated in this study consisted of Caucasian endemic *P. caucasica* Fed. (=*P. communis* subsp. *caucasica* (Fed.) Browicz; N=100), *P. balansae* Decne. (= *P. communis* L.; N=8) and *P. pyraster* (L.) Burgsd. (=*Pyrus communis* L. subsp. *pyraster* (L.) Ehrh.; N=3), which has been obtained from Germany, Hessen, in surrounding of v. Erda. Eighty-one individuals of 26 Georgian local and 22 individuals of 9 introduced cultivars (total 103 individuals of 35 cultivars) have been collected. Some of them are sampled in the collection of the Institute of Horticulture, Viticulture and Oenology, village Skra, Gori district, Georgia; local cultivars were sampled in the collection of aboriginal cultivars of Biological Farming Association Elkana in village Tsnisi, Akhaltsikhe district, Georgia. Many local cultivars are collected in peasant house gardens in different regions of Georgia. The individuals were evaluated by 27 morphological traits, which included one landmark analysis data, 12 leaf and shoot descriptors and 14 fruit descriptors. Morphological characters have been taken as recommended by International Union for the Protection of New Varieties of Plants (UPOV, 2000) for *P. communis* and J. Voltas and colleagues (Voltas et al., 2007), which delimited differences between wild and cultivated taxa of the genus *Pyrus* based on morphometric analysis. In total 21 morphological traits have been analysed by multivariate analysis. The Principal Components Analysis and Hierarchical Cluster Analysis (HCA) methods revealed the distance or similarity measure to be used in clustering with the Ward's method as amalgamation rules. According to HCA's results, pear cultivars, analysed in this study, are clustered into two groups (Fig.6). The first group **A** contains local cultivars related to *P. caucasica* and *P. balansae* by 21 morphological traits of leaves and fruits. Especially close Euclidean similarity distance is revealed between wild Caucasian pear, *Panta* in Georgian and a cultivar named 'Panta mskhali', which confirms etymological and taxonomic similarity within these taxa. *P. balansae* shows very close similarity distance with 'Tsvrili mskhali' and 'Korda'. Very closely related group of local cultivars to wild Georgian pears contains: 'Bebani', 'Samariobo', 'Tavrejuli', 'Kvichicha', 'Khinos mskhali' and 'Akiro'. The other group: 'Shavmskhala', 'Nenes mskhali', 'Borbala', 'Majara', 'Shakara' and 'Kartuli mskhali', is more distanced from wild pears, but located in the same cluster. We assume that these local cultivars must have been originated by direct domestication of wild Caucasian and Balanse's pears in Georgia.

The second cluster B (Fig.6) contains introduced cultivars of pear originated in European countries and some old Georgian cultivars. The group B reveals relationship with wild pear - *P. pyraster,* which is distributed in Europe and does not reach Georgian territory. The area of distribution is up to the middle of Turkey. The most cultivars from intermediate group B are more widespread in Georgia than the local cultivars originated by direct domestication of wild pears from group A. Two local pears 'Gulabi' and 'Khechechuri' are the most widespread among all local pears of Georgia and there are two or more varieties of them in each localities of the country. Moreover, Georgian name of cultivar 'Gulabi', which is also used to be a name of local pear group in classification of N. Khomizurashvili (1973), means pear in Persian. We suggest that local cultivars from cluster B (Fig.6) might have appeared in Georgia very long time ago and were improved by local population using breeding procedures.

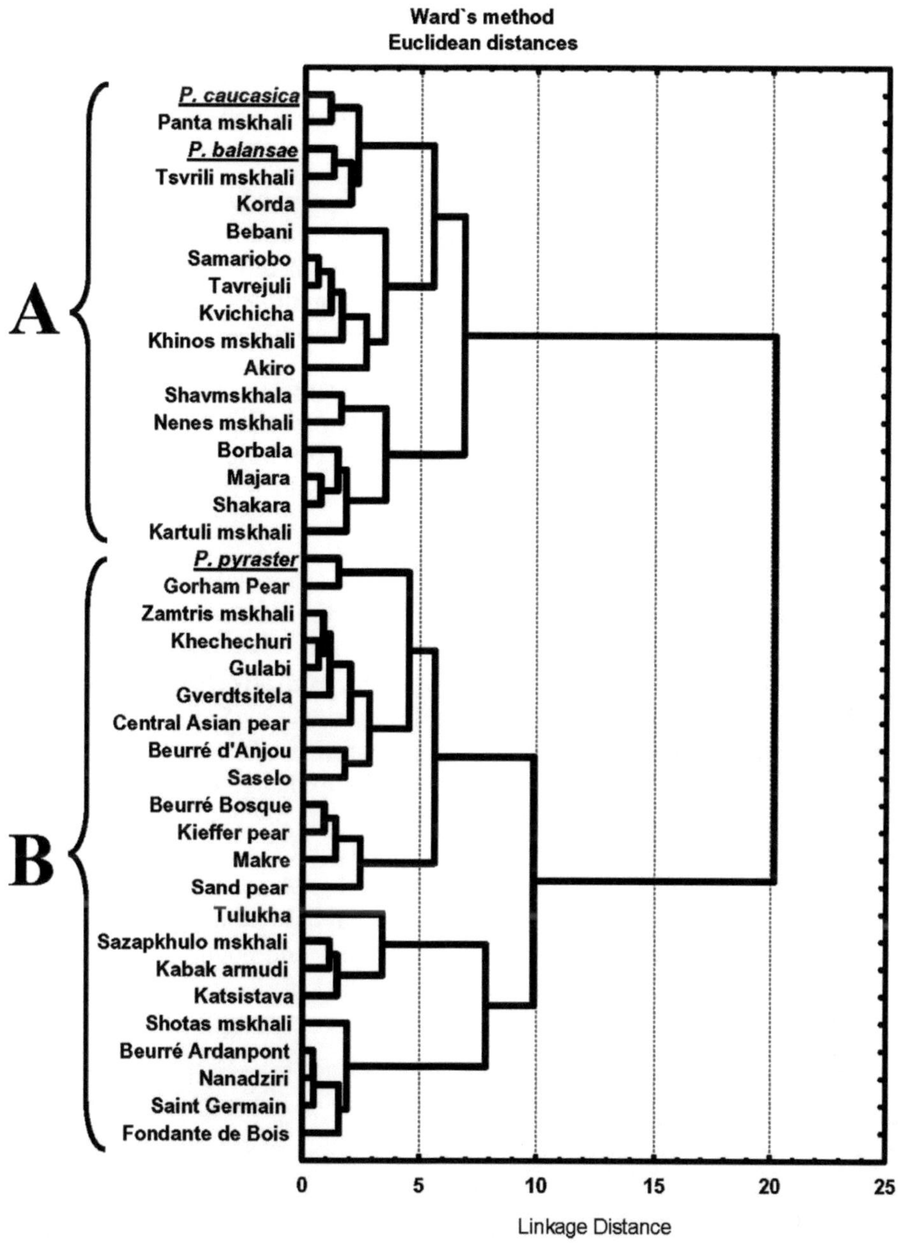

Fig. 6. HCA dendrogram of Euclidean distance with the Ward's method showing the relationships between the 35 cultivars and 3 wild species of pear based on 20 morphological traits of leaf, shoot and fruit and 20 landmark harmonics of mature leaf; The taxa in the dendrogram are clustered into two main groups - A and B. (N=214).

Leaf margin shape is the main morphological trait that differentiates Caucasian pear (*P. caucasica*) from European pear (*P. communis*). Leaf margins are entire in *P. caucasica* and serrate in *P. communis*. This theory was proved by the statistical analysis of the collected samples for the present study. Nowadays, *P. pyraster* is considered as the wild pear of Europe and cultivars are named as *P. communis* (Yamamoto & Chevreau, 2009). 'Communis' or the 'Common pear' group of cultivars has become the name of the cultivated pears of Europe, however, the structure and the diversity of the wild and cultivated pears of this group is not studied in details and needs further genetic and molecular investigations.

Thus, the results of this study have shown that some local cultivars of Georgia are direct domesticated from the native wild pear species - *P. caucasica* and *P. balansae*. The other local cultivars might be obtained due to selective works by breeding of local landraces with introduced cultivars from different countries in historically different periods. The molecular study of these taxa will clear in more details origin of these cultivars.

The results confirm the hypothesis that some local cultivars of Georgia are directly domesticated from the native wild pear species - *P. caucasica* and *P. balansae*. The other local cultivars might be obtained due to selective works by breeding of local landraces with introduced cultivars from different countries in historically different periods.

5. Traditional sustainable use of ancient crops

Since ancient times, agriculture in Georgia has been divided in two zones: lowlands (0-1300 m a.s.l.) and high-mountains (1300-2200 m a.s.l.). This classification was based on production of wine (Javakhishvili, 1987). Winemaking was always the major branch of agriculture in Georgia. Wine was exported from Georgia since ancient times. The vineyards were cut off to reduce income for exporting the wine in neighbour countries during the occupation of the country by the Muslim nations. The exchange of agricultural products took place between lowland and high-mountain regions and not only within the Georgians, but also with North Caucasians. This tradition remained till the end of 20th century when Dagestanian people from Didoeti region visited Kakheti lowland in eastern Georgia in late autumn. They have exchanged agricultural products from high-mountains to lowland crops. In 20th century, they brought from Dagestan cows, cheese and potato and have exchanged them on Kakhetian wine, schnapps from grapevine called *chacha*, bread wheat flour, fruits and vegetables.

A B C

Fig. 7. A - *Satsnekheli* - stone construction for the pressing of grapes in Nekresi monastery (IV century AD), Kakheti; B - Clay vessels for wine storage *Kvevri*, Ikalto monastery (XII century AD), Kakheti; C- Red wine *Saperavi* and *boghlortso* (wheat bread into red wine) in clay cups on Georgian table. Photos by Maia Akhalkatsi.

The information about the traditional use of wine is remained in folklore and ethnographic studies collecting this knowledge. The problem in reducing of written information was again the wars, destruction of settlements and burning of manuscripts during historical times. Therefore, the information on traditional agriculture is based on both literature data and interviews of local people obtained during our field trips in different provinces of Georgia. The ancient stone construction *satsnekheli* for the pressing of grape might be found near many historical monuments (Fig.7A) even in ancient caves of Chachkari near Vardzia monastry complex. Each family in lowland regions have a room for winemaking in houses called *marani* in which the clay vessels are buried in soil named *kvevri* (Fig.7B) where the wine is made and stored by traditional Georgian technology. One of *kvevri* in each *marani* was called *zedashe* and the wine in it might be used only in religious rituals since it belonged to the God. *Zedashe* was filled by schnapps called *Araki* derived from different fruits e.g. wild pear - *P. caucasica,* in high mountain regions and used for rituals (Fig. 8A-C). Wine and bread are ritual accessories of Christian religion and first cross entered in Georgia by St. Nino from Cappadocia in 4th century AD was made from vines. However, the folklore knowledge let us know that grapevine was a ritual plant in ancient religion and represented a tree of the God of sun. The remnants of grapevine images are often demonstrated as stone carvings on Christian churches (Fig.4B) and ancient golden and silver cups and Jewellery. Wine was used not only as alcohol drink but in religious rituals and as traditional food. *Boghlortso* (Fig.7C) was a healthy food prepared by wheat bread placed within a red wine and used by the whole family members including children and chronic invalid and consumptive people.

Bread is served with all Caucasian meals. It is the same ritual food like wine in Georgia. Two landraces of bread wheat - *Triticum aestivum* var. *erythrospermum* and *T. aestivum* var. *lutescens* are used for religious rituals in Svaneti (Girgvliani, 2010). The flour of these cultivars is preserved separately from other reserves of bread wheat flour and used on religious holydays. Milled faba bean and kenaf seeds are added to the bread flour for baking ritual bread. There are barley cultivars: *H. vulgare* var. *pallidum* in Svaneti and *H. vulgare* var. *nutans* in Meskheti, used for traditional bread preparation added to the *T. carthlicum* 'Dika' flour.

A B C

Fig. 8. A- Murtaz Chankseliani collected wild pear (*Pyrus caucasica*) in forest, R. Kheledula gorge, Lower Svaneti; B - Wild Caucasian pear for preparing of alcohol schnapps *Araki;* C - Distillation equipment for preparing of pear schnapps. Photos by Maia Akhalkatsi.

The healthy quality of food in Georgia is connected with usage of fruit and herb sauces for roasted and fried meet. Sour plum sauce (*P. cerasifera* var. *divaricata*) is always added to spit-roasted chicken and pork. Many herbs and spices are added to meet meals. This should be understand to be a modern direction in diet works when an alkaline food is recommended to neutralize acid food, which is everyday meal including meet, sugar and bread. Modern civilization eats considerably more acid-forming foods than alkalizing foods. According to well-known naturopath P. Airola (1984), acidosis, or over-acidity in the body tissues, is one of the basic causes of diseases, especially the arthritic and rheumatic diseases. There is in the internet now a lot of information on alkaline food. The lists of alkaline products show that fruits and vegetables have highest pH most of which are traditional cultivars in Georgia. There are alkaline vegetables such as - alfalfa, barley, beet greens, beets, cabbage, carrot, celery, cucumber, eggplant, garlic, green beans, green peas, herbs, lentils, lettuce, onions, radishes, spices, spinach, watercress, wild greens, etc. which are the traditional cultivars in Georgia. Alkaline traditional fruits are - apple, apricot, berries, cherries, sour cherries, figs, grapes, grapefruit, nectarine, peach, pear, strawberries, watermelon, seeds and nuts, etc. To maintain health, the diet should consist of 60% alkaline forming foods and 40% acid forming foods. To restore health, the diet should consist of 80% alkaline forming foods and 20% acid forming foods (Airola, 1984).

Therefore, we should think about sustainability of traditional use of crops and wild plant species in the past time. The fact that the nature remained undisturbed in the country centuries long and population was characterized by much higher healthy features as current situation exists in Georgia should be explained as occurrence of healthy food and nature tolerant use of the plant resources.

6. Threats

6.1 Threats and conservation of ancient crops

There are several reasons for the genetic erosion of ancient cultivars and the wide distribution of new varieties of introduced crops. First of all, new cultivars have higher yields and are therefore preferred both as a source of food for local people and as a cash crop that determines local income. The second reason why local peasants began to prefer cultivating genetically modified (GM) plants may be explained by introduction of new diseases into Georgian agricultural fields in recent years, causing harm primarily to ancient cereals and vegetables. However, the introduction of new parasites has revealed that endemic forms of Georgian crop plants contain valuable selective disease-resistant material for genetic engineering. The tetraploid and hexaploid endemic wheat species *T. timopheevii* and *T. zhukovskyi*, for example, are characterized by a high level of resistance to a new race (TTKS, commonly known as Ug99) and many other races of *Puccinia graminis* f. sp. *tritici* due to the wheatstem rust resistance gene Sr36 (Tsilo et al., 2008). *T. carthlicum* is characterized by immunity to diseases, a short growing period, and resistance to cold.

Intensive Genetic erosion of ancient crops started in Georgia since 1950s which was also a period of intense selection work in breeding stations in the whole of the Soviet Union, e.g. the highly productive awnless wheat cultivar Bezustaja I developed in Russia has been sown in all wheat fields in Georgia since the 1970s, and this variety eventually replaced Georgian endemic wheat species. Recently new breeder's varieties of wheat and other cereals are introduced from different countries.

The process of genetic erosion of ancient crop varieties has not been a great concern for the mountain areas of Georgia, which until the 1990s acted as a depository of ancient crops. One important consideration that explains why ancient cultivars were conserved longer in mountainous regions than in the lowlands is that the local population preserved their traditional ways of life and socioeconomic structures. The traditional agricultural system is characterized by dependence on local genetic resources and locally developed technologies. Even today, peasants in mountain villages use an ox-drawn sledge made of wood for loading and transporting cereals and a threshing sledge on threshing floors to thresh wheat, oats, rye, and barley. Traditional agricultural equipment makes it possible to cultivate areas even on steep slopes and at high elevations, where modern tractors cannot be used. Moreover, some old landraces of wheat and barley are used to prepare bread and beer for religious rituals. Substitution of these landraces by others would go against centuries-old traditions (Akhalkatsi et al., 2010).

Despite these conditions that support the maintenance of ancient landraces, many endemic and native representatives of crop plants are currently in danger of extinction and face severe problems of genetic erosion in all mountainous regions of Georgia. While agrobiodiversity is declining rapidly in many areas of the world due to anthropogenic pressure (Körner et al., 2007), including population growth, in Georgia the main reason for genetic erosion of ancient crop varieties is demographic decline in mountain regions due to harsh economic conditions and lack of modern infrastructure (Nakhutsrishvili et al., 2009). The shift from ancient cultivars to modern high-yielding crops such as maize and potato, which took place in the lowland areas much earlier, began in mountain villages only in the last 20 years. Greater income from marketing allows families to stay in mountain villages. Moreover, the economic security of the traditional farming systems in these mountain regions appears to be in jeopardy when traditional agriculture is replaced by cattle breeding, which causes abandonment of cultivated fields and their transformation into pastures.

Several research centres maintain *ex situ* germplasm collections of Georgia, such as gene banks and living collections. According to the National Biodiversity Action Plan of Georgia (Jorjadze, 2005), international nature conservation institutions and Georgian scientific and nongovernmental organizations have taken care to preserve the genetic resources of local cultivars. Several gene banks and living collections occur in Georgia. There is one biggest genebank located at the Georgian Institute of Farming established in 2004 through support of International Centre for Agricultural Research in the Dry Areas (ICARDA). They owned a total 3057 accessions of local and introduced cultivars and CWRs in 2010. The other 5 gene banks are located in different research institutes unified with Agrarian University in 2011. Total number of germplasm accessions is 6286 in Georgian gene banks. However, the material kept in *ex situ* collections are not sufficient and need more contribution. Many seed banks worldwide contain about 7000 accessions of germplasm of Georgian cultivars and crop wild relatives. A recently initiated project, "Mountain Biodiversity in the Caucasus and its Functional Significance," supported by the Swiss National Science Foundation Program SCOPES, will build an electronic biodiversity archive for Georgia, and include data on mountain plant biodiversity in Georgia. Because it will be built in compliance with Global Biodiversity Information Facility (GBIF) standards, it will contribute to the Global Mountain Biodiversity Assessment (GMBA) mountain portal with GBIF (www.mountainbiodiversity.org). A research agenda concerned with the use of georeferenced mountain biodiversity data for science and management was developed at a GMBA workshop in Kazbegi, Central Caucasus, in July 2006 (Körner et al., 2007). Such a database of the plant species of the Caucasus will become a prominent entry in the GBIF database and highlight the current status of plant genetic resources in Georgia.

It should be emphasized that establishment and maintenance of *ex situ* collections and databases is just a first step in the conservation process of ancient crop varieties. The next step should be return of conserved seed material to the fields of local farmers. From 2004 to 2009, the Global Environmental Facility/United Nations Development Fund (GEF/UNDP) project "Recovery, Conservation and Sustainable Use of Georgia's Agro-Biodiversity" was carried out with the aim of conservation and sustainable use of threatened local plant genetic resources in the oldest historical mountainous region of Georgia, Samtskhe-Javakheti. This project enabled establishment of sources of primary seed and planting material for threatened crops and fruit varieties, and assisted farmers in accessing markets for organic products from such crops as lentil, grass pea, chickpea, faba bean, common millet, Italian millet, etc. Another project was the return of the Georgian wheat variety *T. aestivum* var. *ferrugineum* 'Akhaltsikhis Tsiteli Dolis Puri' in Meskheti province, where it was sown on 10 ha and produced bread that was introduced in shops featuring organic products in Tbilisi as of 2008. Afterward, this project was supported by the Georgian church, which expressed an interest in cultivating ancient crops on monastery grounds. However, these attempts have been realized only on a small scale and not in larger areas of the country.

In our opinion, the major activity of the corresponding governmental institutions should be directed on supporting local farmers in reintroducing ancient crops on the market and maintain maximum diversity of the target taxon's gene pool. The importance of agricultural achievements not should be oriented only on high yield of crops but the traditional foods to which people have adapted a long time determines their healthy lifestyle. Thus, conservation and reintroduction of ancient cultures to modern agriculture can insure longevity of people.

3.3 Threats and conservation needs of crop wild relatives

The natural populations of many species of CWRs are increasingly at risk. The primary causes of diversity loss of wild plant species are habitat loss, degradation and fragmentation. Many cereal CWRs, including relatives of wheat and millet species, occur in arid or semi-arid lands and are severely affected by over-grazing and desertification. The forest species are affected by habitat disturbances because of illegal forest cutting occurring in 1990s in Georgia. Climate change is having significant impacts of species distributions and survival in a concrete habitat. One of the most important threats to the diversity of CWRs are genetic erosion and pollution. The threat of genetic pollution or introgression, either from genetically modified organisms (GMOs) or from conventionally bred crops, to wild species has become an increasing risk to the in situ genetic conservation of crop wild relatives.

Another problem is that many species of important CWR occur in centres of plant diversity and crop diversity located mainly in developing countries, which often lack resources to invest in the necessary conservation activities. South Caucasus and Georgia, in particular, is the centre of origin and diversity of many of the world's important crop plants. There are several international projects realized by the ICARDA, the International Plant Genetic Resources Institute (IPGRI), US Department of Agriculture (USDA), United Nations Environmental Program (UNEP), etc. contributed in undertaking efforts in monitoring and conservation of plant. Although, additional resources are urgently needed in such areas of high diversity to identify priority species for conservation, determine the necessary conservation activities, monitor the status of key species, improve the use of these valuable resources.

Habitat disturbances are main threats leading to the extinction of rare and endangered plant species. Deforestation took place during last decades in Georgia and caused habitat degradation. The fact detected with the population of wild grapevine has revealed the threat to the riparian forests, which is situated along rivers in very close proximity of settlements and local people uses the resources of this forest in a highest degree. We have detected that some trees were cut representing the support of clambering wild grapevine and the individuals were lying on the earth, which will cause its drying up and death. More great scale cuttings in dark coniferous forests lead to arising of forest openings with high irradiation leading to drying up the underground cover of mosses and lichens, which drastically changes habitat and determines disappearing of natural species adapted to this habitat. Overgrazing of meadows and pastures was a problem in Soviet period, when several million head of sheep were grazing summer pastures of mountainous regions of Georgia. However, now the number of cattle is reduced and does not threaten much the rare species in their natural habitats. In spite of this fact, grazing affect survival of rare species such as *Hordeum spontaneum,* which was found on road side and during the next visit it was grazed completely. Such disturbances as habitat degradation due to road and pipeline construction works threatens the populations but has temporary effect. These types of disturbances are especially threatening the rare and endangered species of high conservation value.

The best way of *in situ* preservation of genetic diversity of valuable plants is creation of nature reserves on the territories, where natural populations of CWR occur. The first nature reserve of Georgia was established in Lagodekhi in 1912. In present, the protected areas occupy 7% of the country's territory, which is equal to 495.892 ha. There are 16 nature reserves, 9 national parks, 12 managed resource protected areas, 14 natural monuments and 2 protected landscapes in Georgia. The problem remains for the species, which are growing in rural habitats and on arable lands mixed with field crops have different assessment to threats. These species are depending in their existence to the monitoring of arable lands, which crop will be sown, how will be transformed filed crop to pasture or hay meadow, or what kind of herbicides and mineral fertilizers will be used in the field. The maintenance of wild populations growing as weeds in cultivated fields depends on sustainable management of agriculture in the region. The governmental institutions should control the processes which might bring to the genetic erosion of CWRs having high value of conservation. In this case the legislation bases should be effective to control local farmers not affect CWRs with ecologically unsuitable for this species actions in the field e.g. use of fertilizers or introduction of new crops leading to changing in technology of field cultivation methodology and leading to disturbances of wild weed species of high conservation value.

Ex situ conservation of the germplasm of CWRs is very valuable material for improvement of crop quality and their resistance against fungal and microbial disease. It will be of interest to collect their seed material and distribute to genbanks, which will contribute to provide necessary germplasm to research centres dealing with the genetic engineering. The Tbilisi Botanical Garden and Institute of Botany has two collections of seeds. One is collection of rare endemic plant seeds, which is collected in the framework of the Millennium project managed by Kew Royal Botanical Garden, UK. The second is collection of aboriginal crop varieties collecting in different regions of Georgia. These program works together with IPK, Gatersleben Germany, where the analogy of the collected material is kept at the gene bank. The living collections of CWRs are very few. Botanical Gardens in Tbilisi and Batumi have some small collections of CWRs collected in the frameworks of international collaborative projects. However, maintenance of the collections after finishing the projects is impossible and they are cancelled in several years. The plant genetic resources documentation in Georgia is mostly computerized. There are several databases, which include all information and passport data for accessions of field crops, but so far they have no free access.

Most threats to biodiversity are the results of human actions, which are expressed in the overuse of natural resources for fuel, fodder, manure, grazing and collecting of ornamental and medicinal plants. This process leads to the loss of genetic diversity including crop wild relatives. The *in situ* protection measures are not easy to implement and, thus, the accent should be directed on *ex situ* conservation.

7. Conclusions

Very old archaeological findings, cultural heritages and so far existing high morphological and genetic diversity of ancient crops and their wild relatives show that Georgia has very old agricultural traditions that have preserved to our times. The fact that large-scale genetic erosion of the ancient landraces in Georgia has reached extreme levels from 1950s and almost all the local varieties of cereals (wheat, barley and millet), legumes (peas, lentils, common vetch and faba bean), and grapes are now disappear from the farms requires special analyses and development of conservation measures. Only the gene banks and living collections hold germplasm of landraces extinct in the farms. An assessment of the effectiveness of current conservation strategies to protect the diversity of ancient crops in Georgia reveals a gap in the reintroduction of conserved germplasm to the fields of local farmers. In our opinion, the corresponding governmental institutions responsible for conservation of biodiversity should refocus the strategy to require complementary *in situ* and *ex situ* conservation actions to maintain maximum diversity of the target taxon's gene pool by supporting local farmers in reintroducing ancient crops on the market and thereby filling this gap. Moreover, at present, neither field crop genebank nor live collections of the permanent crops have sufficient land and equipment in Georgia, as well as funding to carry out *ex situ* conservation at the modern level. Storage of the *in situ* collections should be improved through upgrading the present storage of the field crop genebank facilities.

There is a need to improve public awareness of importance of ex *situ* conservation. Popularity of the data obtained by scientists should be distributed among the local population so that they themselves have contributed to the preservation of national heritages. The results of scientific investigations that some crops represent local cultivars and even domesticated landraces in this area means that this is connected with lifestyle of the local population. The fact that longevity of life in the Caucasus was very high and centenarians lived to 120 years and more should be understand that a healthy diet of mankind is not only amount of calorie but the combination of food of high quality. The modern alkaline diet almost completely coincides with the traditional Georgian cuisine. Therefore, we must appreciate the importance of conservation of local varieties to ensure the health of local people.

The data obtained in our investigations (Ekhvaia &Akhalkatsi, 2010; Ekhvaia et al., 2010, 2011; Asanidze et al., 2011) indicate importance of CWR species in Georgia as many of them represent direct ancestors of local cultivars. The fact that wild grape shows high genetic relation to local varieties of grape indicates that winemaking represents an ancient culture in Georgia, which is expressed even in religious rituals of the nation. Wild grapevine and pear representing the wild ancestors of local varieties are under threat because of wood cutting. Many other CWRs are in the same position. The legislation of species conservation is applied to rare and endemic species and *in situ* conservation is maintained only at protected areas. However there is no legislation that can protect CWRs growing in rural and urban areas and representing weed species. No actions of conservation are undertaken to protect these species. Many CWRs (wild wheat, rye, coriander, etc.) are grown in cultivated lands of local farmers. Many years, wild wheat species were mixed with local varieties of wheat and barley but now they are disappeared. The events which are protecting them are traditional cultivation technology of the landraces to which the local weeds are adapted by their life strategy and propagation character. The threats here will be change of traditional crops to the new varieties, which will need different cultivation events. This might lead to disappearance of the CWRs from the cultivated beds. At present, CWRs *ex situ* collections are almost absent in Georgia.

The problem of genetic erosion of landraces and their wild relatives needs active contributions by national policies and comprehensive measures are urgently needed to avoid the complete loss of ancient crop genetic resources in Georgia. International nature conservation institutions and Georgian scientific and nongovernmental organizations should show more activity to the restoration of ancient crops, which defined the healthy life of Georgians.

8. Acknowledgement

Research on agrobiodiversity in Georgia is supported by grant IZ73Z0_128057, in the framework of the Swiss National Science Foundation (SNSF) program SCOPES (Scientific Cooperation between Eastern Europe and Switzerland). The expeditions were funded by the Global Environmental Facility/United Nations Development Fund (GEF/UNDP) Project GEO/01/G41/A/1G/72. We are grateful to K. Pistrick, R. Fritsch, F. Blattner, and M. Gurushidze from IPK Gatersleben for supporting our investigations.

9. References

Adler, D. S. & Bar-Oz, G. (2009). Seasonal patterns of prey acquisition during the Middle and Upper Palaeolithic of the southern Caucasus. In: *The Evolution of Hominid Diets: Integrating approaches to the study of Palaeolithic subsistence,* Hublin, J.-J. & Richards, M. (Eds), pp. 127-140, Springer, ISBN: 9781402096983, Leipzig.

Airola, P. (1984). *How to Get Well: Handbook of Natural Healing.* Health Plus Pub, (March 1984), ISBN:0932090036, Sherwood, OR.

Akhalkatsi, M., Ekhvaia, J., Mosulishvili, M., Nakhutsrishvili, G., Abdaladze, O. & Batsatsashvili, K. (2010). Reasons and processes leading to the erosion of crop genetic diversity in mountainous regions of Georgia. *Mountain Research and Development,* Vol.30, No.3, (August 2010), pp.304-310. ISSN: 0276-4741.

Asanidze, Z., Akhalkatsi, M. & Gvritishvili, M. (2011). Comparative morphometric study and relationships between the Caucasian species of wild pear (*Pyrus* spp.) and local cultivars in Georgia. *Flora,* Vol.206, No.11, (November 2011), article in press. doi:10.1016/j.flora.2011.04.010.

Badr, A., Müller, K., Schäfer-Pregl, R., El Rabey, H., Effgen, S., Ibrahim, H. H., Pozzi, C., Rohde, W. & Salamini, F. (2000). On the origin and domestication history of barley (*Hordeum vulgare*). *Molecular Biology and Evolution* Vo.17, No.4, (April 2000), pp.499–510. ISSN: 0737-4038.

Bakhtadze, K.E. (1947). *Biologia, selektsia i semenovodstvo chainogo rastenija (Biology, Selection and Seed Productivity of Tea Plants).* Nauka, Moscow. (in Russian).

Batonishvili, V. (1991). *Aghtsera samefosa sakartvelosi (Geographic Description of Georgian Kingdom).* (Ed. 7), Ganatleba, Tbilisi. (In Georgian).

Bedoshvili, D. (2008). *National Report on the State of Plant Genetic Resources for Food and Agriculture in Georgia.* Available from:
http://www.pgrfa.org/gpa/geo/Georgian report on State of PGR Sep 29, 2008.pdf; Accessed on 21 March 2010.

Bockelman, H. E., Erickson, C. A. & Goates, B. J. (2002). National Small Grains Collection: wheat germplasm evaluations. *Annual Wheat Newsletter,* Vol.48, (10 August, 2002), pp.273-286.

Casas, A. M., Yahiaoui, S., Ciudad, F. & Igartua, E. (2005). Distribution of MWG699 polymorphism in Spanish European barleys. *Genome,* Vol.48, No.1, (February 2005), pp.41–45. ISSN: 0831-2796.

Davies, P. (2003). An Historical Perspective from the Green Revolution to the Gene Revolution. *Nutrition Reviews,* Vol.61, No.6, pp.124–134. ISSN: 0029-6643, doi:10.1301/nr.2003.jun.S124-S134.

Dekaprelevich, L. (1947). Saqartvelos martsvlovani kulturebis dziritadi jishebi (Main cultivars of cereals in Georgia). *Proceedings of State Selective Station, Georgia,* Vol.2, No.1, pp.5-47. (In Georgian).

Dekaprelevich, L. & Menabde, V. (1929). K izucheniu polevykh kultur zapadnoi Gruzii. I. Racha. (Study of cereal cultivars in Georgia. I. Racha). *Scientific Papers of the Applied Sections of the Tbilisi Botanical Garden.* Vol.6, No.2, pp. 219-252. (In Russian).

Dvorák, J., Luo, M.-C. & Yang, Z.-L. (1998). Genetic Evidence on the Origin of *Triticum aestivum* L. In: *The Origins of Agriculture and Crop Domestication.* Damania, A.B., Valkoun, J., Willcox, G. & Qualset C. O. (Eds). *Proceedings of the Harlan Symposium,* pp. 254-267, ISBN: 92-9127-084-9. Aleppo, Syria, 10-14 May 1997.

Ekhvaia, J. & Akhalkatsi, M. (2010). Morphological variation and relationships of Georgian populations of *Vitis vinifera* L. subsp. *sylvestris* (C. C. Gmel.) Hegi. *Flora,* Vol.205, No.9, (September 2010), pp.608-617. ISSN: 0367-2530, doi:10.1016/j.flora.2009.08. 002.

Ekhvaia, J., Blattner, F. R. & Akhalkatsi, M. (2010). Genetic diversity and relationships between wild grapevine (*Vitis vinifera* subsp. *sylvestris*) populations and aboriginal cultivars in Georgia. *Proceedings of 33rd World Congress of Vine and Wine, 8th General Assambley of the OIV,* Tbilisi, Georgia, June 2010, Available from:
<http://www.oiv2010.ge/index. php?page=5&lang=0>

Ekhvaia, J., Blattner, F.R., Gurushidze, M. & Akhalkatsi, M. (2011). Relationships between wild grapevine (*Vitis vinifera* subsp. *sylvestris*) populations and native grape cultivars in Georgia. *Proceedings of BioSystematics Berlin 2011*. ISBN: 978-3-921800-68-3, pp.111, Berlin, Germany, 21 - 27 February 2011.

Eliava, G. M. 1992. *Dzvelkolkhuri unikaluri vazis jishta katalogi (Catalog of ancient grape varieties of Colchis)*. (Ed.2), Kutaisi Press N6, Georgia. (In Georgian).

Eritzjan, A. A. (1956). K izucheniu izmenchivosti sostava pchenits v Gruzii (Study of variability of wheat field composition). *Proceedings of Tbilisi Botanical Institute*, Vol.18, No.1. pp.251-277. (In Russian).

Finlayson, C. (2005). Biogeography and evolution of the genus *Homo*. *Trends in Ecology and Evolution*. Vol. 20, No.8, (August 2005), pp.457-463. ISSN: 0169-5347.

Fox, W. Sh. (2004). *Living longer with phytomedicines from the Republic of Georgia*. Woodland Publishing, ISBN-10: 158054388X, Health & Fitness.

Gabunia, L. & Vekua, A. A. (1995). Plio-Pleistocene hominid from Dmanisi, East Georgia, Caucasus. *Nature*, Vol.373, No.6514, (9 February 1995), pp.509-512, ISSN: 0028-0836, doi:10.1038/373509a0.

Gabunia, L., Vekua, A., Swisher, C. C., Ferring, R., Justus, A., Nioradze, M., Ponce de Leon, M., Tappen, M., Tvalchrelidze, M. & Zollikofer Ch. (2000). Earliest Pleistocene hominid cranial remains from Dmanisi, Republic of Georgia: taxonomy, geological setting, and age. *Science*, Vol.288, No. 5578, (July 5, 2002), pp. 85-89. ISSN: *0036-8075*.

Garson, L.K. (1991). The Centenarian Question: Old-Age Mortality in the Soviet Union, 1897 to 1970. *Population Studies*, Vol.45, No.2, pp. 265-278. ISSN: 0032-4728. doi:10.1080/0032472031000145436.

Girgvliani, T. (2010). *Zemo Svanetis khorblis aborigenuli formebis istoria. (The history of aboriginal forms of wheat varieties of the Upper Svaneti)*. Artanuji, ISBN:978-9941-421-20-4, Tbilisi. (In Georgian).

Grant, K., Russel, Ch. & Everill, P. (2009). *Anglo-Georgian Expedition to Nokalakevi. Interim report on excavations in July 2009*. Available from:
http://www.nokalakevi.org/Downloads/Reports/English/Site/2009.pdf.

Grassi, F., Labra, M., Imazio, S., Ocete Rubio, R., Failla, O., Scienza, A. & Sala, F. (2006). Phylogeographical structure and conservation genetics of wild grapevine. *Conservation Genetics*, Vol.7, No.6, (December 2006), pp.837–845. ISSN: 1566-0621.

Heun, M., Schäfer-Pregl, R., Klawan, D., Castagna, R., Accerbi, M., Borghi, B. & Salamini, F. (1997). Site of einkorn wheat domestication identified by DNA fingerprinting. *Science*, Vol.278, No.5341, (14 November 1997), pp.1312-1314. ISSN:0036-8075.

Jackson, J. (2003). *The Biology of Apples and Pears, Series: The Biology of Horticultural Crops*. Cambridge University Press, ISBN: 0521380189, Cambridge.

Javakhishvili, I. (1987). *Sakartvelos ekonomiuri istoria (Economic History of Georgia)*. (Ed. 2), Vol.5., Metsniereba, SB:2927, Tbilisi. (In Georgian).

Jorjadze, M. 2005. Agrobiodiversity, biotechnology and biosafety. In: *National Biodiversity Strategy and Action Plan – Georgia*. NACRES and Fauna and Flora International (FFI). Polygraph, ISBN: 99940-716-6-1. Tbilisi.

Ketskhoveli, N. (1928). *Masalebi kulturul mtsenareta zonalobis shesastsavlad kavkasionze. (Materials on zonal distribution of cultivated plants in the Greater Caucasus)*. Agricultural National Committee Press, Tbilisi. (In Georgian).

Ketskhoveli, N. (1957). *Kulturul mtsenareta zonebi sakartveloshi (Zones of cultivated plants in Georgia)*. Georgian Academy of Sciences Press, Tbilisi. (In Georgian).

Ketskhoveli , N., Ramishvili, M. & Tabidze, D. 1960. Sakartvelos ampelograpia. (Amphelography of Georgia). Georgian Academy of Sciences Press, Tbilisi. (In Georgian).

Khomizurashvili, N. (1973). *Sakartvelos Mekhileoba (Horticulture of Georgia).* Metsniereba, Tbilisi (in Georgian).

Kilian, B., Özkan, H., Kohl, J., von Haeseler, A., Barale, F., Deusch, O., Brandolini, A., Yucel, C., Martin, W. & Salamini, F. (2006). Haplotype structure at seven barley genes: relevance to gene pool bottlenecks, phylogeny of ear type and site of barley domestication. *Molecular Genetics and Genomics,* Vol. 276, No.3, (September, 2006), pp. 230-241, ISSN: 1617-4615. doi 10.1007/s00438-006-0136-6.

Kobakhidze, A. (1965). Sakartvelos samartsvle-parkosani mtsenareebi: *Phaseolus* L. (cereals of Georgia: *Phaseolus* L.). In: *Botanika (Botany),* Ketskhoveli, N. (Ed), pp.87-89, Metsniereba, Tbilisi. (In Georgian).

Kobakhidze, A. (1974). Sakartvelos samartsvle-parkosan mtsenareta botanikur-sistematikuri shestsavlisatvis. (Botanical-systematic study of cereals in Georgia). In: *Botanika (Botany),* Ketskhoveli, N. (Editor), pp. 58--190, Metsniereba, Tbilisi. (In Georgian).

Körner, Ch., Donoghue, M., Fabbro, T., Häuse, Ch., Nogués-Bravo, D., Arroyo, M.T.K., Soberon, J., Speers, L., Spehn, E.M., Hang Sun, Tribsch, A., Tykarski, P. & Zbinden N. (2007). Creative use of Mountain Biodiversity Databases: The Kazbegi Research Agenda of GMBA-DIVERSITAS. *Mountain Research and Development,* Vol.27, No.3, (August 2007), pp.276–281. ISSN: 0276-4741.

Kuthatheladze, Sh. (1980). *Pyrus* L. In: *Sakartvelos Flora (Flora of Georgia),* Ketskhoveli, N. (Ed.), Vol. 6. Metsniereba, ISBN:5-520-00325-7, Tbilisi. (in Georgian).

Kvavadze, E., Bar-Yosef, O., Belfer-Cohen, A., Boaretto, E., Jakeli, N., Matskevich, Z. & Meshveliani, T. (2009). 30,000-Year-Old Wild Flax Fibers. *Science,* Vol.325, No.5946, (11 September 2009), pp.1359, ISSN: 0036-8075.

Laguna, E. (2004). The plant micro-reserve initiative in the Valencian Community (Spain) and its use to conserve populations of crop wild relatives. *Crop Wild Relative,* Issue 2, (July 2004), pp.10-13. ISSN 1742-3627.

Maghradze, D., Failla, O., Bacilieri, R., Imazio, S., Vashakidze, L., Chipashvili, R., Mdinaradze, I., Chkhartishvili, N., This, P. & Scienza, A. (2010). Georgian *Vitis* germplasm: usage, conservation and investigation. *Bulletin de I'OIV,* Vol.83, No. 956-958, (October-December, 2010), pp. 485-496. ISSN: 0029-7127.

Maisaia, I., Shanshiashvili, T. & Rusishvili, N. (2005). *Agriculture of Colchis.* Metsniereba, ISBN:99940-785-3-4, Tbilisi. (In Georgian).

Maxted, N., Ford-Lloyd, B. V., Jury, S., Kell, Sh. & Scholten, M. (2006). Towards a definition of a crop wild relative. *Biodiversity and Conservation.* Vol.15, No.8, (July 2006), pp.2673-2685. ISSN: 0960-3115

Maxted, N., Dulloo, E., Ford-Lloyd, B. V., Iriondo, J. M & Jarvis, A. (2008). Gap analysis: a tool for complementary genetic conservation assessment. *Diversity and Distributions,* Vol.14, No.6, (November 2008), pp.1018–1030. ISSN:1472-4642. doi: 10.1111/j.1472-4642.2008.00512.x.

McGovern, P.E. (2003). *Ancient wine: the search for the origins of viniculture.* Princeton University Press, ISBN: 0691070806, Princeton, New Jersey.

Melikishvili, G. (Ed). (1970). *Sakartvelos istoriis narkvevebi (Historical essays of Georgia).* Sabchota Sakartvelo, Tbilisi. (In Georgian).

Menabde, V. (1938). *Sakartvelos kerebi (Barleys of Georgia).* Georgian Academy of Sciences Press, Tbilisi. (In Georgian).

Menabde, V. (1948). *Pshenitsi Gruzii (Wheats of Georgia).* Georgian Academy of Sciences Press, Tbilisi. (In Russian).

Menabde, V. (1961). Kartuli khorblebi da mati roli khorblis saerto evolutsiashi (Georgian wheat cultivars and their role in wheat evolution). *Proceedings of Tbilisi Botanical Institute,* Vol.21, No.1. pp.229-259. (In Georgian).

Nakhutsrishvili, G. (1999). The vegetation of Georgia (Caucasus). *Braun-Blanquetia,* Vol.15, pp.1-74. ISSN:0393-5434.

Nakhutsrishvili, G., Akhalkatsi, M. & Abdaladze, O. (2009). Main threats to the mountain biodiversity in Georgia (the Caucasus). *Mountain Forum Bulletin,* Vol.9, No.2, (July 2009), pp.18–19. ISSN 1815-2139.

Neidze, V., Bokeria, M., Kharadze, K., Kurtubadze, M. & Garsenishvili, L. (2008). *Sakartvelos geographia (Geography of Georgia).* LogosPress, ISBN: 978-99940-64-48-9, Tbilisi. (In Georgian).

Ovchinnikov, I. V., Gotherstrom, A., Romanova, G. P., Kharitonov, V. M., Liden, K. & Goodwin, W . (2000). Molecular analysis of Neanderthal DNA from the northern Caucasus. *Nature,* Vol.404, No.6777, (30 March 2000), pp.490–493, ISSN: 0028-0836. doi:10.1038/35006625.

Petersen, G., Seberg, O., Yde, M. & Berthelsen, K. (2006). Phylogenetic relationships of *Triticum* and *Aegilops* and evidence for the origin of the A, B, and D genomes of common wheat (*Triticum aestivum*). *Molecular Phylogenetics and Evolution,* Vol.39 , No.1, (April 2006), pp.70–82. ISSN: 1055-7903.

Pistrick, K., Akhalkatsi M., Girgvliani, T. & Shanshiashvili, T. (2009). Collecting plant genetic resources in Upper Svaneti (Georgia, Caucasus Mountains). *Journal of Agriculture and Rural Development in the Tropics and Subtropics,* Supplement 92, pp.127-135. ISSN:1613-8422.

Ramishvili, R., (1988). *Dikorastushchii vinograd Zakavkazia (Wild grape of the South Caucasus).* Ganatleba, ISBN: 5-505-00690-6, Tbilisi. (In Russian).

Sabashvili, M. (1970). *Niadagmtsodneoba (Soil Sciences).* Tbilisi University Press, Tbilisi. (In Georgian).

Schatz, G., Shulkina, T., Nakhutsrishvili, G., Batsatsashvili, K., Tamanyan, K., Ali-zade, V., Kikodze, D., Geltman, D. & Ekim, T. (2009). Development of Plant Red List Assessments for the Caucasus Biodiversity Hotspot. In: *Status and Protection of Globally Threatened Species in the Caucasus.* Zazanashvili, N. & Mallon, D. (Editors). Contour Ltd. pp. 188-192. Tbilisi. ISBN 978-9941-0-2203-6. Available from: <www.assets.panda.org/downloads/ cepf_caucasus_web_1.pdf >

Takhtajan, A. (1986). Floristic Regions of the World. University of California Press, First English Language Edition, (September 1986), ISBN: 0520040279, Berkeley, CA.

This, P., Lacombe, T. & Thomas, M.R. (2006). Historical origins and genetic diversity of wine grapes. *Trends in Genetics* Vol.22, No.9, (September 2006), pp.511–519. ISSN: 0168-9525.

Tsilo, T. J., Yue Jin & Anderson, J. A. (2008). Diagnostic Microsatellite Markers for the Detection of Stem Rust Resistance Gene Sr36 in Diverse Genetic Backgrounds of Wheat. *Crop Science,* Vol.48, No.1, (January–February 2008) pp.253–261. ISSN: 0011-183X.

Tushabramishvili, N. (2011). The main problems of Palaeolithic of Georgia and the complex studies of the Western Georgian cave sites. In: *Khornabuji,* Pitskhelauri, K. (Ed), pp. 39-41, Ilia State University Press, ISBN: 1512-2999, Tbilisi.

UPOV (2000). Guidelines *for the conduct of tests for distinctness, uniformity and stability. Pear* (*Pyrus communis* L.). Geneva. Date of access: 2000-04-05. Available from: <http://www.upov.int/en/publications/tg-rom/tg015/tg_15_3.pdf>

Vavilov, N. I. (1992). *Origin and geography of cultivated plants.* Cambridge University Press, ISBN: 0-521-40427-4, Cambridge.

Volk, G. M., Richards, C. M., Henk, A. D., Reilley, A. A., Bassil, N. V. & Postman, J. D. (2006). Diversity of wild *Pyrus communis* based on microsatellite analyses. *Journal of the American Society for Horticultural Science,* Vol.131, No.3, (May 2006), pp.408–417. ISSN: 0003-1062.

Voltas, J., Pemán, J. & Fusté, F. (2007). Phenotypic diversity and delimitation between wild and cultivated forms of the genus *Pyrus* in North-eastern Spain based on morphometric analyses. *Genetic Resources and Crop Evolution,* Vol.54, No.7, (November 2007), pp.1473–1487. ISSN:0925-9864.

Yamamoto, T. & Chevreau, E. (2009). Pear genomics. In: *Genetics and Genomics of Rosaceae.,* Folta, K.M. & Gardiner, S.E. (Eds.), Springer, ISBN:0387774904, New York.

Zhizhizlashvili, K. & Berishvili, T. (1980). Zemo Svanetis kulturul mtsenareta shestsavlisatvis (Study of cultivated plants in Upper Svaneti). *Bulletin of Georgian Academy of Sciences,* Vol.100, No.2, pp.417-419. ISSN:0132-1447. (In Georgian).

Zohary, D., & Hopf, M. (2000). *Domestication of plants in the Old World.* (3rd edition.) Oxford University Press, ISBN 0-19-850356-3, New York.

4

Metallophytes and Metallicolous Vegetation: Evolutionary Aspects, Taxonomic Changes and Conservational Status in Central Europe

Henryk Baumbach
Friedrich-Schiller-Universität Jena
Germany

1. Introduction

Habitats with particular edaphic conditions such as metalliferous, serpentine or salt sites usually require special morphological, anatomical, and physiological adaptations of colonizing plants. Therefore, they often present specialized flora and vegetation and belong to the ecologically most interesting habitat types of Central Europe.

The vegetation of metalliferous soil sites is legally protected by European law (European Habitats Directive 92/43/EEC, Appendix I, habitat type 6130), national laws (e.g. in Germany by the Federal Nature Conservation Act, BNatSchG, § 30) and several federal state laws.

During the last decades the scientific focus has been on eco-physiological adaptations of metallophytes (for an overwiew see Baker et al., 2010), their community context and syntaxonomical evaluation (e. g. Libbert, 1930; Schwickerath, 1931; Koch, 1932; Schubert, 1953/54; Ernst, 1965, 1966, 1968, 1974; Koch & Kuhn, 1989; Daniëls & Geringhoff, 1994; Pott & Hellwig, 2007), bio-geographical aspects (e. g. Schulz, 1912; Wein, 1926; Heimans, 1936; Schubert 1954a; Ernst, 1974), and their taxonomical status. Since the basic work of Schulz (1912) metallicolous ecotypes of vascular plants have received special attention in numerous publications, and several (endemic) taxa of dubious value have been described. Modern molecular methods (isozymes, DNA fingerprint and sequence analyses) allowed new insights into genetic differentiation of metallicolous populations, micro-evolutionary processes and the taxonomical treatment of several metallophyte taxa. Some scientific studies dealing with genetic differentiation of the character species of metallicolous vegetation (e. g. *Minuartia verna, Silene vulgaris, Armeria maritima, Viola calaminaria, Cardaminopsis halleri, Thlaspi caerulescens*) have shown that the taxonomic rank of some of them should be reconsidered. From this and due to new studies on vegetation composition syntaxonomical changes seem also to be necessary. Remarkable problems for the definition of habitat types may arise from this because the perception of the conservation authorities is focussed on the few character species of metalliferous vegetation.

In the first part of this chapter I will focus on this problem and discuss possible consequences for the conservation of metalliferous sites. In the second part of the chapter I will review the current situation of metalliferous sites and their vegetation in the former copper-shale mining region of Mansfeld and Sangerhausen (the eastern Harz Mountains foothills region, Saxony-Anhalt, Central Germany). Despite the legal protection and several conservation efforts, the number of metalliferous soil sites with their unique flora has decreased in several regions substantially. Therefore, I want to encourage a supra-regional conservation strategy for Central European metalliferous sites.

Fig. 1. Metallicolous vegetation with the character species *Armeria maritima*, *Viola calaminaria*, *Minuartia verna* and *Silene vulgaris* near Rabotrath (Belgium, photograph: H. Baumbach, June 2008).

2. Types of metalliferous soil sites and their distribution in Central Europe

Natural (primary) metalliferous soil sites are relatively rare in Central Europe and confined to the range of the Alps and to only few, small-area localities with particular geological and edaphic conditions in the lowlands (Ellenberg, 1996). Most of the recent metalliferous sites are secondary habitats shaped by ore mining and processing. They are concentrated in some mid-mountain ranges and their forelands in Germany (Saxony-Anhalt and Lower Saxony: Harz Mountains and foothills, North Rhine-Westphalia), and parts of Poland (Legnica, Olkusz), the Czech Republic (Pyšek & Stočes, 1978; Pyšek & Pyšek, 1988, 1989), Slovakia (Banasova, 1980), and Belgium (Simon, 1975; Dejonghe et al., 1993; Graitson et al., 2003). Tertiary sites can be subdivided into those whose genesis is a result of atmospheric deposition in the vicinity of metal smelters (Baumbach et al., 2007; Ernst et al., 2004) or alluvial deposition of metal-enriched substrates by sedimentation in river floodplains and on raised river banks. Well-known examples of this are the sites along the rivers downstream the mines and smelters with their ore crushing facilities in the Harz Mountains (Hellwig, 2002; Knolle, 1989), the Erzgebirge Mountains (Golde, 2011) and the Geul River valley in the Belgium-Netherlands border region (van der Ent, 2007). The flora and vegetation of metalliferous soils in the Alps is beyond the scope of this chapter, for an overview see Punz (1991, 1994, 2008) and Punz et al. (1990).

3. Flora and vegetation

3.1 Flora

3.1.1 Metallophyte classification

A classification of plant species colonizing metalliferous soils was introduced by Lambinon & Auquier (1964). They differentiate between "metallophytes" (taxa found only on metal-contaminated soils) and "pseudo-metallophytes" (taxa found on both contaminated and uncontaminated soils within the same region, e. g. *Silene vulgaris, Agrostis capillaris*). Metallophyte taxa either grow exclusively on metalliferous soils ("absolute" or "eumetallophytes", e. g. *Viola calaminaria, V. guestphalica*) or in phytogeographically distinct areas on uncontaminated soils ("local metallophytes": e. g. *Armeria maritima, Minuartia verna*).

3.1.2 Evolutionary aspects and taxonomic changes

The origin of metallicolous plant taxa, underlying micro-evolutionary processes and taxonomical problems have been controversial topics of the last decades. The most important new findings on genetic differentiation and the resultant taxonomical consequences are given in six metallicolous taxa.

The Spring sandwort (Leadwort), *Minuartia verna* (L.) HIERN (Caryophyllaceae), is a perennial, low-growing cushion-forming species (Fig. 2). It is one of the typical pioneer plants of metalliferous sites in the Central European lowlands and on the British Isles (Garcia-Gonzalez & Clark, 1989). It has a circumpolar distribution and occurs in the arctic regions and in Asian and European high mountains (Meusel et al., 1965), in the Alps particularly on calcareous soils. The subspecies *M. verna* ssp. *hercynica* (WILLK.) O. SCHWARZ has recently been confined to metalliferous soils in the lowland and therefore has a disjunctive range. The Harz Mountains, the east Harz foothills in central Germany as well as the Aachen-Stolberg region in western Germany represent the main areas of distribution today (Benkert et al., 1996; Haeupler & Schönfelder, 1988). Further small-area occurrences are confined to northern Thuringia (Bottendorf Hills), Weserbergland (Blankenrode), the region of Osnabrück (Hasbergen), and eastern Belgium (Liège region). On non-metalliferous sites it is confined to parts of the Franconia Jura mountains (ssp. *verna*; Haeupler & Schönfelder, 1988). The occurrence of *M. verna* on anthropogenic metalliferous sites in the Harz Mountains was first recorded by Thal (1588). This is probably the first recognition of a plant species growing on anthropogenic sites in the scientific literature. Even though the mechanisms of metal-tolerance are not yet completely known, *M. verna* apparently realizes a reduced root uptake and an excreting mechanism of metals through leaf hydathodes (Neumann et al., 1997).

M. verna is confined to open plain sites on raw soils with only a thin cover of fine grained substrate. The dead cushions decompose very slowly and facilitate the initial humus of the raw soils (Schubert, 1953/54). In later successive stages *M. verna* is replaced due to its low shadow tolerance which disables survival in dense grassland vegetation.

Fig. 2. *Minuartia verna*, one of the character species of metallicolous vegetation in Central Europe, is critically endangered in some regions due to the loss of habitats (photograph: H. Baumbach).

Minuartia verna is one of the central European character species of metalliferous soil sites that has been considered as glacial relic and therefore paleo-endemics (Schulz, 1912; Schubert, 1954a; Heimans, 1961; Ernst, 1974). Due to the large morphological variation within the *Minuartia verna*-complex several of the described subspecies and varieties (e. g. Graebner, 1919, Hayek, 1922; Mattfeld, 1922; Halliday, 1964) can only be delineated bio-geographically or edaphically. The necessity of a critical revision of the *Minuartia verna*-complex has already been postulated by Ernst (1974), nevertheless this revision still is lacking.

Molecular data obtained in the only DNA-study (Baumbach, 2005) undertaken support the paleao-endemism hypothesis of the metallicolous lowland forms of *M. verna* which are clearly separated from the alpine taxon (ssp. *gerardii*). One hundred and three individuals from eight metallicolous lowland populations and three non-metallicolous alpine populations have been analysed using AFLP (amplified fragment length polymorphism). Five primer combinations have been used to generate 199 polymorphic loci. Genetic variability within *Minuartia verna*-populations was lower (14 % average heterozygosity; 37 % polymorphic loci) than in metallicolous populations of *Silene vulgaris* (see below) and *Armeria maritima* (Baumbach & Hellwig, 2007). Within non-metallicolous populations, genetic variability was clearly and significantly higher than within metallicolous populations. Geographical and genetic distances were highly positively correlated ($r=0.81$; $p<0.001$). AMOVA revealed a stronger genetic differentiation of the metallicolous populations than of the non-metallicolous populations. Genetic variation within lowland populations was remarkably lower (52 %) than within the alpine populations (86 %). At least some ancestral populations obviously preserved at primary metalliferous sites from the post-glacial until the beginning of ore mining. Furthermore, high genetic differentiation among the metallicolous lowland groups, the low genetic variability within populations and the comparably high ITS variation indicate a long-term isolated evolution within the different regions. This should be kept in mind when maintaining an aggregation of all metallicolous lowland forms to a widely considered ssp. *hercynica* in terms of Schwarz (1949) and Schubert (1954a). Results of an ongoing AFLP- and ITS-study (Baumbach, in prep.) including 30 *M. verna*-populations from 12 geographical regions from central Europe will illuminate further aspects.

Another character species of metalliferous soils is the Bladder Campion, *Silene vulgaris* (MOENCH) GARCKE (Caryophyllaceae). It was reported from nearly all abandoned ore mining sites in Central Europe (compare Ernst, 1974; Pardey, 1999a; Baumbach, 2000; Golde, 2001; and others) where it usually initiates the succession of vascular plant communities, especially on steep slops of spoil-heaps. The morphologically differentiated form growing on mining and smelter heaps was first mentioned by Schulz (1912) who considered it as belonging to the var. *angustifolia* Garcke. It was formally recognised as var. *humilis* by Schubert (1954a). The main motivation for the separation from the var. *angustifolia* was the restriction of the var. *humilis* to metalliferous heaps. Morphologically it was chiefly characterised by a dwarfed, prostrate growth, smaller and narrower leaves, fewer flowers per shoot, and smaller seeds than the normal variety. However, Schubert (1954a) mentioned the large variation of the morphological features found alone in one heap population. Nevertheless, Rothmaler (1963) elevated the status of the var. *humilis* to a subspecies *humilis*. Physiologically plants of the "*humilis*" type were assumed to show higher metal-tolerance (Schubert, 1954a). In contrast, Bröker (1963) and Gries (1966) demonstrated that morphological features and metal-tolerance are not associated in *Silene vulgaris*. This result is consistent with a physiological study (Wierzbicka & Panufnik, 1998) which showed that the *humilis*-form is mainly a morphological response to water stress and nutrient deficiency often occurring on the exposed and coarse-grained mining dumps, but not due to the metalliferous soil conditions themselves. Several recent and previous papers have illuminated particular physiological aspects of the metal-tolerance (for discussion see Baumbach, 2005). Schat et al. (1993) proposed a general model for the genetic control of copper tolerance in *Silene vulgaris*. More important, crossing experiments by Schat et al. (1996) have shown that tolerance loci for zinc, copper, and cadmium in plants of an Irish population from a metalliferous site were identical with those in several German populations from metalliferous sites. They argued that the occurrence of common major genes for tolerance among different geographically isolated populations must have resulted from independent parallel evolution in local non-tolerant ancestral populations. Furthermore Schat & Vooijs (1997) examined the co-segregation of tolerances to Cu, Zn, Ni, Co, and Cd in crosses between distinctly tolerant ecotypes. The results demonstrated a non-pleiotropic genetic control of tolerance to Cu, Zn, and Cd while tolerance to Ni and Co seems to represent the pleiotropic by-product of the tolerance allele of one particular locus for zinc tolerance. As these studies show, there is no all-embracing metal tolerance, rather a population-specific response to the site-specific soil factors. A study using AFLP (Baumbach, 2005) compared the genetic differentiation of *Silene vulgaris*-populations from different regions in eastern Belgium and Germany. One hundred forty individuals out of five pairs of populations (one metallicolous and one non-metallicolous in the surroundings each) and four additional single populations have been sampled. One hundred thirty-four polymorphic and reproducible loci, generated with three primer combinations, have been used. Detected genetic variability within *Silene vulgaris*-populations was comparatively high (20 % average heterozygosity, 64 % polymorphic loci). There was no significant difference in parameters of genetic variability between metallicolous and non-metallicolous populations. An analysis of molecular variance (AMOVA) resulted in a comparatively low genetic differentiation of the populations (Φ_{ST}=0.25). Correspondingly, (hypothetical) gene flow between populations is high (N_em=0.74). Genetic variation within populations (more than 70 %) exceeds variation among them. AMOVA reveals geographical partition of genetic variance while grouping into metallicolous vs. non-metallicolous populations was not supported by the data. The clear geographical grouping of population pairs (metalliferous and normal soil populations each) supports the hypothesis of polytopic and reiterate colonization events of metalliferous sites by tolerant ecotypes from neighbouring populations on normal soils. In conclusion, a *Silene vulgaris* taxon "*humilis*" (subspecies or variety) should be rejected because it is neither characterised by genetic markers, nor by exclusive morphological features or an exclusive metal-tolerance.

One of the most interesting and best studied plant species of metalliferous sites in Central Europe is the Thrift, *Armeria maritima* (MILL.) WILLD. s. l. (Plumbaginaceae), which can also be found in some geographical regions on non-metalliferous soils. The *A. maritima* complex is highly polymorphic and has been subdivided into eight infraspecific taxa (Pinto da Silva, 1972) that can be easily characterized geographically and ecologically but are difficult to distinguish morphologically. The non-metalliferous soil taxa of Central Europe are ssp. *alpina,* which is confined to the subalpine belt of the Alps, ssp. *maritima* (at all Western European coasts), and ssp. *elongata* (mainly on sandy soils in sub continental Europe and at the Baltic Sea coasts). Within the Central European range of ssp. *elongata* we find at least four described taxa of *A. maritima* which are confined to metalliferous soil sites. The status of these metallicolous taxa has been controversial, and has changed several times (Wallroth, 1842; Schulz, 1912; Christiansen, 1931; Schubert, 1954a; Ernst, 1974; Lefebvre, 1985; Wisskirchen & Haeupler, 1998) because the morphological characters used to discriminate these taxa are not always constant at the level of populations. Local populations may contain plants that can be assigned to distinct infraspecific taxa growing side by side. Studies performed with isozymes (Vekemans & Lefebvre, 1997; Vekemans et al., 1992) and RAPD markers (Baumbach & Hellwig, 2003) revealed that the large morphological variation corresponds to large genetic variation occurring within the *A. maritima* complex and strong differentiation at the population level.

Furthermore, results of an AFLP study (Baumbach & Hellwig, 2007) give strong evidence that metallicolous populations have been founded from the ancestral non-metallicolous populations repeatedly and independently in different geographical regions. The metallicolous *Armeria maritima* microendemics *„hornburgensis"*, *„bottendorfensis"*, *„eifeliaca"*, *„calaminaria"* should not be given formal rank as varieties or subspecies within *A. maritima.* Also their treatment as varieties of an *A. maritima* subspecies *halleri* sensu latu (Wisskirchen & Haeupler, 1998) is doubtful because ssp. *halleri* can not be consistently characterized throughout its geographical range and may be an artefact itself. All investigated metallicolous forms show close relations to *Armeria maritima* ssp. *elongata.* If a taxonomical recognition should be considered necessary it is advisable to treat the microendemics as varieties of that subspecies (Baumbach & Hellwig, 2007).

Cardaminopsis halleri (L.) HAYEK (Brassicaceae) is a wide-spread pseudo-metallophyte which occurs at metalliferous and non metalliferous sites (meadows, watersides, rocks). The genetic structure of 28 metallicolous and non-metalicolous populations with a total of 625 individuals was studied by Pauwels et al. (2005) using PCR-RFLP on chloroplast DNA (cpDNA). Eleven distinct chlorotypes were found: five were common to non-metallicolous and metallicolous populations, whereas six were only observed in one edaphic type (five in non-metallicolous and one in metallicolous). No difference in chlorotype diversity between edaphic types was detected. Computed on the basis of chlorotype frequencies, the level of population differentiation was high but remained the same when taking into account levels of molecular divergence between chlorotypes. Isolation by distance was largely responsible for population differentiation. As it was the case in *Silene vulgaris* (Baumbach, 2005) geographically isolated groups of metallicolous populations were more genetically related to their closest non-metallicolous populations than to each other. These results suggest that metallicolous populations have been established separately from distinct non-metallicolous populations without suffering founding events and that the evolution towards increased tolerance observed in the distinct metallicolous population groups occurred independently.

Fig. 3. Old lead-zinc opencast mining near Blankenrode (Germany), type locality of *Viola guestphalica* (small picture), the Westfalian zinc violet (photograph: H. Baumbach, July 2011).

The Alpine penny-cress, *Thlaspi caerulescens* J. PRESL et C. PRESL (Brassicaceae), is a perennial, rosette-forming plant with clusters of white to pink flowers. With occurrences on metalliferous and non-metalliferous soils in Central Europe and on the British Isles the plant is a typical pseudo-metallophyte. The metal-ecotype of *T. caerulescens* was described as subspecies *calaminare* (LEJ.) BÄSSLER. Isozyme-analyses (Koch et al., 1998) have indicated, that a taxonomical subdivision of *T. caerulescens* into a taxon of metalliferous soils (ssp. *calaminare*) and a taxon of non-metalliferous soils is not possible and, furthermore metal tolerance might have evolved twice in populations from different areas.

The Zinc Violet, *Viola calaminaria* (GING.) LEJ., is the characteristic plant species of calamine sites in the Aachen-Stolberg-Liege region (the three border region of West-Germany, East-Belgium, and South-Netherlands). The blue flowering Westphalian zinc violet, *Viola guestphalica* is locally endemic, restricted to the former zinc-lead opencast "Bleikuhle" near Blankenrode (Fig. 3). Taxonomy and biogeography of both species were controversially discussed in numerous publications (e. g. Schulz, 1912; Heimans, 1961; Ernst, 1974; Nauenburg, 1986; and others) within the last decades. A recent study (Hildebrandt et al., 2006) has shown close relationship of both taxa and with *Viola lutea* HUDS. The authors consequently propose the treatment of both taxa as subspecies (ssp. *calaminaria* (LEJ.) ROTHM. and ssp. *westfalica* (W. ERNST) J. HEIMANS) of the alpine Mountain pansy, *Viola lutea*.

3.2 Vegetation

3.2.1 The classical syntaxonomical concept

The metallicolous vegetation of Central Europe is characterized by plant communities of the order Violetalia calaminariae Br.-Bl. et Tx. 1943 nom inval. Ernst (1965, 1974, 1976) proposed the sub-division of this order into the following three alliances:

1. Thlaspion calaminariae Ernst 1965: Metallicolous vegetation of Western-central- and Western Europe with only *Thlaspi caerulescens* as character species of the alliance. It includes the associations of the Violetum calaminariae Schwick. 1931 and the Minuartio-Thlaspietum caerulescens K. Koch 1932. The Violetum calaminariae can be subdivided geographically in eastern and western areas with the blue flowering zinc violet (*Viola guestphalica)* in the Violetum calaminariae westfalicum at Blankenrode (Germany) its only site in the world, and the yellow flowering *Viola lutea* subsp. *calaminaria* in the Violetum calaminariae rhenanicum.
2. Armerion halleri Ernst 1965: Metallicolous vegetation in the more eastern parts of Central Europe with *Armeria halleri* as exclusive character species of the alliance. The Armerion halleri includes the associations of the Armerietum halleri Libb. 1930, the Armerietum hornburgensis Schub. 1974, the Armerietum bottendorfensis Schub. 1953, and the Holco-Cardaminopsietum halleri Hülb. 1980 (Schubert et al., 2001).
3. Galio anisophylli-Minuartion vernae Ernst 1964: Metallicolous vegetation in the Alps, characterized by widely distributed alpine species (*Galium anisophyllum, Poa alpina, Euphrasia salisburgensis, Dianthus sylvestris*) and in the Italian and Austrian Alps with *Thlaspi rotundifolium* ssp. *cepaeifolium* and the endemic *Viola dubyana* (Baker et al., 2010).

3.2.2 Proposed syntaxonomical changes

Despite the fact that this system was accepted for almost four decades, recent publications proposed some new approaches due to new records and a re-visitation of the existing vegetation data. Additionally, some of the taxa whose taxonomical position became uncertain are character species of associations, alliances, order or class (Ernst, 1974; Schubert et al., 2001). This also results in a need for new discussion on the exceptional syntaxonomical position of the calaminarian grasslands.

Baumbach & Schubert (2008) proposed to discuss an assignment of the metallicolous vegetation in a new alliance Minuartion vernae characterized by *Minuartia verna* ssp. *hercynica*. This alliance should be grouped in the vicinity of the Armerion elongatae Krausch 1961 within the order Festuco-Sedetalia acris R. Tx. 1951 and the class Koelerio-Corynephoretea Klika in Klika et Novak 1941.

Further vegetation studies are necessary to illustrate these questions and also the autonomy of the several metallicolous plant associations. Calaminarian grasslands have a specific species composition which is characterized by abundant metal-tolerant taxa, and a lack of non-tolerant grassland species. Their physiognomic structure is shaped by dwarfism and little vegetation cover which allows the growth of a rich flora of mosses and lichens. Naturally, in transition zones between metal-rich and metal-free soils there may occur plant stands that legitimate the assignment to sub-associations of dry grassland communities. A complete negation of discrete communities of metallicolous vegetation was, however, opposed by Baumbach & Schubert (2008).

This is in contrast to Becker et al. (2007) who rejected the Armerietum bottendorfensis. They recognized four communities, in which the metallophytes *Armeria maritima* (ssp. *halleri*) and *Minuartia verna* ssp. *hercynica* occur in high frequency and classified them as special sub-associations (Armerietosum halleri) of four dry grassland associations (Teucrio-Festucetum, Thymo-Festucetum, Filipendulo-Helictotrichetum, Adonido-Brachypodietum).

In the conclusion of a comprehensive study of the metallicolous vegetation of the Western Harz Mountains Dierschke & Becker (2008) recommended the maintenance of an association Armerietum halleri. A synoptic table with inclusion of relevés from the literature in Germany supports the concept of a narrow class Violetea calaminariae with communities poor in species. For the remaining single alliance they propose the name Armerion halleri Ernst 1965 as *nomen conservandum*.

The alliance of the Galio anisophylli-Minuartion vernae is not acceptable in the eastern Alps by the Austrian authors (see Punz, 1991, 1994). Based on new vegetation records and tables, Punz & Mucina (1997) assigned the metallicolous vegetation associations of the Eastern Alps chiefly to the Thlaspietea rotundifoliae (Englisch et al., 1993) and a small part to the Asplenietea trichomanis (Mucina, 1993).

4 Current situation of metalliferous sites in Central Europe

4.1 The situation in the east Harz foothills – a case study in Central Germany

4.1.1 Introduction

The current situation of metalliferous soil sites will be exemplified here by the eastern und south-eastern Harz foothills region (Saxony-Anhalt, Central Germany). For centuries copper-shale mining was active in the Mansfeld (1200-1969) and the Sangerhausen (1382-1990) mining districts.

The metallicolous vegetation of the copper-shale spoil heaps was intensely studied by Schubert (1953/54). Most of these heaps are not only characterized by calaminarian grasslands in different successive stages but also by raw soils, xerothermous vegetation and shrubberies, which occur tessellate and often in very small areas. Therefore, and due to their isolated location within intensive used fields these spoil heaps have a high value for the biodiversity of the landscape.

4.1.2 Metallicolous vegetation

The typical calaminarian grassland of the copper-shale spoil heaps is the Armerietum halleri Libb. 1930. As the Armerietum hornburgensis and the Armerietum bottendorfensis it belongs to the alliance of the Armerion halleri Ernst 1965. When compared with the alliance of the western distributed Thlaspion calaminariae Ernst 1965 it is floristically depleted due to the absence of *Viola calaminaria* and *Thlaspi caerulescens*. However, it is characterized by a greater number of xerothermous grassland species (compare Schubert, 1953/54; Ernst, 1974; Pardey, 1999a; Dierschke & Becker, 2008). *Cardaminopsis halleri* is another typical species of metallicolous vegetation of the Harz Mountains and Western Germany which does not occur in the region.

The character species of the association Armerietum halleri, *Armeria maritima* (subsp. *halleri*), and the character species of the class Violetea calaminariae, *Minuartia verna* (ssp. *hercynica*) and *Silene vulgaris* (var. *humilis*), occur in high frequency (consistency class V). Frequent species are furthermore *Festuca ovina* agg. and *Agrostis capillaris* (consistency class IV) as well as *Achillea millefolium, Asperula cynanchica, Carlina vulgaris, Campanula rotundifolia, Cirsium acaule, Dianthus carthusianorum, Euphorbia cyparissias, Euphrasia officinalis* agg., *Galium verum, Hieracium pilosella, Koeleria macrantha, Pimpinella saxifraga, Potentilla heptaphylla, Potentilla neumanniana, Scabiosa ochroleuca,* and *Thymus praecox* (all consistency class III; Schubert, 2001). Characteristic mosses and lichens of the association are *Ceratodon purpureus, Bryum caespiticium, Cladonia alcicornis, Cladonia chlorophaea,* and *Peltigera rufescens* (all consistency class III). Other typical lichen species occurring at the spoil heaps are *Acarospora bullata, A. sinopica, A. smaragdula, Lecanora stenotropa, L. subaurea, Lecidea inops, Rhizocarpon oederi,* and *Stereocaulon nanodes* (Huneck, 2006). The successive stages of the Armerietum halleri at the copper shale spoil heaps (Minuartia-, Silene-, Euphrasia-, Cladonia-, Armeria-, Festuca- and Brachypodium-stage) have been described by Schubert (1953/54) In the eastern parts of the Mansfeld mining region *Minuartia verna* is replaced by *Alyssum montanum* which obviously fulfils the same ecological niche (Gerth et al., 2011).

4.1.3 Secondary sites

The most extensive occurrences of metallicolous vegetation can be found on the small mining spoil-heaps which arose between 1200 and 1815. The real number of spoil heaps located in the open landscape was controversially discussed in the last decades. In a recent project we mapped about 1500 small spoil heaps and estimated the vegetation cover by Colour-Infrared (CIR) aerial photographs analysis (of 2005) and field surveys of randomly selected heaps. Till now we mapped the floristic inventory of 650 spoil-heaps in the field. For the inventory we distinguished between (I) small spoil heaps, which originated between the 13th century and 1815 (Fig. 4), and (II) large spoil heaps which originated between 1815 and the end of mining. In contrast to the small spoil heaps the date of origin for the large spoil heaps is exactly known from the mining files.

Fig. 4. Spoil heaps of the medieval copper shale mining as secondary metalliferous sites in Central Germany (Dobis, photograph: H. Baumbach, May 2008).

The post-mining landscape consists (state March 2011): in the Mansfeld mining district of 1028 small and 56 large spoil heaps (of which 8 and 12 are slag heaps, respectively), and in the Sangerhausen mining district of 416 small and 9 large spoil heaps (of which 3 and 1 are slag heaps, respectively). Thereby it has to be considered that 60-70 % of the small spoil heaps in the Mansfeld mining district were already destroyed by land burial between 1850 and 1910 (Oertel, 2002). Due to the large amount of metalliferous sites originating in the former copper-mining activities the federal state of Saxony-Anhalt has a particular responsibility towards the preservation of the habitat type 6130. Therefore, three Natura 2000 sites with the habitat type 6130 as the main conservation objective, located in the post-mining landscape of the old mines, have been designated (Tab. 3). The total size of these sites is about 687 ha. From this area 137 hectares (19.9 %) are copper-shale spoil heaps with only 21.8 hectares (3.2 %) of calaminarian grassland. Unfortunately, a comprehensive conservation strategy for these sites is not available to date.

At numerous small spoil heaps the ongoing succession lead to an extensive displacement of the calaminarian grasslands by shrubberies. A total disappearance of the metallicolous vegetation has to be expected on most of the spoil heaps over the long term. The assumption, that metalliferous sites are naturally free of woods (Schubert, 1953/54; Ernst 1974) applies only in early successive stages. In the Sangerhausen mining district more than 75 % of the small spoil heaps are scrubby to more than 80 % (Tab. 2). Only 4 % can still considered as open (scrub encroachment <10 %).

Number of mapped spoil heaps (N=561)	Percentage	Occurrence of the character species	Formation of calaminarian grasslands
222	40 %	three	at least partial optimal
112	20 %	two	fragmentary
98	17 %	one	only initial stages

Table 1. Occurrence of the character species (*Minuartia verna, Silene vulgaris,* and *Armeria maritima*) and formation of calaminarian grasslands (Armerietum halleri) at copper shale spoil heaps (N=561) in the Mansfeld region.

In comparison, the situation in the Mansfeld mining district is much better. Only one third of the small spoil heaps is scrubby to more than 80 %, and 26 % of the spoil heaps can be considered open (scrub encroachment <10 %). At least initial stages or fragmentarily developed calaminarian grasslands can be found on 37 % of the mapped spoil heaps. On 40 % of the spoil heaps optimally developed calaminarian grasslands with all three character species occur at least partially (Tab. 1). When considering only the Natura 2000-sites the situation of scrub encroachment is slightly better (Tab. 3): 18 % of the small spoil heaps are scrubby to more than 80 %, and 31 % can considered as open (scrub encroachment <10 %). Nevertheless, calaminarian grasslands occur at only 62 % of the spoil heaps.

	N	Percentage of spoil heaps per scrub encroachment class							
Mining district		0	1	2	3	4	5	6	7
Mansfeld	996	7.9	17.8	10.9	14.2	8.0	13.0	14.2	14.1
Sangerhausen	408	2.5	1.7	1.2	3.9	4.7	8.8	29.2	48.0
Total	1404	6.3	13.0	8.0	11.1	7.0	11.7	18.4	23.8

Table 2. Scrub encroachment of the spoil heaps in the Mansfeld and Sangerhausen copper shale mining districts. N: total number of spoil heaps. Scrub encroachment classes: 0: 0 %, 1: 1<10 %, 2: 11-20 %, 3: 21-40 %, 4: 41-60 %, 5: 61-80 %, 6: 81-100 %, 7: 100 %.

Assuming, that the large number of sites alone is adequate for the protection of metallicolous vegetation, the need for any restoration measurements was neglected until recently (Baumbach & Schubert, 2008). A rethinking of this position took place with the results of an inventory (Baumbach, 2008) that showed the high degree of scrub encroachment on many spoil heaps. Consequently, a priority list with 50 spoil-heaps which primarily need biotope management was compiled by the nature conservation authority in 2009. The first of the proposed measures (removal of shrubs, cutting of trees and removal of top soil humus layer) were realized in autumn 2010 and spring 2011. As these actions were recent, an evaluation of the project's success is still pending.

Despite natural succession the current main reasons of the decline of metalliferous habitats are conflicts between intensive agriculture (input of fertilizers and herbicides from the surrounding fields, plough up heap margins), removal of material for road construction, motocross and quad activities by juveniles (which is in particular critically for the slow growing cryptogams), and deposition of domestic waste and rubble (especially in the surroundings of villages). Furthermore, the construction of new roads and the extension of housing development led to the loss of numerous spoil heaps within the last 20 years. A serious problem may arise from increasing metal prices which make the metalliferous spoil heaps attractive for extracting their remnant metal contents.

Site Number	6130-area [ha]	N_{total}	6130		Percentage of spoil heaps per scrub encroachment class							
			N	%	0	1	2	3	4	5	6	7
DE4335-301	12,37	273	170	62	18	25	11	15	8	12	7	4
DE4434-302	4,10	84	67	80	6	26	17	23	10	11	4	5
DE4434-303	5,34	104	49	47	0	2	5	12	13	25	30	13
total	**21,81**	**461**	**286**	**62**	**11**	**20**	**11**	**15**	**10**	**15**	**12**	**6**

Table 3. Natura 2000 sites in the Mansfeld region where the spoil heaps with calaminarian grasslands (habitat type 6130) are the main objective. 6130-area: total size of the calaminarian grassland in hectares, N_{total}: total number of spoil heaps in the Natura 2000 site, Number (N) and percentage (%) of spoil heaps with calaminarian grasslands. For definition of scrub encroachment classes see table 2.

The large spoil heaps that arose between 1815 and 1990 dominate the landscape scenery. Nevertheless, they are mainly endangered by production of gravel for road-building. Currently seven large spoil heaps with an original volume of 21 million m^3 and a ground area of 103 hectares are under mining retreat for gravel production. Many other spoil heap surfaces were used as dumping ground for sewage sludge, domestic and industrial waste (e. g. fly ash, filter dust). With regard to metallicolous vegetation this was seen uncritically in the past because two of the character species, *Minuartia verna* and *Armeria maritima*, are either totally absent at the large spoil heaps or have only very small-scale occurrences. Other groups of organisms were more or less neglected. A study by Huneck (2006) revealed a high diversity of lichens (96 recorded species) which are mainly confined to open spoil-heap surfaces. This very fact would be reason enough for a strong conservation of the remnant spoil heaps. Moreover, to date there is no comprehensive knowledge on the importance of the spoil heaps for biodiversity of fauna and fungi.

4.1.4 Primary sites

Within the Mansfeld mining district there are at least three primary metalliferous soil sites which have not been disturbed by mining activities in the past. One of them is the locus classicus of *Armeria maritima* ssp. *hornburgensis* at the Galgenberg Hill near the village of Hornburg. This Natura 2000 site (DE 4535-303) has a total area of 2 hectares from which only 170 m^2 are metallicolous vegetation. Interestingly *Minuartia verna* does not occur there. At the beginning of the 1960s numerous Black pines (*Pinus nigra*) were planted at the hill. Until the end of the 1990s they caused a decline of the site to 10 % of its former size. The population size of *Armeria m.* ssp. *hornburgensis* decreased from still 200 individuals in 1965 to only 100 in 1999 (Baumbach & Volkmann, 2002). In 2001 31 Black pines were completely removed. This and a partial mowing with subsequent removal of the biomass led to a stabilization and extension of the population to 383 in 2006 (Baumbach & Volkmann, 2006). Nevertheless, the situation is furthermore critical due to the tiny area of the site.

Another recently discovered site is located near Mansfeld at the castle mountain. Only a few square metres *Minuartia verna* and *Armeria maritima* grow on solid Rotliegend substrate without a fine soil layer. A detailed study of the vegetation and an analysis of the substrate are still pending.

Fig. 5. Primary metalliferous site near Blankenheim (Saxony-Anhalt, Central Germany). This small site is critically endangered by shading and litter input from the surrounding grassland (photograph: H. Baumbach, June 2006).

A third site, located in the east of the village of Blankenheim, was accidentally discovered in 1999 (Fig. 5). The site is subdivided into 5 subareas with a size of only some square metres each (Tab. 4). The soil layer is only 25 cm thick and covers an outcrop of metalliferous sand-ore. Obviously this site was never disturbed by mining activities because sand-ores were not suitable for smelting. Furthermore, this site is located outside the former active mining area. To date this very small site has no conservational state. It is critically endangered by shading and litter input from the surrounding grassland (Arrhenatheretum) and by accidental destruction. The genotypes of the *Minuartia verna*-population growing at this site have both characters of the alpine populations and characters of the spoil heap populations (Baumbach, 2005). This indicates that the *M. verna*-population has survived on this site since the last ice age and is therefore a real paleo-endemic.

4.1.5 Tertiary sites

The sole tertiary metalliferous soil site of the region is located at the former vineyard in Hettstedt-Burgörner (Baumbach et al., 2007). It originated from decades of emissions from downwind located copper, lead, and zinc smelter Kupferkammerhütte (1493-1990). The first damages to the vegetation resulting from smelter emissions were reported by Freytag (1870). The current vegetation at the vineyard is not only due to metal contamination but also to sulphur dioxide emissions from the former Kupferkammerhütte smelter located in the north and northwest, the smelter Kupfer-Silber-Hütte located in the southwest, and the rolling mill located in the west. The total area influenced by the emissions of the smelters covers approximately 11 hectares. On the most impacted western slope of the vineyard the top soil is covered by a layer of dust up to 25 cm with high amounts of metals and semi-metals.

Subarea	**1**	**2**	**3**	**4**	**5**
Size (m^2)	6	1	3	2	9
Exposition	S	S	S	S	S
Inclination (°)	5	8	8	8	10
Total vegetation cover (%)	98	95	98	100	100
Cover of mosses and lichens (%)	8	2	2	2	5
Number of species (phanerogams)	27	8	8	9	15
Armeria maritima	+		1		1
Minuartia verna	3	+	1	1	r
Silene vulgaris	+	1			
Festuca rupicola	3	3	3	4	4
Galium verum	+	+	1	1	+
Euphorbia cyparissias	+		+	+	+
Hieracium pilosella		1	+	+	r
Rumex acetosa	1			r	+
Potentilla tabernaemontani	+				+
Centaurea stoebe	r				+
Eryngium campestre					r
Achillea pannonica	+			+	r
Dianthus carthusianorum	r				r
Plantago lanceolata	r		r		
Festuca pratensis		+			+
Poa pratensis					+
Leontodon autumnalis					+
Trifolium campestre	+	r			
Tanacetum vulgare	1	r			
Plantago intermedia	r				
Lotus corniculatus	+				
Hieracium lachenalii	r				
Arrhenatherum elatius	1				
Dactylis glomerata	1				
Tragopogon pratensis	+				
Veronica chamaedrys	+				
Brachypodium pinnatum	1				
Agrostis stolonifera				+	
Lathyrus pratensis			+		
Falcaria vulgaris				r	
Daucus carota	+				
Artemisia vulgaris	r				
Betula pendula	1				
Rosa inodora	1				
Rosa elliptica	2				

Table 4. Vegetation relevés of the primary metalliferous site near Blankenheim (Saxony-Anhalt), Braun-Blanquet scale, inventory date: 15th of June, 2006.

Therefore it is characterised by extended metallicolous vegetation (3.5 ha), which is meagre in plant species and dominated by the characteristic species of the Armerietum halleri, *Armeria maritima, Minuartia verna,* and *Silene vulgaris* (all consistency class V). Furthermore, frequently occur *Agrostis stolonifera, Rumex acetosa* and *Festuca rupicola* (V), and with low frequency *Brachypodium pinnatum, Sanguisorba minor, Plantago lanceolata, Pimpinella saxifraga, Ranunculus bulbosus, Daucus carota, Cerastium holosteoides, Artemisia vulgare, Erigeron acre, Convolvulus arvensis, Tussilago farfara, Lonicera xylosteum, Tilia cordata, Ulmus minor, Betula pendula, Acer pseudoplatanus,* and *Acer campestre* (all consistency class I).

Some locally extreme concentrations of metals were found in the soil surface horizon at the middle slope (Tab. 5). Afforestation of the site performed in 1993 was unsuccessful. Nevertheless, parts of the upper slope and the plateau region (7.7 ha) of the vineyard are covered by extended birch pioneer forests as a result of natural succession. Due to shade the metallicolous vegetation disappeared there and was not replaced by other herbaceous vegetation. Therefore the top soil is strongly affected by erosion and solifluction.

SP	N	E	Pb	Zn	Cu	As	Mn	Cd	Co	Ni
1	$^{57}21923$	$^{44}66248$	8294	7482	3121	471	441	116	15.4	40.4
2	$^{57}21904$	$^{44}66277$	16019	14192	3601	1097	574	293	14.8	42.1
3	$^{57}21897$	$^{44}66286$	9664	2537	992	506	155	58	9.6	19.6
4	$^{57}21876$	$^{44}66332$	170	159	146	21	36	2	1.6	8.1

Table 5. Metal mass-concentrations [mg/kg dry weight] in the upper soil layer of four sampling points (SP) on the vineyard. Method: nitric acid pulping (DEV-A32), ICP/MS. Location of the PP: lower (1), middle (2), and upper (3) middle-slope, upper-slope (4). N: northing; E: easting in Gauss-Krüger coordinates (3° wide strips, Bessel ellipsoid, Potsdam datum, Rauenberg central point).

The vineyard will long be characterized by high soil-metal concentrations. With respect to the size of the polluted area a soil exchange or a cover with arable substrates is not possible. Regarding the results of the sequential extraction, the soil and the climate conditions in the region, the danger of medium-term mobilisation of metals via water or air seems to be low. The vineyard is the largest known site with metalliferous vegetation in Central Germany. Therefore, Baumbach et al. (2007) proposed to protect it to allow long-term observations of the regeneration processes in biocenoses at metal-contaminated sites.

4.2 The situation in other regions

4.2.1 Germany

In total, Germany has 36 Natura 2000 sites where the habitat type 6130 occurs. In 21 of them this habitat type is the main reason for site protection. The protection by the European Habitats Directive and the establishment of the Natura 2000 network lead to conservation efforts in several regions.

The only metalliferous site in Thuringia is located at Bottendorf Hill in the lower Unstrut valley (Schubert, 1954b; Becker et al., 2007). It is the type-locality of *Armeria bottendorfensis* and probably the largest primary site in Germany. The primary site on the southern slope and some small copper-shale spoil heaps on top of the hill are part of the Natura 2000 site

(DE4634-303) with a total size of 133 ha. The area is characterized by continental dry-grasslands which are in a very good state of preservation due to sheep pasturing.

The metallicolous vegetation of the West-Harz Mountains (Lower Saxony) was recently revisited by Dierschke & Becker (2008). Although their focus is on subdivision, environmental conditions and a syntaxonomical classification of the Armerietum halleri they also highlight conservational aspects. A comparison of the current number of metalliferous sites with a 1928 Bode list shows a large decline of sites due to human destruction and natural succession. The reasons for declining of (most small area) metalliferous sites and their vegetation are mainly: destruction by building development, flooding by dams, coverage, removal of slag for re-smelting or for road construction, afforestation, use as storage areas for timber or rock waste, motocross activities, trespass and camping, disturbances by mineral collectors (in particular critical for cryptogams), shading and input of litter by surrounding trees (Dierschke & Becker, 2008). There are no (at least no published) experiences with management and restoration of metalliferous soil sites in the Harz Mountains. Remarkably, the largest stands of the Armerietum cladonietosum in the Westharz Mountains can be found in a fenced, undisturbed area in the Oker River valley (Dierschke & Becker, 2008).

More than ten years ago, North Rhine-Westphalia was the first federal state with its own conservation strategy for metalliferous soil sites (Pardey, 1999b). Fortunately, this conservation strategy is not restricted to metallicolous flora and vegetation but also encompasses the fauna of these sites and plans for habitat connectivity in several core areas. To date there are 15 Natura-2000 sites in North Rhine-Westphalia where the habitat type 6130 occurs. In 11 of them it is the main objective for site protection.

First experiments with the restoration of metalliferous sites were made at the end of the 1990s in the Stolberg region. Preliminary results (2003) and a final evaluation (2008) were published by Raskin. Test sites were metalliferous sites with dominating Pine stands and only insular remnants of metallicolous vegetation. Primarily all trees and shrubs were felled and the remnant timber was burned on site in the winter. The next measure was the removal of top soil in strips using muzzle loaders. All organic material (vegetation, humus, litter) was removed from the restoration sites. In the first year six of the seven character species of the calaminarian grasslands colonized on the new raw soil sites. This pioneer vegetation was dominated by *Silene vulgaris* with scattered occurrence of *Minuartia verna*. Only after the 5th year did appear also *Armeria maritima*.

Although the Erzgebirge Mountains (Saxony) were one of the most important ore mining areas in Europe there is meagre knowledge of metallicolous vegetation. Consequentially the first paper on this topic was entitled "Metallicolous vegetation - a so far overlooked habitat type in Saxony" (Golde, 2001). The sites with metallicolous vegetation are fragmented and restricted to the Freiberg region. At these metalliferous sites *Minuartia verna* is totally absent and *Armeria maritima* has only a very small occurrence (Golde, 2001). If metallicolous ecotypes of *Silene vulgaris* und *Thlaspi caerulesens* loose their taxonomical rank the definition of the habitat type 6130 is no longer possible by characteristic angiosperm taxa but only by metallicolous cryptogam communities (mainly the Acorosporetum sinopicae). Two Natura 2000 sites where the habitat type 6130 occurs have been designated. In one of them it is the main objective for conservation (DE4945-303).

4.2.2 Belgium, Netherlands, and Poland

In eastern Belgium most of the 35 metalliferous sites are located in the former mining district of the Vielle Montaigne (Plombieres, Kelmis) close to the German mining district of Aachen-Stolberg and the Liege region (Graitson et al., 2003). In total 76 hectares of the habitat type 6130 are located in Natura 2000 sites.

The only metalliferous site of the Netherlands is that of the tertiary type. It is located in the Geul River floodplains near Epen and has an area of about 0.5 ha. Van der Ent (2007) estimates that in 1925 more than 10 km of the flood plains were dominated by metallicolous vegetation (see also Heimans, 1936). Till now over 99 % has diminished due to a decreased zinc load and increased eutrophication. In 2007/08 first restoration experiments in the field (bulldozing and small scale topsoil removal) were performed.

In Poland metallicolous vegetation of the Armerietum halleri is developed at several spoil heaps in the Olkusz region (Upper Silesia). Two of these sites with a size of 7.4 and 4.9 hectares are Natura 2000 sites (PLH120091, PLH120092). Unfortunately, there is no more (published) information available.

5. Conclusion

The review of new findings outlined in the first part of this paper shows the necessity for a taxonomic re-evaluation of several taxa. So far some of them have been used exclusively for the definition of the conservational value of metalliferous sites by the official nature conservation praxis. In consequence, such a re-evaluation may cause considerable formal problems as the Saxon example shows. One danger of taxonomic changes is the argumentation that with a loss of a certain taxonomic rank of the character species also the conservational value of the metalliferous sites becomes obsolete. Therefore I would like to underline that metalliferous sites and their vegetation deserve protection independent from the taxonomic rank of their character species or the syntaxonomy of their plant communities.

It is beyond a doubt that the particular abiotic factors (metal content, drought, nutrient deficiency) and the isolated geographic occurrence of metalliferous sites facilitate the evolution of special adapted and genetically differentiated plant populations. This genetic differentiation is an ongoing process that can lead to the evolution of neo-endemic species. We are still at the beginning of *Silene vulgaris,* advanced in *Armeria maritima* (centuries to millennia), and very advanced in *Minuartia verna* (since the end of the last ice age). Despite any conventional taxonomical rank these taxa deserve protection due to their unique gene pool and their evolutionary potential. Furthermore, they represent a genetic reservoir which could have prospective importance for the colonization and remediation of polluted anthropogenic sites (Whiting et al., 2004).

Apart from flowering plants which have for a long time been in scientific focus, adapted and genetically differentiated populations have to also be expected in lichens, fungi, bacteria and the soil fauna. Knowledge on biodiversity of metallicolous biocenoses is still insufficient and needs further scientific work. Based on this a comprehensive ecological and conservation biological evaluation should be carried out. Furthermore I propose to protect metalliferous sites as "evolutionary playgrounds", independent of the actual occurrence of rare or legally protected species. This is of course only feasible if there is no danger for human health and the natural environment through causes by the respective sites.

Whereas most of the secondary sites have been well investigated, the importance of primary metalliferous soil sites (for example as source of the palaeo-endemic metallophyte taxa) was underestimated for a long time and the knowledge of these sites is still poor. To date we only know of a few undisturbed, small sized metal outcrops with metallicolous vegetation in the Central European lowland. Most of them are critically endangered. The knowledge of these sites and their management requires further research.

To provide an efficient long-term protection of metalliferous sites and their biocenoses in Central Europe conservation strategies should be implemented both at a regional and a supra-regional scale. A supraregional conservation strategy should develop with participation of all Central European Countries (Germany, Belgium, Poland, Netherlands) and their federal states where metalliferous soil sites occur, respectively. Regional conservation strategies (e. g. Germany: Saxony-Anhalt, Lower Saxony, Thuringia) should consider the specific characteristics of the several regions. Model should be the "Conservation strategy for calaminarian sites in North Rhine-Westphalia" (Pardey , 1999b).

These conservation strategies should account the following main aspects: protection of the remnant metalliferous sites with special regard to the primary ones, population biology of the character species (minimum area, minimum viable population, maximum distance of populations within a given region to still enable gene flow between them), long-term bio-monitoring of representative habitats, information and awareness raising of the community (Haese, 1999), preservation of historic mine buildings and monuments (Wagenbreth, 1973; Philipp, 2000), and an integration of the post-mining landscape in tourist concepts (Slotta, 2003). A good example for such a comprehensive strategy is "the lead legacy" for the Peak District's lead mining heritage (Barnatt & Penny, 2004).

Science still lacks coherent insight into the exact measures for restoration of metallophyte communities. At present it is quite impossible to predict the long-term influence of climate change and the still-rising atmospheric nitrogen deposition on diversity of the biocenoses on metalliferous sites which are limited in water and nutrients.

Research into the geographic distribution, ecological amplitude and niche differentiation of metallophytes, and the impact of ecological management and habitat alteration on metallophyte vegetation, is necessary to facilitate conservation and to develop and manage sites in the future. A successful realization of these points needs an intensive interdisciplinary knowledge-sharing between science and site management. This is realized for example in the "European Heavy Metal Ecology Network" (EHMEN: www.post-mining-ecosystems.org) which is made up of a large group of professionals from science, management and governments across Europe or the "Arbeitsgemeinschaft Bergbaufolgelandschaften e. V." (Working group post-mining landscapes: www.bbfl.de).

6. Acknowledgment

For language correction I would like to thank Janneke Sedlo.

7. References

Baker, A. J. M.; Ernst, W. H. O.; Van der Ent, A.; Malaisse, F. & Ginocchio, R. (2010): Metallophytes: the unique biological resource, its ecology and conservational status in Europe, Central Africa and Latin America, In: *Ecology of Industrial Pollution*, Batty, L. C. & Hallberg, K. B., pp. 7-40, Cambridge, Cambridge University Press, UK

Banasova, V. (1980): Indikationseigenschaften der Kupferhalden-Vegetation in der Slowakei. *Wissenschaftliche Beiträge Martin-Luther-Universität Halle-Wittenberg, Reihe P*, Vol. 27, No.P11, pp. 74-76, ISSN 0440-1298

Barnatt, J. & Penny, R. (2004): *The Lead Legacy. The Prospects for the Peak District's Land Mining Heritage*. Peak District National Park Authority, Buxton

Baumbach, H. (2000): Beitrag zur Flora und Vegetation von Bergbau-, Hütten- und Stollenhalden im Mansfelder und Sangerhäuser Revier. *Schriftenreihe des Mansfeld-Museum (Neue Folge)*, Vol.5, pp. 105-118, ISBN 3-932639-04-9

-(2005): Genetische Differenzierung mitteleuropäischer Schwermetallsippen von *Silene vulgaris*, *Minuartia verna* und *Armeria maritima* unter Berücksichtigung biogeographischer, montanhistorischer und physiologischer Aspekte. *Dissertationes Botanicae* Vol.398, pp. 1-128, ISBN 3-443-64311-6

-(2008): Zur Situation der Schwermetallrasen und ihrer Standorte im östlichen und südöstlichen Harzvorland. *Naturschutz im Land Sachsen-Anhalt* Vol.40, No.2, pp. 3-19

-(in prep.): Genetic differentiation of the central European *Minuartia verna* – complex: molecular evidence for glacial relict endemism of metallicolous populations.

Baumbach, H. & Hellwig, F. H. (2003): Genetic variation within and among metal-tolerant and non-tolerant populations of *Armeria maritima* (Mill.) Willd. s.l. (Plumbaginaceae) in Central and Northeast Germany. *Plant Biology*, Vol.5, No.2, pp. 186-193, ISSN 1435-8603

Baumbach, H. & Hellwig, F. H. (2007): Genetic differentiation of metallicolous and non-metallicolous *Armeria maritima* (Mill.) Willd. taxa (Plumbaginaceae) in Central Europe. *Plant Systematics and Evolution* Vol.269, No.3-4, pp. 245-258, ISSN 0378-2697

Baumbach, H. & Schubert, R. (2008): Neue taxonomische Erkenntnisse zu den Charakterarten der Schwermetallvegetation und mögliche Konsequenzen für den Schutz von Schwermetallstandorten. *Feddes Repertorium* Vol.119, No. 5-6, pp. 543-555, ISSN 0014-8962

Baumbach, H. & Volkmann, H. (2002): Dynamik, genetische Struktur und Schutz kleiner Populationen - das Beispiel von *Armeria maritima* ssp. *hornburgensis*. *Mitteilungen zur floristischen Kartierung in Sachsen-Anhalt* Vol.7, pp. 3-24, ISSN 1432-8038

Baumbach, H. & Volkmann, H. (2006): Zur Situation von *Armeria maritima* ssp. *hornburgensis* - aktuelle Daten zu Populationsgröße, Demographie und Taxonomie. *Mitteilungen zur floristischen Kartierung in Sachsen-Anhalt* Vol.11, pp. 41-50, ISSN 1432-8038

Baumbach, H.; Volkmann, H. & Wolkersdorfer, C. (2007): Schwermetallrasen auf Hüttenstäuben am Weinberg bei Hettstedt-Burgörner (Mansfelder Land) - Ergebnis jahrhundertelanger Kontamination und Herausforderung für den Naturschutz. *Hercynia N. F.* Vol.40, No.2, pp. 87-109, ISSN 0018-0637

Becker, T.; Brändel, M. & Dierschke, H. (2007): Trockenrasen auf schwermetall- und nicht schwermetallhaltigen Böden der Bottendorfer Hügel in Thüringen. *Tuexenia* Vol.27, pp. 255-286, ISSN 1433-8440

Benkert, D.; Fukarek, F. & Korsch, H. (1996): *Verbreitungsatlas der Farn- und Blütenpflanzen Ostdeutschlands*, ISBN 3-437-35066-8, Gustav Fischer Verlag, Jena, Germany

Bode, A. (1928): Reste alter Hüttenbetriebe im West- und Mittelharze. *Jahrbuch der Geographischen Gesellschaft Hannover*, pp. 141-197

Bröker, W. (1963): Genetisch-physiologische Untersuchungen über die Zinkverträglichkeit von *Silene inflata* Sm. *Flora (Jena)*, Vol.153, pp. 122-156

Christiansen, W. (1931): Die mitteldeutschen Formenkreise der Gattung *Armeria*. *Botanisches Archiv*, Vol.31, pp. 247-265

Dejonghe, L.; Ladeuze, F. & Jans, D. (1993): Atlas des Gisements plombo-zinciferes du Synclinorium de Verviers (Est de la Belgique). *Toelichtende Verhandelingen Geologische en Mijnkaarten van België*, Vol.33, pp. 1-483

Daniëls, F. & Geringhoff, H. (1994): Pflanzengesellschaften auf schwermetallreichen Böden der Briloner Hochfläche, Sauerland. *Tuexenia*, Vol.14, pp. 143-150

Dierschke, H. & Becker, T. (2008): Die Schwermetall-Vegetation des Harzes - Gliederung, ökologische Bedingungen und syntaxonomische Einordnung. *Tuexenia*, Vol.28, pp. 185-227, ISSN 1433-8440

Ellenberg, H. (1996): *Vegetation Mitteleuropas mit den Alpen*, 5th edition, Verlag Eugen Ulmer, ISBN 3-8252-8104-3, Stuttgart (Hohenheim), Germany

Englisch, T.; Valachovič, M.; Mucina, L.; Grabherr, G. & Ellmauer, T. (1993): Thlaspietea rotundifolii, In: *Die Pflanzengesellschaften Österreichs. Teil II: Natürliche waldfreie Vegetation*, Grabherr, G. & Mucina, L., pp. 276-342, Gustav Fischer Verlag, Jena, Germany

Ernst, W. (1965): Ökologisch-Soziologische Untersuchungen der Schwermetall-Pflanzengesellschaften Mitteleuropas unter Einschluß der Alpen. *Abhandlungen aus dem Landesmuseum für Naturkunde zu Münster in Westfalen*, Vol.27, No.1, pp. 5-54: -(1966) Ökologisch-soziologische Untersuchungen an Schwermetallpflanzengesellschaften Südfrankreichs und des östlichen Harzvorlandes. *Flora, Abt. B*, Vol.156, pp. 301-318

-(1968): Der Schwermetallrasen von Blankenrode, das Violetum calaminariae westfalicum. *Mitteilungen der floristisch-soziologischen Arbeitsgemeinschaft N. F.*, Vol.13, pp. 261-262

-(1974): *Schwermetallvegetation der Erde*, Gustav Fischer Verlag, ISBN 3-437-30187-X, Stuttgart, Germany

-(1976): Violetea calaminariae, In: *Prodomus of the European Plant Communities. Vol. 3.*, R. Tüxen, pp. 1-132, J. Cramer, Vaduz, Switzerland

Ernst, W.; Knolle, F.; Kratz, S. & Schnug, E. (2004): Aspects of ecotoxicology of heavy metals in the Harz region - a guided excursion. *Landbauforschung Völkenrode*, Vol.54, No.2, pp. 53-71

Freytag, M. (1870): *Wissenschaftliches Gutachten über den Einfluß, welchen die Hüttenwerke der Mansfelder Kupferschieferbauenden Gewerkschaft in dem Wipperthal zwischen Mansfeld und Hettstedt auf die Vegetation der benachbarten Grundstücke und indirect auf Menschen und Thiere ausüben*. Druck von Heinrich Reichardt, Eisleben, Germany

Garcia-Gonzalez, A. & Clark, S. C. (1989): The distribution of *Minuartia verna* and *Thlaspi alpestre* in the British Isles in relation to 13 soil metals. *Vegetatio*, Vol.84, pp. 87-98

Gerth, A.; Merten, D. & Baumbach, H. (2011): Verbreitung, Vergesellschaftung und genetische Populationsdifferenzierung des Berg-Steinkrautes (*Alyssum montanum* L.) auf Schwermetallstandorten im östlichen Harzvorland. *Hercynia N. F.*, Vol.44, No.1, pp. 73-92, ISSN 0018-0637

Golde, A. (2001): Schwermetallfluren - ein in Sachsen bislang verkannter Lebensraumtyp - Überblick über Vorkommen und Ausbildungsformen im Freiberger Bergbaugebiet. *Berichte der Arbeitsgemeinschaft Sächsischer Botaniker N. F.*, Vol.18, pp. 49-60

-(2011): Das Freiberger Bergbaugebiet, In: *Bergbaufolgelandschaften Deutschlands - geobotanische Aspekte und Rekultivierung*, Baumbach, H.; Heinze, M. & Sänger, H., in press, Weissdorn-Verlag, Jena, Germany

Graebner, P. (1919): *Minuartia verna. Synopsis der Mitteleuropäischen Flora*, Vol.5, No. 1, pp. 734-751

Graitson, E.; Melin, E. & Goffin, M. (2003): Inventaire et caractérisation des sites calaminaires en Région wallonne. Société Publique d'Aide à la Qualité de l'Environnement G.I.R.E.A.- Université de Liège, Belgium

Gries, B. (1966): Zellphysiologische Untersuchungen über die Zinkresistenz bei Galmeiformen und Normalformen von *Silene cucubalus* Wib. *Flora, Abt. B*, Vol.156, pp. 271-290

Haese, U. (1999): Der Schutz der Stolberger Schwermetallvegetation im Lichte des öffentlichen Bewußtseins. *LÖBF-Schriftenreihe*, Vol.16, pp. 129-137, ISBN 3-89174-029-8

Haeupler, H. & Schönfelder, P. (1988): *Atlas der Farn- und Blütenpflanzen der Bundesrepublik Deutschland*, Ulmer, ISBN 3-8001-3434-9, Stuttgart, Germany

Halliday, G. (1964): Studies in the *Minuartia verna* complex, I. *Feddes Repertorium*, Vol.69, pp. 8-14

Hayek, A. (1922): Versuch einer natürlichen Gliederung des Formenkreises der *Minuartia verna* (L.) Hiern. *Österreichische Botanische Zeitschrift*, Vol.71, pp. 89-116

Heimans, J. (1936): De Herkomst van de Zinkflora aan de Geul. *Nederlandsch kruidkundig Archief*, Vol.46, pp. 878-897

-(1961): Taxonomic, phytogeographical and ecological problems round *Viola calaminaria*, the zinc violet. *Publicaties van het Natuurhistorisch Genootschap in Limburg*, Vol.XII, pp. 55-71

Hellwig, M. (2002): Die Schwermetallbelastungen und die Schwermetallvegetation im Innerstetal. *Berichte der Naturhistorischen Gesellschaft Hannover*, Vol.144, pp. 3-21

Hildebrandt, U.; Hoef-Emden, K.; Backhausen, S.; Bothe, H.; Bozek, M.; Siuta, A. & Kuta, E. (2006): The rare, endemic zinc violets of Central Europe originate from *Viola lutea* Huds. *Plant Systematics and Evolution*, Vol.257, pp. 205-222, ISSN 0378-2697

Huneck, S. (2006): Die Flechten der Kupferschieferhalden um Eisleben, Mansfeld und Sangerhausen. *Mitteilungen zur floristischen Kartierung in Sachsen-Anhalt*, Sonderheft 4, pp. 1-62, ISSN 1432-8038

Knolle, F. (1989): Harzbürtige Schwermetallkontaminationen in den Flußgebieten von Oker, Innerste, Leine und Aller. *Beiträge zur Naturkunde Niedersachsens*, Vol.42, No. 2, pp. 53-60

Koch, K. (1932): Die Vegetationsverhältnisse des Silberberges im Hüggelgebiet bei Osnabrück. *Mitteilungen des Naturwissenschaftlichen Vereins Osnabrück*, Vol.22, pp. 117-149

Koch, M. & Kuhn, L. (1989): Das Minuartio-Thlaspietum alpestris Koch 1932, eine Pflanzengesellschaft schwermetallhaltiger Böden im Hüggelgebiet, Landkreis Osnabrück. *Osnabrücker naturwissenschaftliche Mitteilungen*, Vol.15, pp. 137-154

Koch, M.; Mummenhoff, K. & Hurka, H. (1998): Systematics and evolutionary history of heavy metal tolerant *Thlaspi caerulescens* in Western Europe: evidence from genetic studies based on isozyme analysis. *Biochemical Systematics and Ecology*, Vol.26, pp. 823-838

Lambinon, J. & Auquier, P. (1964): La flore et la végétation des terrains calaminaires de la Wallonie septentrionale et de la Rhénanie aixoide. *Natura Mosana*, Vol.16, pp. 113-130

Lefebvre, C. (1985): Morphological variation, breeding system and demography at populational and subpopulational levels in *Armeria maritima* (Mill.) Willd. In: *Genetic differentiation and dispersal in plants*, Jacquard, P.; Heim, G. & Antonovics, J., pp. 129-139, Springer, Berlin, Germany

Libbert, W. (1930): Die Vegetation des Fallsteingebietes. *Mitteilungen der floristisch-soziologischen Arbeitsgemeinschaft in Niedersachsen*, Vol.2, pp. 1-66

Mattfeld, J. (1922): Geographisch-genetische Untersuchungen über die Gattung *Minuartia* L. (Hiern). *Feddes Repertorium*, Vol.LVII, No.Beiheft 127

Meusel, H.; Jäger, E. & Weinert, E. (1965): *Vergleichende Chorologie der zentraleuropäischen Flora. Kartenband.* Gustav Fischer, Jena.

Mucina, L. (1993): Aspenietea trichomanis. In: *Die Pflanzengesellschaften Österreichs. Teil II: Natürliche waldfreie Vegetation.* Grabherr, G. & Mucina, L., pp. 241-275, Gustav Fischer Verlag, Jena, Germany

Nauenburg, J. D. (1986): Untersuchungen zur Variabilität, Ökologie und Systematik der *Viola tricolor*-Gruppe in Mitteleuropa. Diss., Math.-Nat. FB, Georg August-Universität, Göttingen, Germany

Neumann, D.; zur Nieden, U.; Schwieger, W.; Leopold, I. & Lichtenberger, O. (1997): Heavy metal tolerance of *Minuartia verna*. *Journal of Plant Physiology*, Vol.151, pp. 101-108.

Oertel, T. (2002): Untersuchung und Bewertung geogener und anthropogener Bodenschwermetallanreicherungen als Basis einer geoökologischen Umweltanalyse im Raum Eisleben-Hettstedt. Diss., Mathematisch-Naturwissenschaftlich-Technische Fakultät, Martin-Luther-Universität Halle-Wittenberg, Germany

Pardey, A. (1999a): Grundlagen des Naturschutzes auf Schwermetallstandorten in NRW. Abiotische Verhältnisse, Flora, Vegetation, Fauna, aktuelle Schutzsituation und zukünftige Zielsetzungen. *LÖBF-Schriftenreihe*, Vol.16, pp. 7-48, ISBN 3-89174-029-8

Pardey, A. (Ed.) (1999b): Naturschutzrahmenkonzeption Galmeifluren NRW. *LÖBF Schriftenreihe*, Vol.16, pp. 1-271, ISBN 3-89174-029-8

Pauwels, M.; Saumitou-Laprade, P.; Holl, A. C.; Petit, D. & Bonnin, I. (2005): Multiple origin of metallicolous populations of the pesudometallophyte *Arabidopsis halleri* (Brassicaceae) in central Europe: the cpDNA testimony. *Molecular Ecology*, Vol.14, pp. 4403-4414

Philipp, R. (2000): Denkmalpflegerische Aspekte der Haldenlandschaft des Kupferschieferbergbaus in Sachsen-Anhalt. *Schriftenreihe des Mansfeld-Museums (Neue Folge)*, Vol.5, pp. 18-24, ISBN 3-932639-04-9

Pinto da Silva, A. R. (1972): *Armeria* Willd. In: *Flora Europaea Vol. 3: Diapensiaceae to Myoporaceae*, Tutin, T. G.; Heywood, V. H.; Burges, N. A.; Moore, D. M.; Valentine, D. H.; Walters, S. M. & Webb D. A., pp. 30-38, Cambridge University Press, ISBN 0-521-08489-X, Cambridge, UK

Pott, R. & Hellwig, M. (2007): Das Armerietum halleri Libbert 1930 aus dem Tal der Innerste am Nordrand des Harzes. *Hercynia N. F.*, Vol.40, pp. 245-255

Punz, W. (1991): Zur Flora und Vegetation über schwermetallhaltigem Substrat im Ostalpenraum - Eine Übersicht. *Verhandlungen der Zoologisch-Botanischen Gesellschaft in Österreich*, Vol.128, pp. 1-18

-(1994): Schwermetallstandorte im mittleren Alpenraum und ihre Vegetation - neue Befunde. *Verhandlungen der Zoologisch-Botanischen Gesellschaft in Österreich*, Vol.131, pp. 1-26

-(2008): Schwermetallstandorte und deren Vegetation im Land Salzburg. *Biotopverbund - Lebensraumvernetzung & Beiträge zum 13. Österreichischen Botanikertreffen*: pp. 375-378

Punz, W. & Mucina, L. (1997): Vegetation on anthropogenic metalliferous soils in the eastern Alps. *Folia Geobot. Phytotax.*, Vol.32, pp. 283-295.

Punz, W.; Schinninger, R. & Engenhart, M. (1990): Floristische Bearbeitungen von Schwermetallstandorten in der Steiermark - Eine Übersicht. *Mitteilungen des naturwissenschafltichen Vereins für Steiermark*, Vol.120, pp. 291-297

Pyšek, A. & Pyšek, P. (1989): Vegetation der Abbaudeponien in Böhmen: Veränderung der Artenzusammensetzung im Verlauf der Vegetationsentwicklung. *Verhandlungen der Gesellschaft für Ökologie*, Vol.XVIII, pp. 37-41.

Pyšek, A. & Stočes, I. (1978): Ein Beitrag zur Kenntnis der Vegetation an blei- und zinkreichen Biotopen in der Erz-Aureole des Massivs von Kladruby in Westböhmen. *Folia Musei Rerum Naturalium Bohemiae Occidentalis, Geologica*, pp. 1-16

Pyšek, P. & Pyšek, A. (1988): Zur spontanen Begrünung der erzhaltigen und erzlosen Abbaudeponien in Böhmen. *Preslia*, Vol.60, pp. 133-155

Raskin, R. (2003): Sind Schwermetallfluren regenerierbar? Erste Ergebnisse einer Erfolgskontrolle im Raum Stolberg. *LÖBF-Mitteilungen* 3/03: pp. 18-22

-(2008): Möglichkeiten und Grenzen der Regeneration von Schwermetallfluren. *Naturschutz und Freizeitgesellschaft*, Vol.8, pp. 60-76, Academia Verlag, Aachen

Rothmaler, W. (1963): *Silene cucubalus* Wibel subsp. *humilis* (R. Schubert) Rothm. *Feddes Repertorium*, Vol.67, No.1-3, p. 4

Schat, H.; Kuiper, E.; Ten-Bookum, W. M. & Vooijs, R. (1993): A general model for the genetic control of copper tolerance in *Silene vulgaris*: Evidence from crosses between plants from different tolerant populations. *Heredity*, Vol.70, No.2, pp. 142-147

Schat, H. & Vooijs, R. (1997): Multiple tolerance and co-tolerance to heavy metals in *Silene vulgaris*: A co-segregation analysis. *New Phytologist*, Vol.136, No.3, pp. 489-496

Schat, H.; Vooijs, R. & Kuiper, E. (1996): Identical major gene loci for heavy metal tolerances that have independently evolved in different local populations and subspecies of *Silene vulgaris*. *Evolution*, Vol.50, No. 5, pp. 1888-1895

Schubert, R. (1953/54): Die Schwermetallpflanzengesellschaften des östlichen Harzvorlandes. *Wissenschaftliche Zeitschrift der Martin-Luther-Universität Halle-Wittenberg, Math.-Nat.*, Vol.3, No.1, pp. 51-70

-(1954a): Zur Systematik und Pflanzengeographie der Charakterpflanzen der Mitteldeutschen Schwermetallpflanzengesellschaften. *Wissenschaftliche Zeitschrift der Martin-Luther-Universität Halle-Wittenberg, Math.-Nat.*, Vol.3, No.4, pp. 863-882

-(1954b): Die Pflanzengesellschaften der Bottendorfer Höhe. *Wissenschaftliche Zeitschrift der Martin-Luther-Universität Halle-Wittenberg, Math.-Nat.*, Vol.4, No.1, pp. 99-120

-(2001): Prodromus der Pflanzengesellschaften Sachsen-Anhalts. *Mitteilungen zur floristischen Kartierung in Sachsen-Anhalt*, Sonderheft 2

Schubert, R.; Hilbig, W. & Klotz, S. (2001): *Bestimmungsbuch der Pflanzengesellschaften Deutschlands*, Spektrum Akademischer Verlag, ISBN 3-8274-0915-2, Heidelberg, Germany

Schulz, A. (1912): Über die auf schwermetallhaltigem Boden wachsenden Phanerogamen Deutschlands. Jahres-Ber. Westfäl. Prov.-Ver. Wiss., Vol.40, pp. 209-227

Schwarz, O. (1949): Beiträge zur Nomenklatur und Systematik der mitteleuropäischen Flora. *Mitteilungen der Thüringischen Botanischen Gesellschaft*, Vol.1, No.1, pp. 98-99

Schwickerath, M. (1931): Das Violetum calaminariae der Zinkböden in der Umgebung Aachens. Eine pflanzensoziologische Studie. *Beiträge zur Naturdenkmalpflege*, Vol.14, pp. 463-503

Simon, E. (1975): La dynamique de la végétation de quelques sites métallifères dans les régions d'Eupen et d'Aix-la-Chapelle en relation avec les facteurs édaphiques. *Bulletin de la Société Royale de Botanique de Belgique*, Vol.108, pp. 273-386, ISSN 0583-8819

Slotta, R. (2003): Die Mansfelder Kupfer-Straße - Chance oder Utopie? *Der Anschnitt*, Vol.55, No.3-5, pp. 224-235, ISSN 0003-5238

Thal, J. (1588): *Sylva Hercynia*. Frankfurt am Main.

van der Ent, A. (2007): Kansen voor herstel van zinkflora in het boven-Geuldal. *De Levende Natuur*, Vol.108, No.1, pp. 14-19

Vekemans, X.; Lambert, A. & Lefebvre, C. (1992): Isozyme variation at the populational level in *Armeria maritima*. *Belgian Journal of Botany*, Vol.125, No.2, pp. 270-275

Vekemans, X. & Lefebvre, C. (1997): On the evolution of heavy-metal tolerant populations in *Armeria maritima*: Evidence from allozyme variation and reproductive barriers. *Journal of Evolutionary Biology*, Vol.10, No.2, pp. 175-191

Wagenbreth, O. (1973): Zur landeskulturellen Erhaltung von Bergbauhalden. *Geographische Berichte*, Vol.68, No.3, pp. 196-205

Wallroth, F. W. (1842): Monographischer Versuch über die Gewächs-Gattung *Armeria* Willd. *Beiträge zur Botanik*, Vol.1, No.1, pp. 168-218

Wein, K. (1926): Die Beziehungen zwischen Erzvorkommen und Pflanzenverbreitung in Deutschland. *Der Naturforscher*, Vol.3, No.5, pp. 240-243

Whiting, S. N.; Reeves, R. D.; Richards, D.; Johnson, M. S.; Cooke, J. A.; Malaisse, F.; Paton, A.; Smith, J. A. C.; Angle, J. S.; Chaney, R. L.; Ginocchio, R.; Jaffre, T.; Johns, R.; McIntyre, T.; Purvis, O. W.; Salt, D. E.; Schat, H.; Zhao, F. J. & Baker, A. J. M. (2004): Research priorities for conservation of metallophyte biodiversity and their potential for restoration and site remediation. *Restoration Ecology*, Vol.12, No.1, pp. 106-116

Wierzbicka, M. & Panufnik, D. (1998): The adaptation of *Silene vulgaris* to growth on a calamine waste heap (S. Poland). *Environmental Pollution*, Vol.101, No. 3, pp. 415-426

Wisskirchen, R. & Haeupler, H. (1998): *Standardliste der Farn- und Blütenpflanzen Deutschlands*, Verlag Eugen Ulmer, ISBN 3-8001-3360-1, Stuttgart (Hohenheim), Germany

Part 3

Landscape Metrics and Nature Conservation

5

The Application of Landscape Indices in Landscape Ecology

Péter Csorba and Szilárd Szabó
Institute of Geosciences, University of Debrecen, Debrecen
Hungary

1. Introduction

Landscape ecology is essentially the ecological approach of landscape research, the theoretical background that facilitates landscape use and planning in practice (Mander & Jongman, 2000; Richling et al., 1994; Wu & Hobbs, 2007; Szilassi et al., 2010). It is a study of spatial variation in landscapes, research of the biophysical and societal causes and consequences of landscape heterogeneity. In order to plan the proper landscape use according to the constantly changing social requirements, we should know the actual condition of the landscape, its function, its spatial structure and history (Tress et al., 2003).

In undisturbed landscapes, landscape function indicates the analysis of the complex relations of several landscape factors - relief, climate, hydrogeology, soil etc. In the large part of our planet, the function of the environment is not natural. Anthropogenic effects deformed it (Odum, 1963). It is more difficult to understand the function of this anthropogenic-technical (Ellenberg, 1983), biogeocybernetician (Naveh & Lieberman, 1984) system. Additionally, in countries with different level of development, the social requirements are also different. In some countries the primary objective is the excavation of raw materials or the maximum use of soil potential, while in other places the sustenance of a landscape beauty and augmentation of its role in personal, local, national identities are inevitable requirements (Pedroli, 2000; Wöbse, 2002).

Globally, population increase results in intensified landscape use. The execution of social objectives is often observable - for example increased water demand, larger construction, infrastructural or recreation areas, etc. - but is ultimately limited by landscape characteristics and capacities. The multifunctional use of landscapes (Antrop & van Eetvelde, 2000, Brandt et al., 2000; MacFarlane, 2007) requires social compromise and landscape ecology has an important role in this process (see: landscape management; Jongman, 1995). Landscape use should adapt to existing land use or to the consequences of the former land uses (Wascher, 2005) in several places. Therefore, one of the main objectives of practical landscape research is the examination of the landscape's capacity, sensitivity, vulnerability, susceptibility, resistibility and elasticity (Haines-Young, 2005; Wiens & Moss, 2005; Wu & Hobbs, 2007).

The analysis of landscape function deals with small pilot areas. However, the interpolation of the results is significantly facilitated by remote sensing. The technical quality and the resolution of satellite images are sufficient to recognize several ecological processes. A new field of science called landscape metrics (Blaschke, 2000; Klopatek & Gardner, 1999; Farina, 1998; Ingegnoli, 2003) studies the ecological consequences of landscape patterns and landscape mosaics (Forman, 1995).

Much of the actual research of the International Association for Landscape Ecology is related to landscape pattern. The core themes are:

- the spatial patterns or structures of landscapes, ranging from wilderness to cities;
- the relationships between patterns and process in landscapes;
- the relationships of human activities to landscape pattern, process and change;
- the effects of scale and disturbance on the landscape (www.landscape-ecology.org).

The spatial arrangement of the patch-matrix-corridor tells a lot about the ecological function, the condition of the landscape and the landscape planning possibilities (Benson & Roe, 2007; Leitão, et al., 2006). Based on the spatial configuration and the temporal change of the landscape mosaic we can draw conclusions about the isolation-migration conditions (Opdam, 1991), the fragmentation-metapopulation processes (Hanski, 1999) and the spatial distribution of the nutrients. In situ control is always necessary since we can identify the ecological materials and energy processes only indirectly from landscape patterns.

2. Landscape metrics and landscape ecological background

2.1 Landscape pattern as ecological frame

There are fewer sharp boundaries in the nature than those found in anthropogenic land uses. For example, there is the wide zone of dwarf mountain pine between the forest associations of the mountains and the alpine grasslands. There is also a wide zone between taiga and tundra. Relatively sharp boundary can be found between the terrestrial and aquatic ecosystems, but even near the seaside or along riverbanks there is a narrower-wider transitional habitat (e. g. the brackish water zone of the Wadden-see). The examination of these transitional zones (ecotones) is an actual ecological or landscape ecological topic, since these zones become more and more important parts of our environment and the material and energy processes (Burel & Baudry, 2005; Hansen et al., 1988).

Sharp land use boundaries (i.e. the large ecotone contrast) reduce the stability of the populations in protected areas. Every natural and semi-natural association has a core area and a boundary zone, and if this boundary or transitional zone is lacking, the core zone will shrink so a transitional ecotone appears spontaneously within the area (Pietrzak, 2001).

In the cases of the intensively used cultural landscapes in developed European countries, the protected areas often directly border on areas with higher hemeroby values (Krzywinski et al., 2009). It is not rare that - due to obvious causes - a protected wetland, a forest patch or a valuable sandy grassland is surrounded by inhabited, industrial or mining areas. In many cases the buffering zone is not assured and within meters from the protected patch there is a motorway, an industrial area or a mine. Sharp boundary lines characterize these landscapes in the hemeroby map that shows the intensity of the anthropogenic effects; there is a large difference between the levels of naturalness of adjacent patches.

Regarding landscapes strongly influenced by anthropogenic effects, the success of nature protection significantly depends on the land use pattern of the wider surroundings. Nowadays, area isolation or the artificial fragmentation of habitats is considered to be the process that most endangers the ecological value of cultural landscapes.

In the landscape ecology literature a common question regards the size of an area necessary for a particular species. There are data for several animal species. For example it is known that a field cricket (*Gryllus campestris*) population needs 3 hectares for long-term viability, but a minimum of 100-150 km^2 area is necessary for a brown bear (*Ursus arctos*) population (Farina, 1998; Fahrig, 1997; Margóczi, 2005; Šimanauskienė et al., 2008). These data are very useful, but since the protection of nature has evolved from species from species protection towards association protection and more complex landscape protection over the last few decades. It is better to examine associations or landscape types. The minimum area requirements for some associations are also well-known. For example, in the cases of softwood forests in the Carpathian Basin, 30-40 hectare patches are considered to be viable.

As nature conservation and landscape protection approach each other, it is a more and more urgent task to determine the minimum area needed for the most important landscape types that survive as a fragment of a given landscape. In the case of European landscape types, the minimum extent is thought to be 10-50 km^2, supposing well-functioning natural or semi-natural landscapes. Unfortunately, few European landscapes have truly natural material and energy processes, thus in our opinion, European landscape protection should aim to protect 50-100 km^2 continuous landscape types. The more severely antrophogenic influences affect a region, the larger the area needed a proper frame of a given biogeographic unit in terms of viable population and complete associations and regarding landscape aesthetics. A 150-200 km^2 large landscape can probably represent the given ecological landscape type in the case of an average European cultural landscape - such as the Breton bocage, the Tuscan coltura promiscua - that is moderately altered by anthropogenic effects (meso- or euhemerob level). An area of this size corresponds to the category of microchore, namely a "Kleinlandschaft" in the European landscape hierarchy (Bastian, 2000).

Optimal area size is not necessary in every case to maintain the functioningof the landscape. Depending on the landscape type, not only one but several smaller and connected patches might also proper for a healthy landscape. Since the 1980s landscape ecology has intensively examined the linkage of habitats, landscape patches, and ecological corridors for landscapes (Brandt, 1995; Csorba, 1996; Linehan et al., 1995; Konkoly-Gyuró, 2005; Schreiber, 1988). Through researche the nomenclature of these linkage elements has been specialized significantly. Several terms have appeared: ecological networks, riparian buffers, greenways, ecological infrastructure, and others. An excellent summary of this can be found in Ahern (1995). We use the most prevailing term: ecological corridors.

Although ecological corridors is not controversial in nature conservation, assessment of their effectiveness varies (Wrbka et al., 2005). It has been proven that ecological corridors can be very useful for certain components of the biosphere - such as small mammals, reptilians, amphibians - and often help to avoid isolation and the "extinction vortex". However, it is obvious that complete associations in landscape ecological corridors do not function with similar efficiency. There are positive experiences with conserved or created corridors (riparian buffers) along rivers. But for other associations, the fragmentation-compensating effect of the corridors is only partly proven. It is obvious that the often artificially created "way of escape" is improper for some species within the associations, the linkage do not function with adequate efficiency.

There is scant data regarding the function of landscape linkage elements and landscape corridors; they would be useful for the landscape nature protection. Unfortunately, in intensive cultural landscape environments (i.e. most of Europe), the conservation or the creation of this system is difficult. Due to the built-up area of the continent, and particularly the density of highways, it is impossible to connect the isolated landscape types with wide landscape corridors. On the other hand, it is also true that artificial linkages, for instance, of the forests of two adjacent mountains, would be unnatural since they were not connected originally. Otherwise, migration through such ecological corridors can be harmful, for example when the habitat is special, or if the immigration of the affinitive species and the genetic combination change the species composition. In these circumstances the level of genetic variation decreases between populations (this is so called outbreeding depression, Fenster et al., 1999). What one has managed to save artificially is not the same association as it was before. In this case the only solution would be to decrease fragmentation isolation, specifically by increasing the size of the habitat.

Unconnected landscape ecological corridors, the so-called stepping stone corridors can function well regarding certain species and can compensate for habitat fragmentation (Opdam, 1991). This, however, is only species protection. It does not strengthen the functioning of associations or landscape types.

2.2 Fundamentals of landscape metrics

Landscape metrics are used in research of landscape structure to analyse the spatial heterogeneity of landscapes and to understand the ecological function of the landscape. In landscape metric studies patches are examined since their geometric characteristics (area, perimeter, shape, etc.) and relative spatial situation (such as the proximity of patches, connectivity) can be easily calculated in mathematically (Voss et al., 2001).

At present, the indices not only describe the characteristics of a unique patch but remain an important part of landscape analysis. Landscape indices can be determined at 3 levels: (1) standard patch level, (2) class level and (3) landscape level. At patch level the indices give the area, the perimeter, the area/perimeter ratio etc. of the patches. Class-level indices describe the simple or weighted mean of the aggregated characteristics of the patches that belong to the same category, or can also be considered in terms of characteristics to allow discernment of their spatial patterns (spatial distribution, proximity, connectivity). At the landscape level, indices are calculated using the characteristics of all the patches in the landscape (Cushman et al., 2008; Forman & Godron, 1986; McGarigal & Marks, 1995; Szabó et al., 2008).

In the 1980s many landscape indices were developed (Lóczy, 2002). Their use was significantly facilitated by the appearance and spread of GIS software, and relatively cheaper and more readily available aerial photos as well as satellite imagery (Szabó, 2010).

Mathematically, landscape indices quantify the landscape characteristics in reproducible i.e. controllable ways. The shape and the spatial configuration of the patches and the situation, isolation and connectivity of the patch types can be determined using these indices. Based on these indices we can determine how suitable a given patch or patch network is for the presence or survival of species that have narrow tolerance limits or are sensitive to anthropogenic disturbance. Botanical and zoological assessments determine the species composition of habitats; the landscape metric completes the studies so that the processes of the landscape can be identified, their existence can be confirmed and the direction of the expected changes can be predicted. Their use was significantly facilitated by the appearance and spread of GIS software, and the cheap and easily available aerial photos as well as satellite images.

There is a lot of overlap among indices and they often have high correlations. Attempts to eliminate redundancies have been made; the most well-known is the work of Riitters et al. (1995) in which 55 landscape metric indices calculated from 85 maps were evaluated. The number of variables was reduced to 26 due to the contravention of the strong cross-correlations and the normality conditions. The 26 metrics were integrated to 6 main components: average patch compactness; patch texture; average patch shape; the number of the attribute classes and the perimeter/area fractal dimension of the large patches. It is also important that the authors doubted the significance of the results since they were able to identify only one research area (and it was an extreme value). The factor structure is not constant; based on our earlier results (Szabó et al., 2008; Szabó, 2010). It seems that the significance of the cross-correlations depends on the metrics used, the characteristics of the unit examined and the design of the map used. Accordingly, it is impossible to determine all the indices that can generally be used anywhere in the landscape analysis since the aforementioned indices correlation of the indices is not constant and statistical analyses do not produce a consistent value.

2.3 Landscape metric software

Landscape indices can be extracted from land use maps derived from aerial photographs, satellite images or field surveys. Base data can be in raster or vector format, depending on the requirements of the landscape metric software.

Software using categorical raster maps include FRAGSTATS, Leap II, and we have to mention the Pattern Module (Eastman, 2003) of the Idrisi. The most popular of these is FRAGSTATS (McGarigal & Marks, 1995), because it was the first and its output easily can be easily imported into other statistical software for further analysis. These calculations are significantly biased by the "grain size" of the raster map. The coarser the grain, the more error and uncertainty is in the results (Szabó, 2009).

Calculation of landscape indices based on vector maps is more precise, faster and more reliable. The results do not depend on resolution, but the number of indices that can be calculated are limited (some metrics can be determined only from rasters). The vLATE (Lang & Tiede, 2003) and the Patch Analyst Extension of ArcGIS 9.3 (Rempel, 2010) are well-known vector based landscape metric software. In addition, there is a software package for habitat-patch connection evaluating called Conefor Sensinode that determines the probability that a focal species will be able to reach the next habitat patch (Saura & Torné, 2009).

2.4 Limits of landcape metrics

During the processes of landscape measurement we must accept some limiting ecological-landscape ecological fundamentals (MacArthur & Wilson, 1967; Harris, 1984). These factors influence the interpretation of the results landscape metrics. This is explained by computer analysis limitations to be precise: sometimes the methods have not properly designed.

The first important limiting factor is our inability to differentiate the linear and the patch-like landscape elements; every unit that is different from its surrounding environment is considered to be a patch.

Patches usually border on the matrix, but the matrix often consists of patches so they are treated like patches during the processing. Since there is no difference between the matrix and the patch with regarding their geometry and topology, two actions can be taken: either (1) accept this fact and evaluate the matrix as a patch; or (2) leave the matrix out of the calculation. The latter one is recommended only when the matrix consists of one or only some large patches as it would confuse the calculations (especially at the class and landscape levels).

The distances between the patches are usually determined using Euclidean distance (in the present case the shortest distance between habitats) and ecological distance (using a so called friction surface [Gonzalez et al., 2008; Gyenizse, 2009] that reflects the movement characteristics of the focal species). Both solutions are inaccurate. Ecological distances are more realistic but during the definition of the friction surface we are not able to consider every influential factor.

Most of the software does not deal with the anthropogenic barrier. This was prepared for only the effective mesh size index elaborated by Jaeger (2000), due to Girvetz et al. (2008). In several cases we can calculate the ecological distances descibed above instead of the Euclidean distence, but they may not properly reflect the linear anthropogenic elements (linear elements appear as pixels bordering one another in a raster and there can be permeability if the line is diagonal; with low resolution, a 2-pixel-thick implementation over-represents the width of the linear element). Therefore, during the landscape metric analyses using Euclidean distances, fragmentation, isolation and connectivity pertain to natural landscapes that are free from anthropogenic ecological barriers. Thus fragmentation actually refers only to the mosaics of the natural landscapes. While interpreting the results, if the conceptional background is explained, the user will know whether the results refer to anthropogenic or natural landscapes.

In landscape ecology, we can talk about the connectivity of the matrix or the corridors (Forman, 1995; Jordan et al., 2003), but using software of landscape metrics these distinctions cannot be separated. Thus, in this case, connectivity is based on the corridor system of patches (stepping stones). Connectivity refers to the potential for species to reach landscape elements (Báldi, 1998; Whitakker & Fernández-Palácios, 2007). If we define the minimum distance that a species unconditionally covers to get to the next patch, and if the patches are within this distance, then we can call this connectivity (this is actually a nomenclatural problem since in the cases of discontinuous corridors the species must also cover the same distance to move on).

3. Landscape indices and landscape ecology

3.1 Evaluation of habitat size and habitat availability

Mosaics are the product of heterogeneity. They are the characteristics of natural landscapes and should not be confused with fragmentation: fragmentation is the breaking-up of the habitat patches or land use units, while mosaics are the form of appearance of the landscape elements (Forman, 1995). A population or an association are not necessarily only associated with large and coherent areas, but can also be found in networks of smaller patches (such as mosaic loess oakwoods with open grasslands, G. Fekete et al., 2000). If this pattern had evolved naturally, it would be considered a mosaic; but if it were anthropogenically produced then it is considered a fragment (Bissonette & Storch; 2003; Kerényi, 2007).

In the process of fragmentation there is an initial condition (a large patch that is not disturbed) and disturbance (breaking the large patch into smaller patches). In order to model the fragmentation process, Jaeger identified 6 phases:

- perforation,
- incision,
- dissection,
- dissipation,
- shrinkage,
- attrition (Jaeger, 2000, 2002).

These phases also correspond to population-degradation stages and the degradation of the quality of habitats.

Where the average land use patch size of the landscape do not reach 0.5 km^2 (i.e. approximately 200-300 m diameter units form the main part of the landscape), habitat fragmentation is probably significant there. This fragmentation datum is useful for comparison of adjacent landscapes, and can be informative for practical landscape planning and regional design in that it helps technical experts understand landscape functions.

The landscape ecology interpretation of the patch size significantly depends on what kind of patches to which the tierce or the half km^2 average refers. It is obvious that an upland grassland area of 0.5 km^2 is ecologically more valuable than a mono-cropped farm field of the same size. This difference can be well represented if fragmentation is determined with a hemeroby index.

The minimum habitat size depends on the requirements of the organisms, but the environmental factors can modify it (for example either adjacent to a busy road or distant from any anthropogenic disturbance). The characteristics of the adjacent patches, the quality of the matrix and the distance of the closest identical patch are also important determinants (Pope et al., 2000; Sisk & Haddad, 2002). Generally we can say that the smallest necessary patch size is adequate for the species that need the largest area among all local species and also takes into consideration possible disturbances (Margóczi, 2005). Thus, knowing only the patch size is insufficient to determine the probabiliy of a species' survival, but it is a good start for its assessment.

3.2 The deviation of patch size, the SLOSS debate: Single Large or Several Small?

The deviation of landscape patches shows whether there are significant differences in landscape patch pattern within the landscape in terms of scenery and landscape functioning. If the variability of the patch size is small, the landscape consists of land use units of more identical size.

Thus, patch size alone is also important information; its differential interpretation resulted in the SLOSS (Single Large or Several Small) debate in the 1970s. It has also been a controversial issue recently: can several small patches provide for as much species richness as a large patch (or some large patches)? The answer is not obvious. Both have advantages and disadvantages (Margóczi, 2005; Simberloff & Abele, 1981; Whitakker & Fernández-Palácios, 2007). Studies have shown that in several cases (butterflies, birds and forest associations) that small fragments result in higher species richness than a single large patch of the same size (Baz & Garcia- Boyero, 1996; Bánszegi et al., 2000; Tscharntke et al., 2002). This does not mean that large patches are not valuable, they are significant for the preservation of specialist species.

Several ecologists feel that neither of the two landscape structures is favourable. There is no conclusive evidence as to the most stabile eco-geometrical pattern for landscape structure. Thus, a mosaic of 2-3 medium-sized and some small patches can be very advantageous. This is a landscape structure that is the least vulnerable to external forces, such as natural catastrophes or anthropogenic threats.

3.3 Connectedness

The SLOSS debate brings us to the problem of patch availability. Small patches form a functional system only if the distance between them is small enough that resident species can reach them. Every system of habitat patches is stable only if there is functional connectedness between patches. Landscape connectivity indices numerically determine whether the landscape structure facilitates or impedes the movement of the animals.

The numerical definition and the interpretation of the results are not easy, however. There are several questions about what ensures connectivity. Is the distance between the patches sufficient? What effects does the matrix have on species movement? For example, birds require no connection between habitat patches since they can easily fly over long distances, but a small reptile or amphibian require connections. Not only the the distance is important, but the characteristics and quality of the matrix between the habitat patches are also important factors. Therefore, connectivity cannot be determined by a single number. Species-specific calculations are necessary (Crooks & Sanjayan, 2006). There is often little information about the movement of species, we can usually only identify the distance that a given animal can move (a threshold distance), and connectedness is calculated based on this value.

3.4 Shape

The determination of the area/perimeter ratio can be an indicator for landscape metric analyses. This method appeared in the earliest studies (Jurko, 1987) and eventually became a prevalent index (Farina, 1998; Fekete & Fekete 1998; Turner & Gardner, 1991).

If the area/perimeter ratio of the patches is high, the patches have short perimeter on average, i.e. zigzag patch shapes do not dominate. In terms of ecology, these landscapes have less ecotone than other areas of a similar size, and they are more stable ecologically. Patches encompassed by long border zones (low area/perimeter ratio), are unstable ecologically are more likely to be threatened by external forces.

Obviously, there are communities (such as those surrounding a hot spring) that require only an area of 0.5 km^2, long, narrow habitat patch providing stable living conditions, but probably a large, homogeneous habitat geometrical characteristic that have extensive, inner patch section is necessary for most ecological patches.

The shape indices of patches determine the compactness and help to indicate the sensitivity of the patches. The more the border line of the patch is similar to a straight shoreline (with few "bays" or "peninsulas"), the more likely the patch is ecologically stable. Peninsulas reduce the area in which disturbance-sensitive species can reside as matrix interaction is more intensive.

In the course of our studies of shape indices (Szabó, 2009) we examined the dependence of the indices on the patch size, pixel count (in the case of raster analysis), orientation (whether rotation of the patches affects the values) and uniqueness (whether the value of a given index belongs to only one patch shape). Accordingly, the results are that:

- the patch size does not or only slightly influences the Shape Index, Related Circumscribing Circle;
- the number of pixels does not or only slightly influences Area, Perimeter, Radius of Gyration, Perimeter/Area, Fractal Dimension, Related Circumscribing Circle, Shape Index;
- the azimuth of the object does not or only slightly influences: Radius of Gyration, Fractal Dimension, Related Circumscribing Circle, Contiguity Index;
- unique, namely does not or only slightly correlates with the other indices: Radius if Gyration, Related Circumscribing Circle;
- their value is definitely true for only one form (patch shape): Shape Index, Related Circumscribing Circle.

Accordingly, the most useful index is the Related Circumscribing Circle, but its use is not prevalent in practice. The area/perimeter ratio of the patches is often applied, as it was mentioned above, to characterize the eco-geographic stability since both its calculation and interpretation is easy. At the same time, it can be deceptive because the value of the index also changes in the case of the patches of the same shape but different size, and the ratio can be the same regarding the patches of different shape. In terms of the use in landscape ecology, other authors emphasize Fractal Dimension and indices related to a reference shape e. g. Shape Index (Krummel et al., 1987; Mezősi & Fejes, 2004). Leitão et al. (2006) do not recommend the use of the Shape Index in patch morphology examinations since the values can be the same if patches have identical perimeters and areas.

There are several other indices that are not implemented in landscape metric software, but they can contribute to the identification of sensitive patches, such as the elongation and the convolution index suggested by Forman (1995). Among the metrics, the Radius of Gyration is the most favourable to estimate the elongation: of two case of the same size, the patch of having a higher value is the elongated patch (Leitão et al., 2006). This can be useful in morphometric analyses, but size dependence of Radius of Gyration must be considered; the comparison is not clear - patches in the nature are usually of different sizes.

3.5 Patch gradients

The system of landscape ecology corridors with stepping stone habitat patches can sustain the oligo- or mesohemerobe level of the cultural landscape. However, this complete ecological system provides sufficient linkage to sustain and stabilize neither populations of less mobile animals nor of the whole ecosystem. The only migration option for them is movement toward a patch that is only slightly different from their original natural habitat (forest, scrub forest, or grassland), when possible. For example, if a shrubby-grassy connection exists between two isolated grass patches, then this direction is a real migration possibility for the species. If the grass patch is surrounded by ploughed farmlands or plantations, and there is only one way to reach adjacent grassland through a single shrubby-grassy patch, then this connection is strategically important. Based on the landscape types of Central Europe, the members of the patch gradient can consist of the following land use patches:

forest → scrub forest → meadow/grassland → fallow → ploughland →
vineyard/plantation → built-up areas

These categories can also be separated well according to the land use patch type of the CORINE. Naturally, no such (complete) patch series spectrum is found in most of the landscapes or landscape fractions. Most areas consists of only 2-4 patch types, such as:

forest → scrub forest → meadow/grassland → fallow

scrub forest → meadow/grassland → fallow → vineyard

fallow → ploughland → built-up areas

In our opinion, ecology-based landscape planning should increase the consideration these kinds of landscape patterns. The patch series where members are the most similar (thus they are the smallest barrier against migration) can be easily marked on detailed landscape maps.

The method can be the practical application of the landscape structure concept (McGarigal & Cushman, 2005). Instead of using increasing number of general landscape indices:

„We would better served by quantifying the local landscape pattern across space as it may be experienced by the organisms of interest, given their perceptual abilities."
(McGarigal & Cushman, 2005)

„We advocate the expansion of the paradigm to include a gradient-based concept of landscape structure that subsumes the patch-mosaic model as a special case. The gradient approach we advocate allows for a more realistic representation of landscape heterogeneity by not presupposing discrete structures, facilitates multivariate representations of heterogeneity compatible with advanced statistical and modelling techniques used in other disciplines, and provides a flexible framework for accommodating organism-centred analyses." (McGarigal & Cushman, 2005)

According to the authors, the moving-window analysis of FRAGSTATS software is appropriate for such analyses.

Determination of patch gradient is very similar to the McGarigal & Cushman (2005) method. It provides valuable background information for planning to achieve protection of species and ecosystems. It is obvious that these patch pattern gradients "generally" provide less useful alternatives for species. But if the aim of nature protection is to enable a species or species group to broaden their narrow or endangered habitat, this method can be helpful. There is also a chance that one could use patch gradient maps to create strategic goals for nature protection.

3.6 Landscape patches and built-up areas, traffic infrastructure

Serious ecological consequences from road construction are obvious, so when investments have reached the social threshold, the public often tries to force decision makers to find the most environmentally friendly alternatives (Lodé, 2000).

The harmful effects of busy roads have been studied since the rise of mass adoption of motorized transportation (Clarke, 1930). Beck (1956) called roads ecological infection canal ("Infektionkanal"). In the 1970s it was proven with efficient experiments that roads act as ecological barriers (Mader, 1979). Seven hundred and forty-two ground-beetles (*Abax ater*) were marked near a not-too-busy upland road in Germany and only two beetles were able to cross it. The rest of beetles turned back from the edge of the concrete. They were not killed by the cars, but the strange surface and open space prevented the movement of the animals. Mader also proved that only a few ground beetles surmounted a moderately busy road wider than 2.5 m. Less than 10% of spiders and small mammals overcome the road barrier (Mader, 1984).

Other authors reported similar results with regard to the habitat fragmenting of roads and urban areas (Forman, 1997, Harris, 1984; Schreiber 1988; Tóthmérész et al., 2011). Indeed, one study analysed the effects of landscape fragmentation on the function of the entire food chain (Polis et al., 2004).

Despite the vast number of animals that are killed along the roads (Erritzoe et al. 2003, Forman, 1997; Langgermach et al., 2006; van der Zande et al., 1980; Voss & Opdam, 1993), the ecological balance is not impacted to the extent that one might suspect. Even in the cases of busy roads, the reduction of the ecological diversity and population stability can be proven clearly only in the 200-500 m wide roadside zone (Farina, 1998; Forman & Alexander, 1998).

There are few studies that demonstrate that the fragmenting effects of the built-up areas play important roles in ecological fragmentation and the habitat shrinkage (Mühlenberg & Slowik, 1997; Paddison, 2001). Settlement ecology examines settlements rather than living creatures settling in unique habitats or being crowded from them, and it considers how the settlements surround the special habitats, how sensitive ecotopes are isolated, or how the settlements impede species movement. Settlements are less permeable than a road. A narrower village is a wider physical barrier than a highway with 3x3 lanes. However, comparing the strength of the ecological barriers it is not certain that the migration-preventing effect of a small village is more significant than a highway enclosed within a fence.

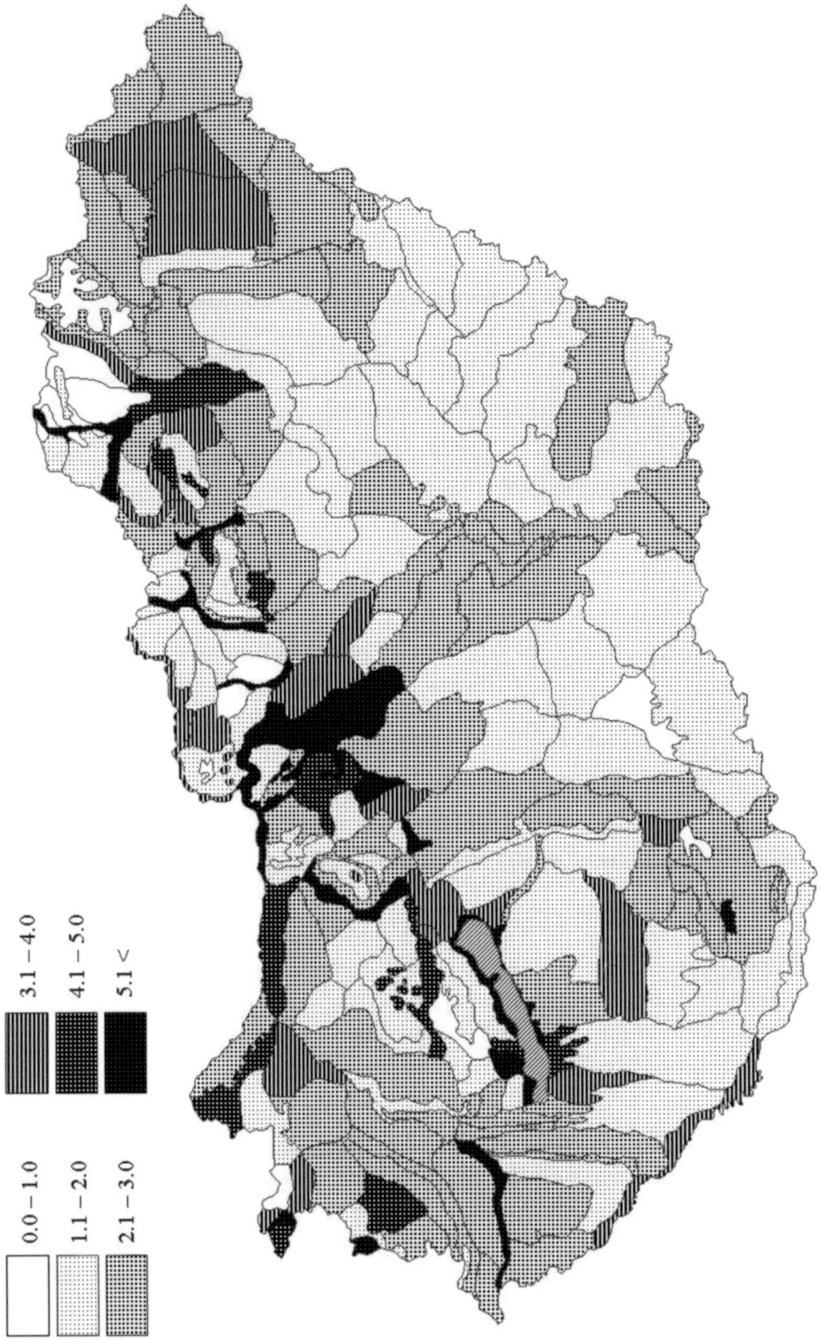

Fig. 1. Weighted and summarized data for the fragmentation effect of settlements and traffic infrastructure in Hungary (modified data km/km^2)

Regarding biodiversity, a settlement landscape can exceed natural species richness. Although ecosystems are certainly less diverse along roads, species diversity of roadside weed vegetation is generally very small.

The degree of ecological fragmentation of landscapes is a useful index for landscape protection and planning. An index was determined using the 1:250 000 scale maps of the Cartographia Road Atlas of Hungary (Csorba, 2008). Based on the Cadastral of Microregions of Hungary, the boundaries of the microregions were drawn into the maps, and then within those fixed boundaries the greatest diameter of small settlements and the length of roads and railways were measured. In the case of large settlements the extent of inner parts and traffic intensities of the roads were considered. In the case of railways, whether railway lines are single- or double-tracked was considered. Results were modified using weighting, where the locations of protected natural areas compared to locations of settlements and then roads and railways were incorporated. The agglomeration processes of large settlements were also considered as they may too restrict the ecological gates and corridors along the migration routes of plant and animal species. Values of the fragmentation index were mapped in km/km^2 for the 230 microregions of Hungary, but in the present study the values distorted by weighting were presented (Fig. 1). In 23 microregions, the fragmentation index is higher than 5. The strongest ecological barriers are found in microregions in valleys and small basins within mountain regions of medium height and their environs.

4. Landscape metrics in practice

The results and the methods of landscape metric research are incorporated into practical applications relatively slowly (Jongman, 1995, Marsh, 1997). The two most significant uses are represented through the following examples.

The Forestry Commission of Scotland and Scottish Natural Heritage planned afforestation sites using BEETLE (an ESRI ArcView module) that examines landscape pattern in terms of a given species' needs. The Commission attempted to ensure landscape connectivity in the new system of forest fragments and in newly planted forest patches (Stone, 2007). In the programme (Cairngorms Forest and Woodland Framework - CFWF, 1999) an attempt to increase the habitat of corn bunting (*Miliaria calandra*) using landscape ecology principles was undertaken with the help of the local land-owners. The owners were encouraged to plant forests in the designated sites to build up the ecological system of the Scottish Highlands. The programme was not successful since the timeframe was short and the interest of local land owners was often not oriented toward afforestation. However, two separated forest patch in the Beinn Eighe National Nature Reserve were planted with 20 ha of forest. Drawing the lesson from the failure of the CFWF program, the Forest Commission elaborated a support system for owners who planted forests. They were not told where to plant forest, but they planted where they wanted to plant. Then the connectivity was tested using BEETLE model. The rate of the support depended on whether doing so would improve the connectivity of the plantation. During 5 months 1.6 million pounds were distributed to create 400 ha of new forest and 41 landscape ecological corridors that made a total area of 8500 ha more functional (Stone, 2007). Considering money as motivation, and the use owner's own ideas, it succeeded in improving the environment in terms for nature conservation; this is what all decision-maker should notice.

The second example is practical use of MESH index elaborated by Jaeger (2000). The German Federal Environment Agency made recommendations to reduce fragmentation based on the effective mesh density (m_{eff}). Accordingly, large habitat patches must be conserved or increased, and by 2015 the ratios of the area with the certain 2002 mesh density values (m_{eff}) must be reduced: 1.9% of patches had areas less than 10 km^2, 2.4% ranged from 10-20 km^2 areas, and 2.8% were 20-35 km^2 areas (Penn-Bressel, 2005). Limiting construction of roads, preventing territorial expansion of cities, encouraging in-fill developments, and long-term regional development plans were necessary tools. The effective mesh size as an indicator of the ecological condition of the cantons (Jaeger et al., 2008) was also used in the Swiss Monitoring System of Sustainable Development (MONET).

The effective mesh size is a popular metric not just in Europe but has also been used in the USA: in California, m_{eff} as the index of fragmentation (Jaeger et al., 2008) is used for the optimization of transport lines and in nature conservation and landscape protection decisions. This is explained by the modified transport rules: nature conservation must be considered during planning (Safe, Accountable, Flexible, Efficient Transportation Equity Act [SAFETEA-LU]). Accordingly, the possible effects of transportation line construction on habitats and methods to reduction these impacts must be identified before the construction of the transport lines. CALTRANS (California Department of Transportation) entrusted the research crew with effect analysis and thus the m_{eff} index came to the front (Girvetz et al., 2008; Thorne et al., 2009).

5. Acknowledgment

This paper was supported by the János Bolyai Research Scholarship of the Hungarian Academy of Sciences (bo_391_09) and the Meridian Foundation.

6. References

Ahern, J. (1995). Greenways as a planning strategy. *Landscape and Urban Planning*. Vol.33, pp. 131-155, ISSN 0169-2046

Antrop, M. & van Eetvelde, V. (2000). Holistic aspects of suburban landscapes: visual image interpretation and landscape metrics. *Landscape and Urban Planning,* Vol.50, No.1, pp. 43-58, ISSN 0169-2046

Bastian, O. (2000). Landscape classification in Saxony (Germany) – a tool for holistic regional planning. *Landscape and Urban Planning,* Vol. 50, No.1, pp. 145-155, ISSN 0169-2046, Elsevier, The Netherlands

Burel, F. & Baudry, J. (2005). *Landscape Ecology. Concepts, Methods and Applications,* Science Publishers Inc., 362 p, ISBN 1-57808-214-5, Enfield (USA) - Plymouth (UK)

Baz, A. & Garcia-Boyero, A. (1996). The SLOSS dilemma: a butterfly case study. *Biodiversity and Conservation,* Vol.5, No.4, pp. 493-502, ISSN 1572-9710

Báldi A. (1998). Theoretical considerations in the design of ecological networks (in Hungarian). *Állattani Közlemények,* Vol.83, pp. 29-40, ISSN 0002-5658

Bánszegi Z.; Báldi A. & Bankovics A. (2000). Species-area relationship and nestedness in nature reserves. *Ornis Hungarica,* Vol.10, pp. 17-26, ISSN 2061-9588

Beck, O. (1956). Oasen der Ruhe für den Wanderer. *Veröffentlichtungen Landesanstalt Naturschutz und Landschaftspflege* Baden-Württemberg, 24. pp. 72-79.

Benson, J.F. & Roe, M (Eds.) (2007). *Landscape and sustainability* Routledge, 320 p. ISBN 0-415-40443-6, London and New York, UK & USA

Bissonette, J. & Storch, I. (2003). Understanding Fragmentation: Getting Closer to 42. *Conservation Ecology* Vol.7, No.2, r5, 15.07.2011, Available from URL: http://www.consecol.org/vol7/iss2/resp5/, ISSN 1708-3087

Blaschke, T. (2000). Landscape Metrics: Konzepte eines jungen Ansatzes der Landschaftsökologie und Anwendungen in Naturschutz und Landschaftsforschung. *Archiv für Naturschutz und Landschaftsforschung*, Vol.39, pp. 267-299, ISSN 0003-9306

Brandt, J. (1995). Ecological networks in Danish planning. *Landschap*, Vol.12, No. 3, pp. 63-76, ISSN 01696300

Brandt, J.; Tress, B. & Tress, G. (Eds.) (2000). Multifunctional Landscapes: Interdisciplinary Approaches to Landscape Research and Management, Conference material for the international conference on "*Multifunctional Landscapes: Interdisciplinary Approaches to Landscape Research and Management,*"Centre for Landscape Research, University of Roskilde, Denmark October 18-21, 2000, Centre for Landscape Research Roskilde

Clarke, H. (1930). Birds killed by automobiles. *Bird-Lore*, Vol.32, No.4, p. 271

Crooks, K.R. & Sanjayan, M. (2006). *Connectivity Conservation*. Cambridge University Press, 712 p, ISBN 9780521857062, New York, USA,

Csorba, P. (1996). *Landscape ecological corridors on the East Foothills of the Tokaj Mountains (Hungary),* Arbeiten aus dem Institut für Landschaftsökologie, Westfälische Wilhelms Universität, Vol.2, pp. 217-229, ISSN 1431-1313, Münster, Germany

Csorba, P. (2008). Landscape ecological fragmentation of the small landscape units (Microregions) of Hungary based on the settlement network and traffic infrastructure. *Ekologia (Bratislava),* Vol.27, No.1, pp. 99-116, ISSN 1337-947X

Cusman, S.A.; McGarigal, K. & Neel, M.C. (2008). Parsimony in landscape metrics: Strenth, universality and consistency. *Ecological Indicators,* Vol.8, pp. 691-703, ISSN 1470-160X

Eastman, J. R. (2003). *IDRISI Kilimanjaro: Guide to GIS and Image Processing,* Clark Laboratories, Clark University, Worcester, USA.

Ellenberg, H. (1983). *Ökologische Beiträge zur Umweltgestaltung*. Ulmer Verlag, Stuttgart

Erritzoe, J., Mazgajski, T.D. & Rejt, L. (2003). Bird casualities on European roads. *Acta Ornithologica,* Vol.38, No.2, pp. 77-93, ISSN 1734-8471

Fahrig, L. (1997). Relative effects of habitat loss and fragmentation on population extinction. *Journal of Wildlife Management,* Vol.61, No.3, pp. 603-610, ISSN: 1937-2817

Farina, A. (1998). *Principles and methods in landscape ecology: Towards a science of the landscape,* Springer-Verlag, 235 p, ISBN 0412730405,

Fekete G. & Fekete Z. (1998). Distance distribution between patch systems: a new method to analyse community mosaics. *Abstracts Botany,* Vol 22, pp. 29-35, ISSN 0133 6215

Fekete G.; Virágh K.; Aszalós R. & Précsényi I. (2000). Static and dynamic approaches to landscape heterogeneity in the Hungarian forest-steppe zone. *Journal of Vegetation Science,* Vol.11, pp. 375-382, ISSN 1100-9233

Fenster, C.B. & Galloway, L.F. (1999). Inbreeding and outbreeding in natural populations of Chamaecrista fasciculate (Fabaceae). *Conservation Biology,* Vol.14, No.5, pp. 1406-1412, ISSN 0006–3207

Forman, R.T.T. (1995). *Land Mosaics – The Ecology of Landscape and Regions,* Cambridge University Press, 632 p, ISBN 9780521479806, Cambridge, UK

Forman, R.T.T. (1997). *Ecological effects of roads: Toward three summary indices and an overview for North America. Habitat fragmentation and infrastructure,* In: Proceedings of the International. Conference on Habitat Fragmentation Infrastructure and the Role of Ecological Engineering, Canters, K. (Ed.), Maastricht, 17-21 Sept. 1995, pp. 40-54.

Forman, R.T.T. & Alexander, L.E. (1998). Roads and their major ecological effects. *Annual Review Ecology and Systematics,* Vol.29, pp. 207-231, ISSN 1545-2069

Forman, R.T.T. & Godron, M. (1986). *Landscape Ecology,* John Wiley and Sons, 619 p.

Fukamachi, K.; Iida, S. & Nakashizuka, T. 1996. Landscape patterns and plant species diversity of forest reserves in the Kanto region, Japan. *Vegetatio,* Vol.124, No.1, pp. 107-114, ISSN 1385-0237

Girvetz, E.H.; Thorne, J.H. & Jaeger, J. (2007). *Integrating habitat fragmentation analysis into transportation planning using the effective mesh size landscape metric.* The 2007 International Conference on Ecology & Transportation „Bridging the Gaps, Naturally", Transportation and Conservation Planning, Little Rock, Arkansas, pp. 281-293.

Girvetz, E.H.; Thorne, J.H.; Berry, A.M. & Jaeger, J. (2008). Integration of landscape fragmentation analysis into regional planning: A statewide multi-scale case study from California, USA. *Landscape an Urban Planning,* Vol.86, pp. 205-218, ISSN 0169-2046

Gyenizse, P. (2009). *Geinformation analysis in Pécs,* Geographica Pannonica Nova Series, No.7, HU ISSN 1789-0527, Pécs, Hungary

Gonzalez, J.R.; del Barrio, G. & Duguy, B. (2008). Assessing functional landscape connectivity for disturbance propagation on regional scales – A cost surface model approach applied to surface fire spread. Ecological Modelling Vol.211, pp. 121-141, ISSN 0304-3800

Haines-Young, R. (2005). *Landscape pattern context and process,* In: Issues and Perspectives in Landscape Ecology, Wiens, J. & Moss, M. (Ed.), Cambridge Studies in Landscape Ecology, Cambridge University Press, Cambridge, pp. 103-111, ISBN 10-0-521-53754-1

Hansen, A.J., di Castri, F. & Naiman, R.J. (1988). Ecotones: what and why? *Biology International,* Vol.17, No. Special Issue, pp. 9-46, ISSN 0253-2069

Hanski, I. (1999). *Metapopulation ecology,* Oxford Series in Ecology and Evolution, Oxford University Press, 313 p, ISBN 0198540655, New York (USA) – Oxford, UK

Harris, L.D. (1984). *The fragmented forest: Island biogeography theory and the preservation of biotic diversity,* Univ. of Chicago Press, 211 p, ISBN 3 540 42743, Chicago, IL

Ingegnoli, V. (2003). *Landscape ecology: A Widening foundation,* Springer Verlag, 357 p, ISBN-10 3540427430, New York (USA), Berlin, Heidelberg (Germany)

Jaeger, A.G.J. (2000). Landscape division, splitting index, and effective mesh size: new measures of landscape fragmentation. *Landscape Ecology* Vol.15, pp. 115-130. ISSN 0921-2973

Jaeger, J. (2002). *Landschaftszerschneidung,* Ulmer Verlag, 447 p, ISBN-10: 3800136708, Stuttgart, Germany

Jaeger, J.; Bertiller, R.; Schwick, C.; Müller, K.; Steinmeier, C.; Ewald, C.K. & Ghazoul, J. (2008). Implementing Landscape Fragmentation as an Indicator in the Swiss

Monitoring System of Sustainable Development (MONET). *Journal of Environmental Management,* Vol.88 No.4, pp. 737-751, ISSN 0301-4797

Jordan, F.; Báldi, A.; Orczi, K-M.; Rácz, I. & Varga, Z. (2003). Characterizing the importance of habitat patches and corridors in maintaining the landscape connectivity of a Pholidoptera transsylvanica (Orthoptera) metapopulation. *Landscape Ecology,* Vol.8, pp. 83-92, ISSN 0921-2973

Jongman, R. (1995). Nature conservation planning in Europe: developing ecological networks. *Landscape and Urban Planning,* 32: 169-183, ISSN: 0169-2046

Jurko, A. (1987). Constancy and resistence as basic components of stability in landscape ecology: a case study on vegetation in East Slowakian Lowland. *Ekológia (CSSR),*Vol.6, No.4, pp. 417-438. ISSN 1337-947X, Bratislava, Czech Republic

Kerényi, A. (2007). *Landscape Preservation* (in Hungarian), Pedellus Tankönyvkiadó. 184 p, ISBN 9639612545, Debrecen, Hungary

Klopatek, J.M. & Gardner, R.H. (eds.) (1999). *Landscape ecological analysis, issues and applications,* Springer Verlag, 400 p, ISBN 0387983252, New York (USA), Berlin, Heidelberg, Germany

Konkolyné Gyuró É. (2005). *The Role of Green Corridors in Landscape Restoration.* In: Konkoly Gyuró É. (ed.) Greenways. Conference Presentations on Ecological Corridors, Green Corridors, Sopron, Hungary, pp. 1-21.

Krummel, J.R.; Gardner, R.H.; Sugihara, G.; O'Neill, R.V. & Colemann, P.R. (1987). Landscape Patterns in a Disturbed Environment. *Oikos,* Vol.48, pp. 321-324, ISSN 1600-0706

Krzywinski, K.; O'Connell, M. & Küster, H. (2009). *Europäische Kulturlandschaften,* Aschenbeck Media, ISBN 978-3-941624-70-2, Bremen, Germany

Lang, S. & Tiede, D. (2003). *vLATE Extension für ArcGIS - vektorbasiertes Tool zur quantitativen Landschaftsstrukturanalyse,* ESRI Anwenderkonferenz 2003, Innsbruck, Austria

Langgermach, T.; Kenntner, N.; Krone, O.; Müller, K. & Sömmer, P. 2006: Anmerkungen zur Bleivergiftung von Seadlern (Haliaeetus albicilla). *Natur und Landschaft,* Vol.81, No.6, pp. 320-326, ISSN 0028-0615

Leitão, A.B.; Miller, J.; Ahern, J. & McGarigal, K. (2006). *Measuring Landscapes: A Planner's Handbook,* Island Press, 272 p.

Linehan, J.; Gross, M. & Finn, J. (1995). Grenneway planning: developing a landscape ecological network approach. *Landscape and Urban Planning,* 33. pp. 179-193, ISSN 0169-2046

Lóczy, D. (2002). *Landscape evaluation, land evaluation* (in Hungarian), Dialóg Campus, 308 p, ISBN 9789639310278, Pécs, Hungary

Lodé, T. (2000). Effects of a motorway on mortality and isolation of wildlife populations. *Ambio,* Vol.29, pp. 163-166, ISSN 0044-7447

Mader, H-J. (1979). *Die Isolationswirkung von Verkehrsstrassen auf Tierpopulationen untersucht am Beispiel von Arthropoden und Kleinsaugern der Waldbiozönose,* Schriftenreihe für Landschaftspflege und Naturschutz, Bonn-Bad Godesberg, Heft 19, 126 p.

Mader, H-J. (1984). Animal habitat isolation by roads and agricultural fields. *Biological Conservation,* Vol.29, pp. 81-96, ISSN 0006-3207

Mander, Ü. & Jongman, R. (eds), (2000). *Consequences of land use changes,* In: Advances in Ecological Sciences Series 5, WIT Press, pp. 11-38. ISSN: 1369-8273, Southampton, Boston, USA

Marsh, W.M. (1997). *Landscape Planning: Environmental Applications,* John Wiley and Sons, 434 p, ISBN 0471854344, New York, USA

Margóczi K. (2005). *Biological Aspects of Landscape Protection (in Hungarian)* JATEPress, 108 p, Szeged, Hungary, ISBN 963 482 228 2

MacArthur, R.H. & Wilson, E.O. (1967). *The theory of island biogeography,* Princeton University Press, 203 p, ISBN 0691088365, Princeton, USA

MacFarlane, R. (2007). Multi-functional landscapes: Conceptual and planning issues for the countryside. In: Benson, J.F. & Roe, M. (Eds.) *Landscape and sustainability* pp. 138-166. Routledge, 320 p. ISBN 0-415-40443-6, London and New York, UK & USA

McGarigal, K.; Cushman, S.A. (2005). *The gradient concept of landscape pattern,* In: Issues and Perspectives in Landscape Ecology, Wiens, J. & Moss, M. (Ed.), Cambridge Studies in Landscape Ecology, Cambridge University Press, pp. 112-119, ISBN 10-0-521-53754-1, Cambridge, UK

McGarigal, K. & Marks, B.J. (1995). *FRAGSTATS: spatial pattern analysis program for quantifying landscape structure.* USDA For. Serv. Gen. Tech. Rep. PNW-351. 141 p.

Mezősi, G. & Fejes, Cs. (2004). *Landscape metrics* (in Hungarian), In: Táj és Környezet, MTA FKI, pp. 229-242, ISBN:963-9545-04-X, Budapest, Hungary

Moser, D.; Zechmeister, H.G.; Plutzar, C.; Sauberer, N.; Wrbka, T. & Grabherr, G. (2002). Landscape patch shape complexitiy as an effective measure for plant species richness in rural landscapes. *Landscape Ecology,* Vol.17, pp. 657-669, ISSN 0921-2973

Mühlenberg, M. & Slowik, J. (1997). *Kulturlandschaft als Lebensraum.* UTB 1947, Quelle und Meyer, 321 p. ISBN 382521947X, Wiesbaden, Germany

Naveh, Z. & Lieberman, A.S. (1984). *Landscape Ecology Theory and Application.* Springer Verlag, 341 p. ISBN 3-540-97169-5, New York, USA

Odum, E.P. (1963). *Fundamentals of Ecology.* Saunders, Philadelphia 574 p.

Opdam, P. (1991). Metapopulation theory and habitat fragmentation: a review of holarctic breeding bird studies. *Landscape Ecology,* Vol. 5, No.2, pp. 93-106, ISSN 0921-2973

Paddison, R. (ed.) (2001). *Handbook of Urban Studies.* Sage, London, XVII, 494 p.

Pedroli, B. (ed.) (2000). *Landscape – Our Home,* Lebensraum Landschaft. Indigo, Zeist, 222 p. ISBN 90 6038 4903, Stuttgart, Germany

Penn-Bressel, G. (2005). Begrenzung der Landschaftszerschneidung bei der Planung von Verkehrswegen. *GAIA,* Vol.14, No.2, pp. 130–134, ISSN 0940-5550

Pietrzak, M. (ed.) (2001). *Landscape Boundaries* (in Polish). Problemy Ekologii Krajobrazu, Vol. 7, 197 p. ISBN 83-89105-92-6, Sosnowiec, Poland

Polis, G. A. & Power, M. E. ,& Huxel, G. R. (eds.) (2004). *Food Webs at the Landscape Level.* The University Chicago Press, XV. 548 p.

Pope, S.E. & Fahrig, L. & Merriam, G. (2000). Landscape complementation and metapopulation effects on leopard frog populations. *Ecology,* Vol.81, pp. 2498-2508, ISSN 1365-2745

Richling, A.; Malinowska, E.; Lechnio, J. (eds.) (1994). *Landscape Research and its Applications in Environmental Management,* IALE Polisch Assoc., 289 p. ISBN 83-85785-11-6, Warsaw, Poland

Riitters, K.H.; O'Neill, R.V.; Hunsaker, C.V.; Wickham, J.D.; Yankee, D.H.; Timmins, S.P.; Jones, K.B.; Jackson, B.L. (1995). A factor analysis of landscape pattern and structure matrics. *Lansdcape Ecology,* Vol.10, No.1, pp. 23-40, ISSN 0921-2973

Rempel, R. (2010). *Patch Analyst for ArcGIS*, Centre for Northern Forest Ecosystems Research, 07.15.2011. Available from URL: http://flash.lakeheadu.ca/~rrempel/patch/index.html

Saura, S. & Torné, J. (2009). Conefor Sensinode 2.2: a software package for quantifying the importance of habitat patches for landscape connectivity. *Environmental Modelling & Software*, Vol.24, pp. 135-139, ISSN 1364-8152

Schreiber, K-F. (Ed.) (1988). *Connectivity in landscape ecology,* Proceedings of the 2nd International Seminar of the IALE in Münster. Münsterische Geographische Arbeiten, 29, 255 p, ISBN 3506732293, Schöningh, Paderborn, Germany

Šimanauskienė, R.; Linkecienė, R. & Taminskas, J. (2008). *Spatial analysis of forest fragmentation (the case study of Lithuania).* In: Implementation of LAndscpe Ecology in New and Changing Condtions, Boltoziar, M. (Ed.), Institute of Landscape Ecology, Slovak Academy of Sciences, ISBN 978-80-89325-03-0, Nitra, Slovakia

Simberloff, D. & Abele, L.G. (1981). Refuge design and island biogeographic theory: effects of fragmentation. *The American Naturalist,* Vol.120, No.1, pp. 41-50, ISSN 0003-0147

Sisk, T.D. & Haddad, N.M. (2002). *Incorporating the effects of habitat edge into landscape models: Effective area models for cross-boundary management,* In: Integrating landscape Ecology into Natural Resource Management, Lui, J. & Taylor, W. (Ed.), Cambridge University Press, pp. 208-240, ISBN 9780521784337, Cambridge, UK

Smith de, M.J.; Goodchild, M.F. & Longley, P.A. (2006). *Geospatial Analysis: A Comprehensive Guide to Principles, Techniques and Software Tools,* Troubador Publishing Ltd., 416 p., ISBN 13: 9781848761582

Stone, D. (2007). Big plans and little plans: delivering land use change designed by landscape ecology. *Acta Geographica Debrecina Landscape & Environment,* Vol.1, No.1, pp. 68-74, ISSN 1789-7556

Szabó, Sz.; Csorba, P. & Varga, K. (2008). *Landscape indices and landuse – Tools for landscape managment.* In: Methods of landscape research, Plit, J. & Andreychouk, V. (Ed.), Dissertations Commission of Cultural Landscape, Vol. 8, pp. 7-20, ISSN 1896-1460, Sosnowiec, Poland

Szabó, Sz. (2009). Examination of application possibilities of landscape metrics in landscape analysis (in Hungarian), Habilitation thesis, University of Debrecen, Debrecen, Hungary, 31.07.2011. Available from URL: http://landscape.geo.klte.hu/pdf/szszabo_habil.pdf (manuscript)

Szabó, Sz. (2010). Comparison of CLC50 and CLC2000 databases with landscape metrics (in Hungarian). *Tájökológiai Lapok,* Vol. 8, No. 1, pp. 13-23, ISSN 1589-4673

Szilassi, P.; Jordán, Gy.; Kovács, F.; van Rompaey, & van Dessel, W. (2010). Investigating the link between soil quality and agricultural land use change. A case study in the lake Balaton catchment, Hungary. *Carpathian Journal of Earth and Environmental Sciences,* Vol.5, No.2, pp. 61-70, ISSN 1844 - 489X

Thorne, J.H.; Huber, P.R.; Girvetz, E.H.; Quinn, J. & McCoy, M.C. (2009). Integration of regional mitigation assessment and conservation planning. *Ecology and Society,* Vol.14, No.1, 47, 15.07.2011. Available from URL: http://www.ecologyandsociety.org/vol14/iss1/art47/, ISSN 1708-3087

Tischendorf, L.; Grez, A.; Zaviezo, T.; Fahrig, L. (2005). Mechanisms affecting population density in fragmented habitat. *Ecology and Society,* Vol.10, No.1, 7, 15.07.2011.

Available from URL: http://www.ecologyandsociety.org/vol10/iss1/art7/, ISSN 1708-3087

Tscharntke, T.; Steffan-Dewenter, I.; Kruess, A. & Thies, C. (2002). Contribution of small habitat fragments to conservation of insect communities of grassland-cropland landscapes. *Ecological Applications,* Vol.12, No.2, pp. 354-363, ISSN 1051-0761

Tóthmérész, B.; Máthé I.; Balázs E. & Magura, T. (2011). Responses of carabid beetles to urbanization in Transylvania (Romania). *Landscape and Urban Planning,* Vol.101, No.4, pp. 330-337, ISSN 0169-2046

Tress, B.; Tress G.; van der Valk, A. & Fry, G. (2003). Interdisciplinary and Transdisciplonary landscape Studies: Potential and Limitations. *Delta Series* Vol.2, 192 p, ISBN 90 807637 13, Wageningen, The Netherlands

Turner, M.G. & Gardner, R.H. (Eds.) (1991). *Quantitative Methods in Landscape Ecology: the analysis and interpretation of landscape heterogeneity.* Springer Verlag, XV. 536 p, ISBN 0387942416

van der Zande, A.N.; ter Kreurs, J. & van der Weijden, W.J. (1980). The impact of roads on the densities of four bird species in an open field habitat - evidence of a long distance effect. *Biological Conservation,* Vol.18, pp. 299-321, ISSN 0006–3207

Voss, C.C. & Opdam, P. (Eds.), (1993). *Landscape ecology of a stressed environment.* Chapman and Hall, 310 p. ISBN 0412448203, London, UK

Voss C.C.; Verboom, J.; Opdam, P. F.M. & Ter Braak, C.J.F. (2001). Toward Ecologically Scaled Landscape Indices. *The American Naturalist,* Vol.183, No.1, pp. 24-41, ISSN 0003-0147

Wascher, D. (ed.) (2005). *European Landscape Character Areas.* Alterra Report No. 1254, 160 p.

Whitakker, R.J. & Fernández-Palácios, J.M. (2007). *Island Biogeography. Ecology, evolution, conservation.* Oxford University Press, 392 p, ISBN 978-0-19-856612-0, Oxford (UK) – New York, USA

Wiens, J. & Moss, M. (Eds.) (2005). *Issues and Perspectives in Landscape Ecology.* Cambridge University Press, 403 p, ISBN 10-0-521-53754-1, Cambridge, UK

Wöbse, H.H. (2002). *Landschaftsästhetik.* Ulmer Verlag, 304 p. ISBN-10: 3800132176

Wrbka, T.; Stocker-Kiss. A.; Schmitzberger, I. & Peterseil, J. (2005). *Network Landscapes of Austria – Identification and Ecological Assessment.* In: Konkoly Gyuró É. (ed.) Greenways. Conference Presentations on Ecological Corridors, Green Corridors, Sopron, Hungary, pp. 47-62.

Wu, J. & Hobbs, R. (2002). Key Issues and research priorities in landscape ecology: an idiosycratic synthesis. *Landscape Ecology* Vol.17, pp. 355-365, ISSN 0921-2973

Wu, J. & Hobbs, R.J. (Eds.) (2007). *Key Topics in Landscape Ecology.* Cambridge Studies in Landscape Ecology, Cambridge University Press, 297 p, ISBN-13 978-0-521-61644-7, Cambridge, UK

6

Impacts of Woodland Fragmentation on Species' Occurrences – The Combination of a Habitat Model with Landscape Metrics

Britta Eggers and Raul Köhler
Institute for World Forestry (vTI), Hamburg
Germany

1. Introduction

Habitat fragmentation can be defined as the modification of once natural habitats into smaller, isolated subareas surrounded by other types of habitat (more or less hostile; the matrix; e. g. Valladares et al., 2006). Fragmentation includes both, the separation of habitats and habitat loss, but the most dramatic and consistently negative effects on biodiversity can be attributed to habitat loss (Fahrig, 2003). In our central European man-made landscape more and more habitats become destroyed or fragmented because of the increasing anthropogenic need of available land and despite the growing knowledge about the problem. Roads, residential, and industrial areas separate formerly connected habitats into small remnants and thus create small subpopulations. Accordingly, habitat fragmentation and the associated effects like (1) the biodiversity decline of stenotopic species in smaller habitats (Desender et al., 1999; Magura et al., 2001), (2) the loss of genetic diversity and variability (Debinski & Holt, 2000; Keller & Largiader, 2003), and (3) a higher probability of inbreeding in smaller populations, belong to the main reasons for extinction of species (Groom et al., 2006). Especially stenotopic species with low dispersal power are endangered because the exchange of specimen between different habitat patches is reduced or entirely inhibited (Hanski et al., 1995). Thus, standardised, comprehensible quantification methods of fragmentation are greatly important for the development of management and conservation plans for habitat networks.

Species distribution models have become more and more important tools to analyse species-habitat relations in ecological and conservation research (Guisan & Thuiller, 2005). The principal aim of such habitat models is to predict habitat suitability for the species as a function of the given environmental variables (Basille et al., 2008). Based on biotic and abiotic key factors habitat models allow the quantification of habitat quality for selected species (Kleyer et al., 1999). For this reason they can help to predict the occurrence of rare and often hidden species (Pearce et al., 2001). Furthermore, the model predictions can be used to estimate influences of landscape changes on different species, to find habitats for resettling species and, to identify potential conflicts concerning anthropogenic activities (Klar et al., 2008; Kramer-Schadt et al., 2007).

Given the conservation status of many species, it is becoming increasingly important to understand the relationship between patterns of species' occurrences and the landscape environment. In fact, landscape approaches are relevant for conservation management because landscape planning and management is conducted on wide scales (Franklin, 1993). Landscape condition, which includes properties of landscape pattern as fragmentation, isolation, and patch size, is often not easy to understand, and yet this is very important for deciding how to manage problems like habitat loss, reconstruction and/or habitat (re-) connections (Mortelliti et al., 2010; 2011).

Before the interactions between landscape patterns and ecological processes can be understood the landscape structure must be quantified in meaningful ways (Turner, 1989). The condition and changes of spatial pattern of landscape can be quantified by statistical measurements, so called landscape metrics or indices (Gustafson, 1998; McGarigal & Marks, 1995; O'Neill et al., 1988). In this study we investigate the shape, pattern and the fragmentation status of the German woodlands by means of landscape metrics and analyse the impacts of fragmentation on different FFH species' occurrences.

The results can be used (1) to estimate the level of fragmentation of the studied woodlands, (2) to predict potential occurrences of the species in other woodland areas, and (3) to analyse the suitability of the calculated landscape metrics to predict the occurrence of threatened species. Accordingly, linking landscape metrics with species' occurrences can be an important approach to support the development of management plans in nature conservation.

2. Methods

2.1 Landscape metrics as measure of degree of fragmentation

Jaeger (2003) describes the fragmentation of landscape elements as a process with regard to the change of landscape structure, whereas the degree of fragmentation represents the state of the landscape at a specific time. Landscape metrics can be successfully used to characterise the condition of a landscape at different time points, they are useful tools for the comparison of different landscapes (Turner et al., 2001). Landscape patterns can be differentiated by means of landscape indices from GIS databases, which contain e. g. classifications of remote sensing data like aerial photographs or satellite imagery. An advantage is the efficient analysis of large areas with standardized methodological approaches, which allows generalization at large spatial and temporal scales.

We selected 19 landscape indices that can be used to quantify fragmentation of woodlands through an assessment of the literature (see annex for details).

2.2 Data and preparation

The basis for an adequate use of landscape metrics is the complete mapping of full coverage of the land use and land cover respectively with non-overlapping, integrated, and unambiguous landscape objects. If thematic maps fulfil these prerequisites, geographic information systems (GIS) can be used to analyse the maps with regard to shape, structure, and distribution of landscape elements.

The digital landscape model (DLM) from ATKIS[1]-data (year 2008) was the basis for our calculation of landscape metrics. The DLM provides a topographic description of the landscape of the Federal Republic of Germany in a vector format on a scale of 1:25000. Currently, it contains the most detailed spatial data in Germany which provide a comprehensive description of shape, location, and distribution of woodland patches. One decisive factor for the choice of these data is the inclusion of linear elements with potential separating impact on forested areas.

The DLM describes objects in the landscape as a result of definitions of the ATKIS-based feature type catalogue (ATKIS-OK), which allows the classification of more than one single feature type at one spatial point/area. Non-redundant binary maps of woodland and non-woodland were created by aggregating feature classes (see steps in figure 1). At this, the thematic feature types 'forest' and 'grove' were combined to the new class 'woodland'. Polygons, which describe urban, transportation, water, and vegetation areas (except forests and groves) were aggregated to the class 'non-woodland'. 'Woodland' geometries were overlaid by 'non-woodland' and by buffered forest-separating linear elements (e. g. roads, rivers, railways, etc.). The selection of these lines follows the approach described in Jaeger et al. (2001). The following linear features classes were transformed to polygons by buffer procedure:

- motorways, streets
- (active) railways
- water courses with a width > 6 m or used for shipping

The individual buffer size was determined from the recorded width or rather width class of line elements. Missing information was replaced by mean width of the specific feature type.

The degree of woodland fragmentation was calculated for the reference level 'Borders of topographic map sheets on a scale of 1:25000' (TK25). A binary 'woodland vs. non-woodland'-map was generated for each map sheet (2947 units) by the cutting-out procedure, as polygons were directly intersected with the reference level data (compare with Moser et al., 2007). All 19 landscape metrics were calculated for each map sheet.

In addition to woodland geometries for calculation of landscape indices we required occurrence data of the species, which could be used as input for the habitat modelling. We were granted access to presence data of the species which are based on map sheets of the topographic map on a scale of 1:25000 (TK25) from the German Federal Agency of Nature Conservation (BfN).

Data preparations were done by means of GIS software ArcGIS™ Desktop (ESRI) and PostGIS tools (Refractions Research). A PostgreSQL database was used for data management and indices calculation.

---

[1]Amtlich Topographisch-Kartographisches Informationssystem (the official digital topographic maps of Germany)

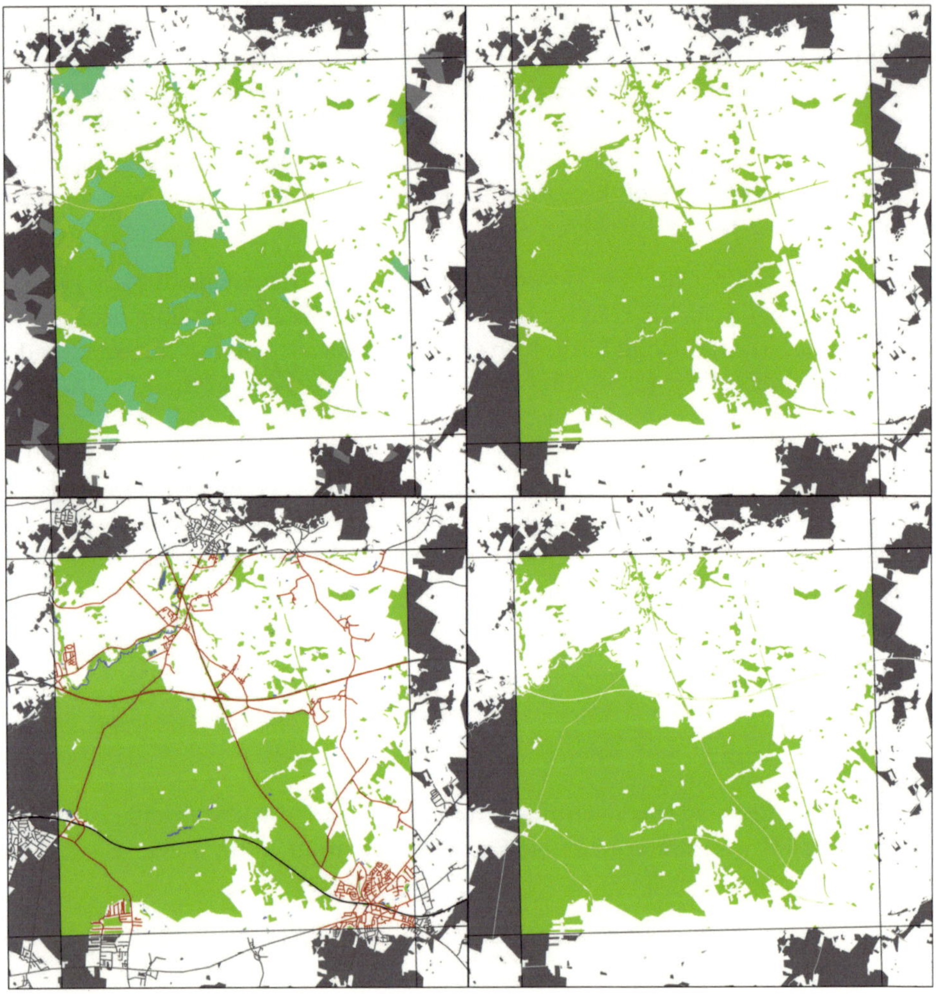

Fig. 1. All pictures show landscape elements in map sheet TK2428. top left: forest types. top right: aggregated forest. lower left: aggregated forest and transportation lines. lower right: binary forest map with separated forest areas.

2.3 Selected species

For the habitat models we chose the barbastelle bat *Barbastella barbastellus*, the black stork *Ciconia nigra*, the European wildcat *Felis silvestris*, Bechstein's bat *Myotis bechsteinii*, and the stag beetle *Lucanus cervus*. The selected species are all of high interest for nature conservation as species of the EU Habitat Directive (FFH) or the Conservation of Wild Birds Directive (The Council of the European Communities, 2004). They all occur in woodlands with special structures and have very different dispersal abilities.

B. barbastellus can rarely be found in woodland areas smaller than 1 km^2. Its home range sizes have strong variations (1.25-25 km^2: Hillen et al., 2009; home range diameter 4-5 km: Russo et al., 2004; Steinhauser, 2002), but compared to *M. bechsteinii* it is able to cross motorways (Kerth & Melber, 2009).

M. bechsteinii occurs in mature, natural woodland areas of at least 2.5-3 km^2 (Kanuch et al., 2008). Home range sizes vary from 1 km^2 (Kerth & Morf, 2004; Kerth et al., 2001) to diameters of 3 km (Steinhauser, 2002). In fragmented woodland areas home range sizes of *M. bechsteinii* increase, which leads to a lower probability of survival (Norberg & Rayner, 1987; Siemers & Swift, 2006).

The highly specialised species *C. nigra* occurs only in undisturbed deciduous woodlands (> 80 % deciduous trees). Additionally, the distance to the next water body should not exceed 1 km (Augutis & Sinkevicius, 2005). The main threat to this species is habitat degradation because of deforestation (particularly the destruction of large traditional nesting trees), the rapid development of industry and farming, and the building of dams and lake drainage for irrigation and hydroelectric power production (BirdLife International, 2009; Rosenvald & Lohmus, 2003). The absence of disturbances is one of the main predictions for the occurrence of *C. nigra*.

The European wildcat (*F. silvestris*) only occurs in large and undisturbed areas. Home range sizes fluctuate from 0.7-14 km^2 (females) to 2-50 km^2 (males; Monterroso et al., 2009). *F. silvestris* avoids approaching areas of settlement, e. g. it is known to make detours of a distance up to 900 m for a small village and 200 m for a single house (Klar et al., 2008).

L. cervus lives in large, unbroken oak woodlands. Its occurrence depends mainly on the presence of dead wood (Pratt, 2000). Home range sizes vary from 0.2 ha (females) to 1 ha (males) and cover distances of 1 km (females) to 3 km (males; Rink & Sinsch, 2007; Sprecher-Uebersax, 2003).

2.4 Habitat models

Niche models of these species were calculated to show the impacts of woodland fragmentation on different woodland species. These models can be used to expose potential habitats that are not yet colonised.

We analysed the suitability of the abovementioned landscape metrics to predict the occurrence of threatened species. We used the selected landscape metrics as environmental variables for the habitat models. Accordingly, 19 landscape metrics were calculated for each of the 2947 map sheets covering the territory of the Federal Republic of Germany. To generate the habitat models a correlation analysis is necessary to exclude variables with an absolute value of the correlation coefficient higher than 0.7 (Fielding & Haworth, 1995; Schröder & Reineking, 2004).

Predictive distribution modelling was done with the software program Maxent (Vers. 3.3.3). Maxent was developed by the machine learning community and uses a statistical technique called maximum entropy that generates a prediction from incomplete information (Phillips et al., 2006; Phillips et al., 2004). Adopting this method, 100 iterative models per species were created. In each cycle, Maxent was configured in such a way that 75 % of the total presence records were used to train the models and 25 % were reserved to test the resulting models. Logistic output format was chosen.

All other statistical analysis was done using R (R Development Core Team, 2010) and SPSS 17 (IBM).

3. Results

3.1 Landscape indices

Besides the geo data and thematic input layers the results of the indices calculation are stored in a PostgreSQL-database. Unambiguous keys enable us to link the results with the reference level 'TK25' map sheets.

The calculated landscape indices for each single map sheet are the input layers for the habitat models. The distribution of the landscape metrics at reference level 'TK25' is described by the means of descriptive statistical quantities. Mean, standard deviation, minimum and maximum values are shown in table 1. These values identify the statistical distribution of indices calculated from the binary 'woodland/non-woodland'-landscapes in 2947 map sheets.

Landscape metrics	MEAN	SD	MIN	MAX
TA (km^2)	129.73	7.05	7.90	176.92
CA_f (km^2)	38.57	26.40	0.00	129.60
$PLAND_f$	0.29	0.20	0.00	0.96
NP_f	250.90	139.09	0	832
PD_f ($1/km^2$)	1.94	1.07	0.01	6.44
PD_{ff} ($1/km^2$)	15.60	36.34	0.05	1257.26
MPS_f (km^2)	0.22	0.44	0.00	19.15
LPI_f	0.08	0.09	0.00	0.66
MESH (km^2)	2.71	5.59	0.00	61.39
MPAR (km/km^2)	104.68	129.05	26.64	5431.75
MSI	1.81	0.23	1.12	3.59
AWMSI	0.93	0.85	0.00	7.60
LSI	9.10	4.40	0.00	25.92
MPFD	1.41	0.11	1.22	5.05
TE (km)	368.92	180.70	0.00	1056.38
ED (km/km^2)	2.82	1.35	0.00	8.09
MENN (km)	0.11	0.81	0.00	43.51
TLD_S (km/km^2)	1.92	1.30	0.00	12.87
TLD_R (km/km^2)	0.08	0.17	0.00	2.25

Table 1. Descriptive statistics of selected landscape indices at reference level 'TK25'

Total area (TA) and forest area (CA_f) represent the reference values for several other landscape metrics calculated for each map sheet. The fragmentation composition can be described with following metrics: number of patches (NP_f), mean patch size (MPS_f), largest patch index (LPI_f), patch density (PD_f, PD_{ff}), and effective mesh size (MESH). Other indices, like shape index (MSI, AWMSI, LSI), mean area-perimeter-ratio (MPAR), or mean patch fractal dimension (MPFD), characterise the shape of fragmentation. Patch fragmentation indices, like distance to nearest neighbour (MENN), describe connectivity/isolation of patches in the map sheets. Transport line density (TLD_S, TLD_R) was derived from the length of linear elements and was not directly comparable with edge metrics (TE, ED) which include all borders of patches (for specified descriptions see annex).

The following figures show selected landscape indices as spatial distribution grid maps, as measures are based on comparable sizes of reference units (map sheets).

Figure 2 represents a typical forest map of Germany illustrating the spatial extent of the index 'percentage of forest in landscape' ($PLAND_f$).

Figure 3 allows comparison with the landscape index 'effective mesh size' (MESH) for woodlands in Germany. MESH provides a relative measure of patch structure and is interpreted as the size of the areas when the region under investigation is subdivided into areas of the same size and with the same degree of landscape division (Jaeger, 2000; McGarigal & Marks, 1995). High values (maximum about 61 km^2) describe large, contiguous forest areas (e. g. in the Alps or the low mountain ranges), whereas small measures identify a low percentage of forested area and highly fragmented woodlands respectively (e. g. coastal regions or the central lowland in Saxony-Anhalt).

Fig. 2. Spatial distribution of 'percentage of forest area' ($PLAND_f$) in Germany

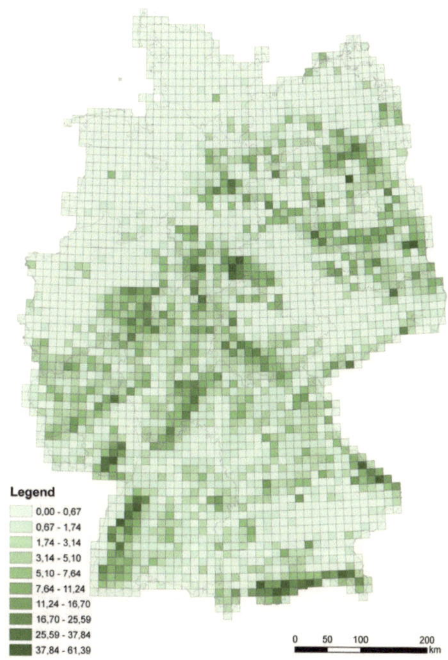

Fig. 3. Spatial distribution of 'effective mesh size' (MESH in km^2) of woodland in Germany

The 'mean shape index' (MSI), a simple and straightforward measure of overall shape complexity, is shown in figure 4. The value increases as the mean patch shape becomes more irregular (McGarigal & Marks, 1995), e. g. in some regions in Baden-Württemberg or North Rhine-Westphalia.

Figure 5 displays the 'mean Euclidean distance to the nearest neighbour' (MENN) of woodland patches at reference level 'TK25'. In regions with small percentages of forest area,

e. g. lowlands in central Germany or the North Sea coast, MENN extends to a maximum of 43 km. At the other hand, in heavily forested areas, e. g. low mountain ranges, only marginal values < 0.1 km can be observed.

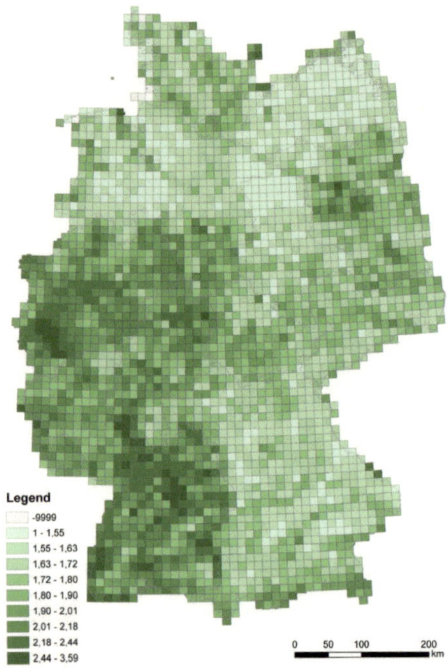

Fig. 4. Spatial distribution of 'mean shape index' (MSI) of woodland patches in Germany

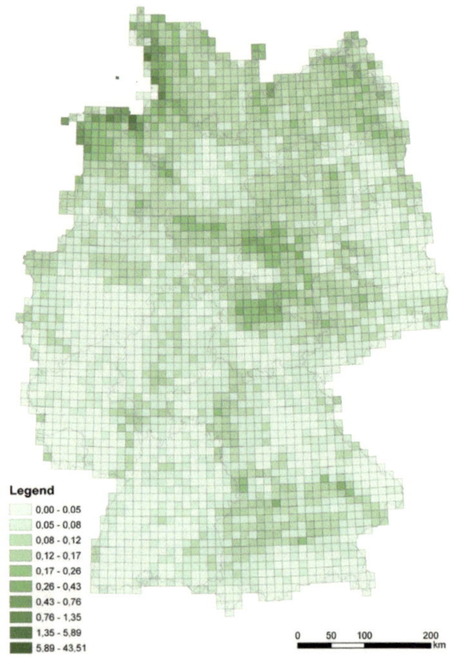

Fig. 5. Spatial distribution of 'mean Euclidean distance to nearest neighbour' (MENN in km) of woodland patches in Germany

3.2 Model results

After the correlation analysis we calculated a model with the following landscape metrics: MENN, MESH, MPAR, MPFD, MSI, NP_f, TLD_S, und TLD_R. Table 2 shows a summary of the habitat model results of the analysed species.

species	pattern quality (AUC) ± standard deviation	most important variable in the model	second important variable in the model
Barbastella barbastellus	0.764 ± 0.011	MESH	MSI
Ciconia nigra	0.832 ± 0.011	MESH	TLD_S
Felis silvestris	0.889 ± 0.008	MESH	MSI
Lucanus cervus	0.777 ± 0.009	MSI	MESH
Myotis bechsteinii	0.761 ± 0.009	MESH	MENN

Table 2. Summary of the model results of the different species.

The results demonstrate that there is a relationship between species' occurrences and calculated landscape metrics. Fragmentation indices like the 'effective mesh size' and the 'mean Euclidean distance' to the next woodland patch and metrics concerning the shape of woodland patches (MSI) are particularly predictive of species occurrences. The occurrence probability of *C. nigra* also depends on the 'transportation line density' of streets (TLD_S). The best occurrence probability was calculated in the models of *C. nigra* and *F. silvestris*. Both model results show higher AUC-values (area under the curve) than 0.8, which signifies models with 'good' predictive power (Reineking & Schröder, 2004).

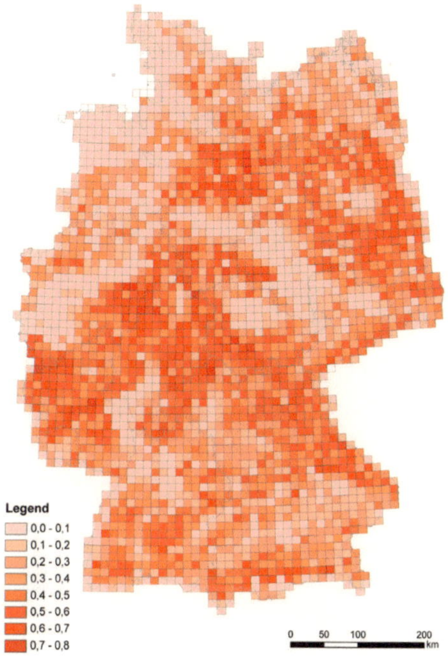

Fig. 6. Habitat model results of *C. nigra*.

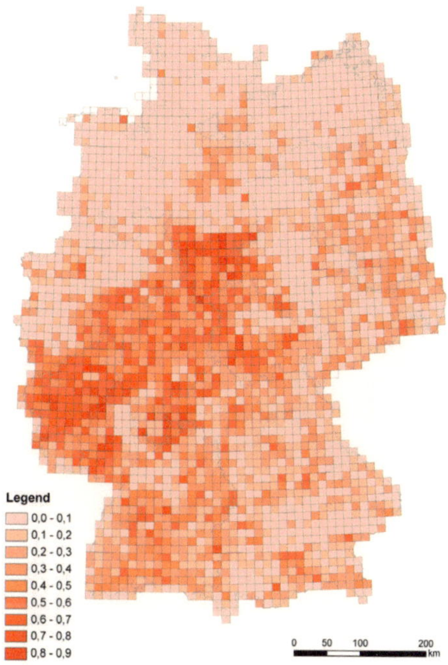

Fig. 7. Habitat model results of *F. silvestris*.

The map of the model results of *C. nigra* (figure 6) shows that the highest occurrence probability is correlated with densely wooded areas in Germany. Following the map, the black stork occurs in all lower mountain ranges (in central Germany) and in sparsely populated areas (in Mecklenburg-West Pomerania and Brandenburg).

The map of the wildcat's model results (figure 7) shows that occurrence probability is low in the northern parts of Germany and highest in the low mountain ranges in central Germany.

The model results of the three other species have been 'acceptable' indicated by AUC-values above 0.7 (Reineking & Schröder, 2004). The maps of the models reveal less obvious results than the results of *C. nigra* and *F. silvestris* present. These model results are shown in figure 8 - 10. Compared to the Bechstein's bat and the stag beetle, the barbastelle bat is more orientated to the eastern part of Germany. The Bechstein's bat and the stag beetle show higher occurrence probabilities in the southwest.

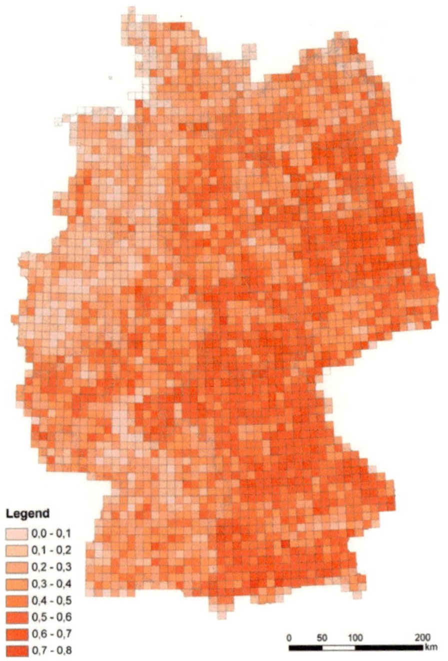

Fig. 8. Habitat model results of *B. barbastellus*.

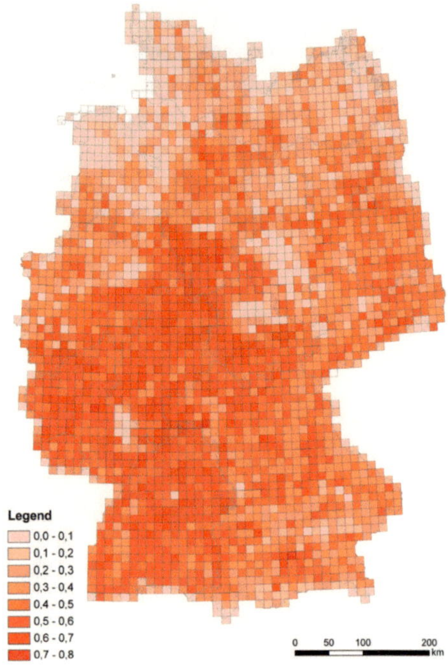

Fig. 9. Habitat model results of *M. bechsteinii*.

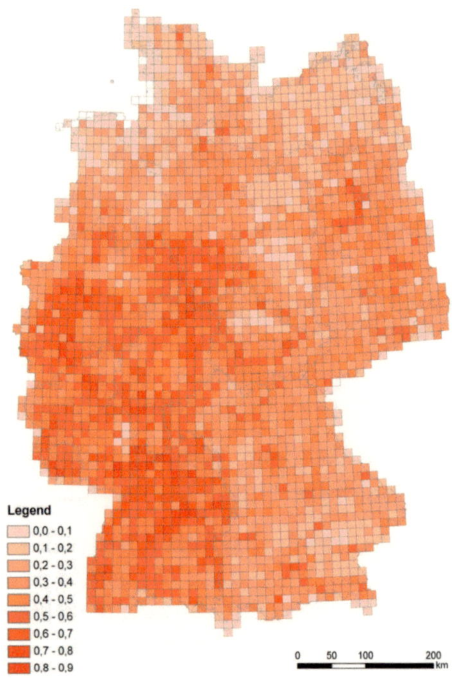

Fig. 10. Habitat model results of *L. cervus*.

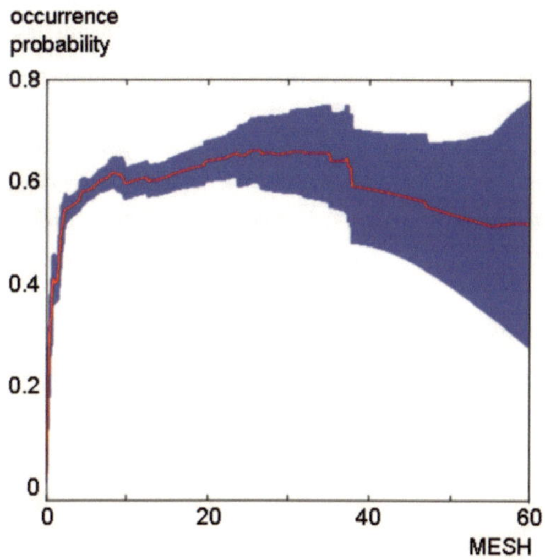

Fig. 11. Response of the occurrence of *F. silvestris* dependent to the 'effective mesh size' (MESH in km²; standard deviation is coloured in blue).

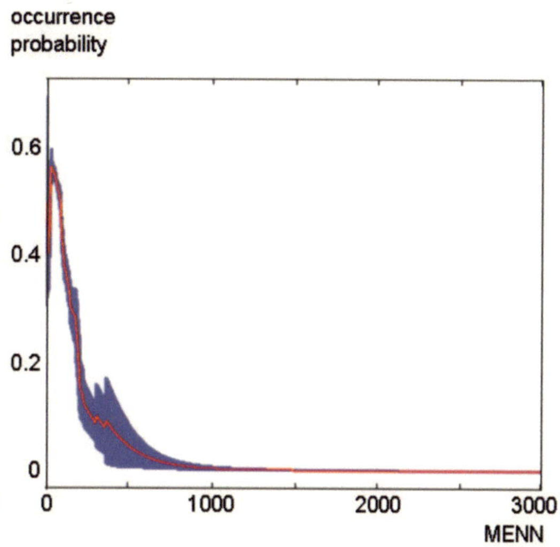

Fig. 12. Response of the occurrence of *M. bechsteinii* dependent to the'mean Euclidean distance to nearest neighbour' (MENN in m) of woodland patches (standard deviation is coloured in blue).

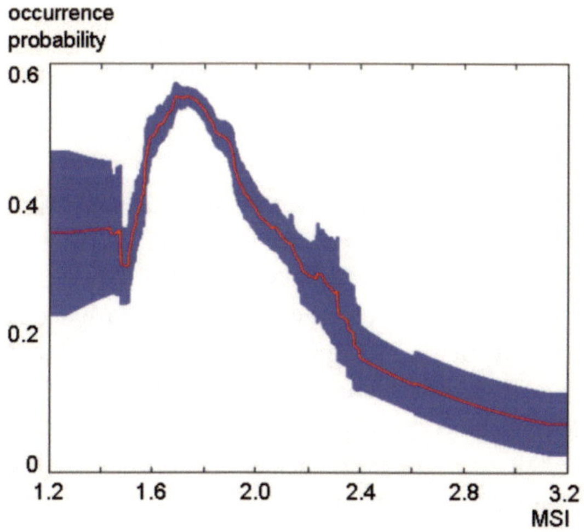

Fig. 13. Response of the occurrence of *B. barbastellus* dependent to the 'mean shape index' (MSI; standard deviation is coloured in blue).

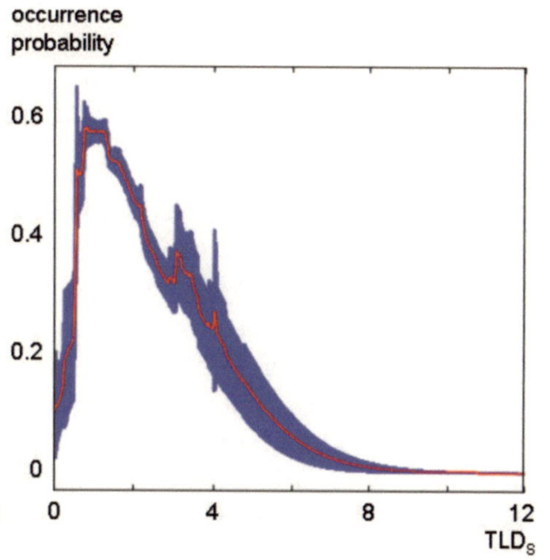

Fig. 14. Response of the occurrence of *C. nigra* dependent to the 'transportation line density of streets' (TLD_S in km/km²; standard deviation is coloured in blue).

The landscape metrics with the highest influence in the model have also been comparable: The occurrence probabilities of *F. silvestris, M. bechsteinii, C. nigra,* and *L. cervus* were highest if MESH reaches mean values (compare results of *F. silvestris* in figure 11). The index 'mean Euclidean distance to nearest neighbour' of woodland patches. is also critical for the occurrence probability of the species (compare figure 12). The closer the neighbouring woodland patches (= the lower the MENN values), the higher is for example the occurrence probability of *M. bechsteinii*. The occurrence probabilities of *B. barbastellus, L. cervus,* and *F. silvestris* concerning the woodland structure are high if the 'mean shape index' (MSI) indicates medium values (for *B. barbastellus* see figure 13). Furthermore, the response of the occurrence probability of *C. nigra* to the 'transportation line density of the streets' shows that the probability declines with increasing density of streets (figure 14).

4. Discussion

In this study we investigated the shape, pattern, and the fragmentation status of the German woodlands by landscape metrics and analysed the impacts of fragmentation on different FFH species' occurrences.

4.1 Landscape metrics for predicting species occurrences

The results of our habitat models show that it is possible to predict species occurrences with measures of landscape structure concerning habitat fragmentation.

The presence of the selected species can be described by different landscape metrics like MESH, MENN, MSI and TLD_S. The contribution to habitat suitability of the predictors in the final model reflects the understanding of the ecology of our target species. As expected, the presence of forest habitats was an important determinant of suitability for the selected target species, but considering and comparing the different species we can assume that not only large woodland areas are essential for their occurrences. Comparing the maps of the habitat models with the map of the woodland area (figure 2) there is only the model of *C. nigra* which shows higher congruencies.

F. silvestris depends on large undisturbed areas but it also needs special woodland structures (mean MSI results in high occurrence probabilities) because low distances to e. g. woodland edges and rivers result in higher densities of prey (Doyle, 1990; Gomez & Anthony, 1998; Osbourne et al., 2005). The same holds true for species like the barbastelle bat. It also depends on large woodland areas but has its foraging grounds at woodland edges (Kerth & Melber, 2009; Norberg & Rayner, 1987; Steinhauser, 2002). *L. cervus* has less power of dispersal than the other selected species. It needs large amounts of dead wood in more or less undisturbed woodlands (Pratt, 2000) and covers distances up to 3 km (males; Rink & Sinsch, 2007). The model results of *L. cervus* are to be considered with care, because in fact data of smaller scales (presence data and landscape metrics) is necessary to calculate more significant models (compare Garcia-Gigorro & Saura, 2005; Wu, 2004).

The importance of fragmentation indices (like MESH, MENN and TLD_S) in the model results shows the general sensitivity of the analysed species to fragmentation due to their ecological traits (like high trophic level, large home range sizes or low dispersal ability). Like other studies show, we can confirm that simple fragmentation components have higher influences on species occurrences than more complex ones (compare Fischer et al., 2004; McGarigal & McComb, 1995; Rutledge & Miller, 2006).

Landscape metrics are important tools to quantify the fragmentation, but most of them have only limited explanatory power for ecological processes because of their potential for inconsistent and ambiguous statistical relationships with response variables of ecological processes (Tischendorf, 2001). However, spatial pattern analysis should be used to explain structural changes in landscape and consider ecological processes (Li & Wu, 2004). Linking fragmentation metrics with the occurrence of species is one important step, but we need additional data, like information about habitat quality, to improve the understanding of spreading of animals, pattern quality, and the model's explanatory power. Consequently, as accentuated by our results for our target species, the landscape structure is an important aspect to be considered in defining patterns of habitat suitability.

Our methods could be used as a framework and forms a basic concept for further research. Furthermore, our approach could be extended to other fragmentation-sensitive species.

4.2 Possibilities for nature conservation

The combination of landscape metrics and habitat suitability for species results in a methodology that improves the potential for understanding patterns of species distribution.

There is no doubt that preservation and restoration of large undisturbed areas are priorities for conservation of many species (e. g. Drees et al., 2011). Mortelliti et al. (2011) confirm this and add that structural connectivity is necessary and should not be regarded without the amount of available habitat in landscapes. However, structural connectivity should also be considered in conjunction with habitat quality, which is strongly driven by species-specific determinants (Mortelliti et al., 2010).

Accordingly, the main aims should be to map potential habitats that are adapted to different scales (depending on species selection) in combination with habitat quality to develop a connected habitat system. This is especially true because global warming contributes to fragmentation of landscapes. Species that are unable to disperse will become extinct because they may live in restricted geographical ranges and are unable to reach disconnected ecological niches or other intact habitats (compare Habel et al., 2010).

Linking landscape metrics with species' occurrences can be an important step to support the development of management plans for conservation.

5. Conclusions

The behaviour of species concerning landscape configuration is generally very complex. Consequently, estimating and calculating patterns based on models are quite difficult.

However, combining landscape structure with the presence of species could be a feasible approach for the quantification of landscape-species relationships and thus, could provide a foundation for studies on regional and/or local scales. Additionally, this methodology may give new insight into nature conservation and landscape management practices.

6. Acknowledgments

We are grateful to the Federal Agency of Nature Conservation (BfN) for granting access to the presence data of the FFH species for the habitat models. We also thank André Iost for not becoming desperate after reading our manuscript, for numerous discussions, and support. We were financially supported by the Federal Ministry of Food, Agriculture and Consumer Protection (BMELV).

7. Annex

Landscape metric	equation	Result space/unit
Total area (reference unit)	$TA = \frac{A}{1000000}$	$TA > 0$, km²
Total class area (forest)	$CA_f = \frac{\sum_{j=1}^{n} a_{ij}}{1000000}$	$CA_f \geq 0$, km²
Percentage of landscape (forest)	$PLAND_f = \frac{CA_f}{TA}$	$0 \leq PLAND_f \leq 1$
Number of patches (forest)	$NP_f = n_i$	$NP_f \geq 0$
Patch density (reference to total area)	$PD_f = \frac{n_i}{TA}$	$PD_f \geq 0$, n/km²
Patch density (reference to total class area)	$PD_{ff} = \frac{n_i}{CA_f}$	$PD_{ff} > 0$, n/km²
Mean patch size (forest)	$MPS_f = \frac{CA_f}{n_i}$	$MPS_f > 0$, km²
Largest patch index (forest)	$LPI_f = \frac{\max_{j=1}^{n}(a_{ij})}{1000000 \times TA}$	$0 \leq LPI_f \leq 1$
Effective mesh size (forest)	$MESH = \frac{\sum_{j=1}^{n} \frac{a_{ij}^2}{1000000}}{TA}$	$0 \leq MESH \leq TA$, km²
Mean perimeter-area ratio (forest)	$MPAR = \frac{\sum_{j=1}^{n} \frac{p_{ij}}{a_{ij}} \times 1000}{n_i}$	$MPAR > 0$, km/km²

Landscape metric	equation	Result space/unit
Mean shape index (forest)	$MSI = \frac{\sum_{j=1}^{n}\left(\frac{p_{ij}}{2\sqrt{\pi * a_{ij}}}\right)}{n_i}$	$MSI \geq 1$
Area-weighted mean shape index (forest)	$AWMSI = \sum_{j=1}^{n}\left[\left(\frac{p_{ij}}{2\sqrt{\pi * a_{ij}}}\right)\left(\frac{a_{ij}}{\sum_{j=1}^{n} a_{ij}}\right)\right]$	$AWMSI \geq 1$
Landscape shape index	$LSI = \frac{\sum_{k=1}^{m} e_{ik}}{1000 \times 2\sqrt{\pi * TA}}$	$LSI \geq 1$
Mean patch fractal dimension	$MPFD = \frac{\sum_{j=1}^{n}\left(\frac{2 \times \ln(p_{ij})}{\ln(a_{ij})}\right)}{n_i}$	$1 \leq MPFD \leq 2$
Total edge (without landscape boundary)	$TE = \frac{\sum_{k=1}^{m} e_{ik}}{1000}$	$TE \geq 0$, km
Edge density (without landscape boundary)	$ED = \frac{\sum_{k=1}^{m} e_{ik}}{1000 \times TA}$	$ED \geq 0$, km/km²
Mean nearest-neighbour distance (forest)	$MENN = \frac{\sum_{j=1}^{n} \min d_{jg}}{1000 \times n_i}$	$MENN \geq 0$, km
Transportation line density (streets/railways)	$TLD = \frac{\sum_{k=1}^{m} l_{ik}}{1000 \times TA}$	$TLD \geq 0$, km/km²

Table 3. Landscape metrics – equations and result space (Jaeger, 2000; McGarigal & Marks, 1995).

Notation:

A	total area (m^2) of landscape (TK25 map sheet)
a_{ij}	area (m^2) of patch i_j
d_{jg}	distance (m) between patch j and g of class i
e_{ik}	total length (m) of edge in landscape involving patch type (class) i
l	length (m) of line elements of class i (streets/railways)
n_i	number of patches in landscape of class i
p_{ij}	perimeter (m) of patch i_j

8. References

Augutis, D. & Sinkevicius, S. (2005). Application of geographic information system (GIS) and technologies in identification of potential nesting habitats of black stork (*Ciconia nigra*), *Acta Zoologica Lituanica,* Vol. 15 No. 1, pp. 3-12

Basille, M., Calenge, C., Marboutin, E., Andersen, R. & Gaillard, J. M. (2008). Assessing habitat selection using multivariate statistics: Some refinements of the ecological-niche factor analysis, *Ecological Modelling,* Vol. 211 No. 1-2, pp. 233-240

BirdLife International. (2009). *Ciconia nigra,* In: *IUCN Red List of Threatened Species,* Version 2011.1, available at: www.iucnredlist.org (Downloaded on 28 July 2011)

Debinski, D. M. & Holt, R. D. (2000). A survey and overview of habitat fragmentation experiments, *Conservation Biology,* Vol. 14 No. 2, pp. 342-355

Desender, K., Ervynck, A. & Tack, G. (1999). Beetle diversity and historical ecology of woodland in flanders, *Belgian Journal of Zoology,* Vol. 129 No. 1, pp. 139-155

Doyle, A. T. (1990). Use of riparian and upland habitats by small mammals, *Journal of Mammalogy,* Vol. 71 No. 1, pp. 14-23

Drees, C., de Vries, H., Härdtle, W., Matern, A., Persigehl, M. & Assmann, T. (2011). Genetic erosion in a stenotopic heathland ground beetle (Coleoptera: Carabidae): a matter of habitat size?, *Conservation Genetics,* Vol. 12 No. 1, pp. 105-117

Fahrig, L. (2003). Effects of habitat fragmentation on biodiversity, *Annual Review of Ecology Evolution and Systematics,* Vol. 34, pp. 487-515

Fielding, A. H. & Haworth, P. F. (1995). Testing the generality of bird-habitat models, *Conservation Biology,* Vol. 9 No. 6, pp. 1466-1481

Fischer, J., Lindenmayer, D. B. & Fazey, I. (2004). Appreciating ecological complexity: Habitat contours as a conceptual landscape model, *Conservation Biology,* Vol. 18 No. 5, pp. 1245-1253

Franklin, J. F. (1993). Preserving biodiversity - species, ecosystems, or landscapes, *Ecological Applications,* Vol. 3 No. 2, pp. 202-205

Garcia-Gigorro, S. & Saura, S. (2005). Forest fragmentation estimated from remotely sensed data: Comparison across scales possible?, *Forest Science,* Vol. 51 No. 1, pp. 51-63

Gomez, D. M. & Anthony, R. G. (1998). Small mammal abundance in riparian and upland areas of five seral stages in western Oregon, *Northwest Science,* Vol. 72 No. 4, pp. 293-302

Groom, M. J., Meffe, G. K. & Carroll, R. (2006). *Principles of conservation biology,* Sinauer Associates, 0-87893-518-5, Sunderland, Massachusetts

Guisan, A. & Thuiller, W. (2005). Predicting species distribution: offering more than simple habitat models, *Ecology Letters,* Vol. 8 No. 9, pp. 993-1009

Gustafson, E. J. (1998). Quantifying landscape spatial pattern: What is the state of the art?, *Ecosystems,* Vol. 1 No. 2, pp. 143-156

Habel, J. C., Schmitt, T., Meyer, M., Finger, A., Roedder, D., Assmann, T. & Zachos, F. E. (2010). Biogeography meets conservation: the genetic structure of the endangered lycaenid butterfly *Lycaena helle* (Denis & Schiffermuller, 1775), *Biological Journal of the Linnean Society,* Vol. 101 No. 1, pp. 155-168

Hanski, I., Pakkala, T., Kuussaari, M. & Lei, G. C. (1995). Metapopulation persistence of an endangered butterfly in a fragmented landscape, *Oikos,* Vol. 72 No. 1, pp. 21-28

Hillen, J., Kiefer, A. & Veith, M. (2009). Foraging site fidelity shapes the spatial organisation of a population of female western barbastelle bats, *Biological Conservation,* Vol. 142 No. 4, pp. 817-823

Jaeger, J. (2003). Landschaftszerschneidung, In: *Handbuch Naturschutz und Landschaftspflege,* Konold, W., Böcker, R. and Hampicke, U. (Eds.), pp. 1999-2028, Ecomed-Verlag, Landsberg

Jaeger, J., Esswein, H., Schwarz-von Raumer, H.-G. & Mueller, M. (2001). Quantitative analysis of landscape fragmentation in Baden-Wuerttemberg, *Naturschutz und Landschaftsplanung,* Vol. 33 No. 10, pp. 305-317

Jaeger, J. A. G. (2000). Landscape division, splitting index, and effective mesh size: new measures of landscape fragmentation, *Landscape Ecology,* Vol. 15 No. 2, pp. 115-130

Kanuch, P., Danko, S., Celuch, M., Kristin, A., Pjencak, P., Matis, S. & Smidt, J. (2008). Relating bat species presence to habitat features in natural forests of Slovakia (Central Europe), *Mammalian Biology,* Vol. 73 No. 2, pp. 147-155

Keller, I. & Largiader, C. R. (2003). Recent habitat fragmentation caused by major roads leads to reduction of gene flow and loss of genetic variability in ground beetles, *Proceedings Of The Royal Society Of London Series B-Biological Sciences,* Vol. 270 No. 1513, pp. 417-423

Kerth, G. & Melber, M. (2009). Species-specific barrier effects of a motorway on the habitat use of two threatened forest-living bat species, *Biological Conservation,* Vol. 142 No. 2, pp. 270-279

Kerth, G. & Morf, L. (2004). Behavioural and genetic data suggest that Bechstein's bats predominantly mate outside the breeding habitat, *Ethology,* Vol. 110 No. 12, pp. 987-999

Kerth, G., Wagner, M. & Konig, B. (2001). Roosting together, foraging apart: information transfer about food is unlikely to explain sociality in female Bechstein's bats (*Myotis bechsteini*), *Behavioral Ecology and Sociobiology,* Vol. 50 No. 3, pp. 283-291

Klar, N., Fernandez, N., Kramer-Schadt, S., Herrmann, M., Trinzen, M., Buttner, I. & Niemitz, C. (2008). Habitat selection models for European wildcat conservation, *Biological Conservation,* Vol. 141 No. 1, pp. 308-319

Kleyer, M., Kratz, R., Lutze, G. & Schröder, B. (1999). Habitatmodelle für Tierarten: Entwicklung, Methoden und Perspektiven für die Anwendung, *Zeitschrift für Ökologie und Naturschutz,* Vol. 8, pp. 177-194

Kramer-Schadt, S., Revilla, E., Wiegand, T. & Grimm, V. (2007). Patterns for parameters in simulation models, *Ecological Modelling,* Vol. 204 No. 3-4, pp. 553-556

Li, H. B. & Wu, J. G. (2004). Use and misuse of landscape indices, *Landscape Ecology,* Vol. 19 No. 4, pp. 389-399

Magura, T., Ködöböcz, V. & Tóthmérész, B. (2001). Effects of habitat fragmentation on carabids in forest patches, *Journal of Biogeography,* Vol. 28, pp. 129-138

McGarigal, K. & Marks, B. J. (1995). FRAGSTATS: Spatial pattern analysis program for quantifying landscape structure, *U S Forest Service General Technical Report PNW,* Vol. 0 No. 351, pp. 1-122

McGarigal, K. & McComb, W. C. (1995). Relationships between landscape structure and breeding birds in the Oregon coast range, *Ecological Monographs,* Vol. 65 No. 3, pp. 235-260

Monterroso, P., Brito, J. C., Ferreras, P. & Alves, P. C. (2009). Spatial ecology of the European wildcat in a Mediterranean ecosystem: dealing with small radio-tracking datasets in species conservation, *Journal of Zoology,* Vol. 279 No. 1, pp. 27-35

Mortelliti, A., Amori, G. & Boitani, L. (2010). The role of habitat quality in fragmented landscapes: a conceptual overview and prospectus for future research, *Oecologia,* Vol. 163 No. 2, pp. 535-547

Mortelliti, A., Amori, G., Capizzi, D., Cervone, C., Fagiani, S., Pollini, B. & Boitani, L. (2011). Independent effects of habitat loss, habitat fragmentation and structural connectivity on the distribution of two arboreal rodents, *Journal of Applied Ecology,* Vol. 48 No. 1, pp. 153-162

Moser, B., Jaeger, J., Tappeiner, U., Tasser, E. & Eiselt, B. (2007). Modification of the effective mesh size for measuring landscape fragmentation to solve the boundary problem, *Landscape Ecology,* Vol. 22 No. 3, pp. 447-459

Norberg, U. M. & Rayner, J. M. V. (1987). Ecological morphology and flight in bats (Mammalia, Chiroptera) - wing adaptations flight performance foraging strategy and echolocation, *Philosophical Transactions of the Royal Society of London B Biological Sciences,* Vol. 316 No. 1179, pp. 335-427

O'Neill, R. V., Krummel, J. R., Gardner, R. H., Sugihara, G., Jackson, B., DeAngelis, D. L., Milne, B. T., Turner, M. G., Zygmunt, B., Christensen, S. W., Dale, V. H. & Graham, R. L. (1988). Indices of landscape pattern, *Landscape Ecology,* Vol. 1 No. 3, pp. 153-162

Osbourne, J. D., Anderson, J. T. & Spurgeon, A. B. (2005). Effects of habitat on small-mammal diversity and abundance in West Virginia, *Wildlife Society Bulletin,* Vol. 33 No. 3, pp. 814-822

Pearce, J., Ferrier, S. & Scotts, D. (2001). An evaluation of the predictive performance of distributional models for flora and fauna in north-east New South Wales, *Journal of environmental management,* Vol. 62, pp. 171-184

Phillips, S. J., Anderson, R. P. & Schapire, R. E. (2006). Maximum entropy modeling of species geographic distributions, *Ecological Modelling,* Vol. 190 No. 3-4, pp. 231-259

Phillips, S. J., Dudik, M. & Schapire, R. E. (2004). A maximum entropy approach to species distribution modeling, In: *Proceedings of the twenty-first international conference on Machine learning* pp. 83-90, Banff, Alberta, Canada

Pratt, C. R. (2000). An investigation into the status history of the stag beetle *Lucanus cervus* Linnaeus (Lucanidae) in Sussex, *Coleopterist,* Vol. 9 No. 2, pp. 75-90

R Development Core Team. (2010). R: A Language and Environment for Statistical Computing, In: *Computing, R. F. f. S.*, www.R-project.org, Vienna, Austria

Reineking, B. & Schröder, B. (2004). Gütemaße für Habitatmodelle, In: *Habitatmodelle - Methodik, Anwendung, Nutzen.Tagungsband zum Workshop vom 8.-10.Oktober 2003 am UFZ Leipzig,* Dormann, C. F., Lausch, A., Blaschke, T., Söndgerath, D. and Schröder, B. (Eds.), pp. 27-37, UFZ, 0948-9452, Leipzig

Rink, M. & Sinsch, U. (2007). Radio-telemetric monitoring of dispersing stag beetles: implications for conservation, *Journal of Zoology,* Vol. 272 No. 3, pp. 235-243

Rosenvald, R. & Lohmus, A. (2003). Nesting of the black stork (*Ciconia nigra*) and white-tailed eagle (*Haliaeetus albicilla*) in relation to forest management, *Forest Ecology and Management,* Vol. 185 No. 3, pp. 217-223

Russo, D., Cistrone, L., Jones, G. & Mazzoleni, S. (2004). Roost selection by barbastelle bats (*Barbastella barbastellus,* Chiroptera : Vespertilionidae) in beech woodlands of central Italy: consequences for conservation, *Biological Conservation,* Vol. 117 No. 1, pp. 73-81

Rutledge, D. T. & Miller, C. J. (2006). The use of landscape indices in studies of the effects of habitat loss and fragmentation - a survey of 566 peer-reviewed studies, *Naturschutz und Landschaftsplanung,* Vol. 38 No. 10-11, pp. 300-307

Schröder, B. & Reineking, B. (2004). Modellierung der Art-Habitat-Beziehung - ein Überblick über die Verfahren der Habitatmodellierung, In: *Habitatmodelle - Methodik, Anwendung, Nutzen.Tagungsband zum Workshop vom 8.-10.Oktober 2003 am UFZ Leipzig,* Dormann, C. F., Lausch, A., Blaschke, T., Söndgerath, D. and Schröder, B. (Eds.), pp. 5-25, UFZ, 0948-9452, Leipzig

Siemers, B. M. & Swift, S. M. (2006). Differences in sensory ecology contribute to resource partitioning in the bats *Myotis bechsteinii* and *Myotis nattereri* (Chiroptera : Vespertilionidae), *Behavioral Ecology and Sociobiology,* Vol. 59 No. 3, pp. 373-380

Sprecher-Uebersax, E. (2003). Faunistical and taxonomical notes about Lucanidae in Nepal (Insecta: Coleoptera), In: *Biodiversität und Naturausstattung im Himalaya,* Hartmann, M. (Ed.), pp. 205-209, Verein der Freunde und Förderer des Naturkundemuseums Erfurt e. V., Erfurt

Steinhauser, D. (2002). Investigations on the ecology of the barbastelle, *Barbastella barbastellus* (Schreber, 1774), and the Bechstein's bat, *Myotis bechsteinii* (Kuhl, 1817) in southern Brandenburg, Germany, In: *Ökologie, Wanderungen und Genetik von Fledermäusen in Wäldern. Untersuchungen als Grundlage für den Fledermausschutz.* pp. 81-98, Bundesamt für Naturschutz (BfN), 3-7843-3612-4 Bonn

The Council of the European Communities. (2004), Council Directive 92/43/EEC of 21 May 1992 on the conservation of natural habitats and of wild fauna and flora, pp. 1-57, Office for Official Publications of the European Communities, Brussels

Tischendorf, L. (2001). Can landscape indices pedict ecological processes consistently?, *Landscape Ecology,* Vol. 16 No. 3, pp. 235-254

Turner, M. G. (1989). Landscape ecology - the effect of pattern on process, *Annual Review of Ecology and Systematics,* Vol. 20, pp. 171-197

Turner, M. G., Gardner, R. H. & O'Neill, R. V. (2001). *Landscape ecology in theory and practice. Edition 1. Pattern and process*, Springer-Verlag, 0-387-95122-9, Berlin, New York

Valladares, G., Salvo, A. & Cagnolo, L. (2006). Habitat fragmentation effects on trophic processes of insect-plant food webs, *Conservation Biology*, Vol. 20 No. 1, pp. 212-217

Wu, J. G. (2004). Effects of changing scale on landscape pattern analysis: scaling relations, *Landscape Ecology*, Vol. 19 No. 2, pp. 125-138

7

The Role of Landscape in Contact Zones of Sister Species of Lizards

Gabriela Cardozo[1], Sergio Naretto[1], Marcelo Zak[2] and Margarita Chiaraviglio[1]
[1]*Laboratorio de Biología del Comportamiento, Universidad Nacional de Córdoba*
[2]*Departamento de Geografía, Universidad Nacional de Córdoba*
Argentina

1. Introduction

Elucidating the factors regulating the spatial distribution of ecologically similar species is a key pursuit (Dammhahn and Kappeler, 2008; Peres-Neto, 2006). Many biotic and abiotic variables might influence species distribution and determine allopatry or simpatry (Di Cola & Chiaraviglio, 2010). Moreover, species patterns are strongly associated with habitat variables at different spatial scales (Hatten & Paradzick, 2003). However, the role of landscapes in contact zones is not completely understood.

Contact zones have long been recognized as natural laboratories of evolution (Bridle et al., 2001). The geographic structure of contact zones determines dynamic evolutionary processes; however, since landscape structure influences population processes (Cardozo et al, 2007; Cardozo & Chiaraviglio, 2008) the maintenance of contact zones is likely to depend on landscape patterns.

Morphologically similar species are more likely to interact than morphologically dissimilar ones simply because a major portion of the behavioral and ecological activities of animals is associated with morphology (Losos, 1990; Pianka, 1986). Morphological similarity among coexisting animal species induces potential interactions that may lead to niche segregation (Huey, 1974; Huey & Pianka 1977). It is widely accepted that niche differentiation is often the basis for the coexistence of competitors (MacArthur & Levins, 1967; Roughgarden, 1979); however, how the coexisting species use landscape-scale resources is not clear. Interspecific competition might favour niche differentiation between competitors because it may optimise their behaviour in different ways (Law et al., 1997; Maynard Smith & Parker, 1976). Thus, niche differentiation in ecologically similar species might induce divergence of landscape-scale habitat use.

In this work, we focused on two closely related lizard species: *Tupinambis merianae* and *Tupinambis rufescens*; they are particularly interesting because they occupy the southernmost area of *Tupinambis* group distribution in South America (Peters & Donoso-Barros, 1986). *T. rufescens* would be restricted to the dry Chaco whereas *T. merianae* would occur in diverse regions (Cei, 1993; Colli et al., 1998; Lopes & Abe, 1999) from southern Amazonia to northern Patagonia (Carvalho et al., 2006). The contact zone of the two lizards coincides principally with the arid South American Gran Chaco.

The species have similar body size and external morphological traits, as well as overlapping macro-habitat use and general foraging mode (Castro & Galetti, 2004; Williams et al., 1993). Therefore, a potentially extensive interespecific interaction would represent a significant pressure in sympatric areas. The combination of morphological similarity, typically terrestrial habits and territoriality renders *Tupinambis* lizards ideal models for examining differential use of resources in sympatric areas based on landscape structure.

Considering that habitat loss is a serious environmental problem in many ecosystems (Ishwar et al., 2003; Luiselli & Capizzi, 1997; Mac Nally & Brown, 2001), the conservation status of landscapes in key wildlife habitats, such as contact zones, becomes strikingly relevant for species conservation. Numerous research works indicate that several species are globally threatened by habitat loss, and how changes in spatial patterns influence ecological processes has received great attention. For instance, Cardozo & Chiaraviglio (2008) found that landscape influences life history parameters and spatial distribution of reproductive individuals in snakes, leading to geographical variations in mating systems and therefore variations in reproductive potential. Furthermore, Cardozo et al. (2007) showed that landscape fragmentation affects dispersal patterns, reducing gene flow.

Investigations on landscape-scale Squamata habitats may provide essential knowledge to understand interspecific interactions and to implement measures for the conservation of herpetological communities (Filippi & Luiselli, 2006). Nevertheless, not only does habitat loss pose a threat to individual species but also landscape modifications could affect species interactions. Thus, the understanding of the associations between landscape conservation status and the distribution of sister species in contact zones could be useful to design conservation plans not only for individual species but also for ecological systems.

We examined landscape-scale habitat use in contact and allopatric zones between the two teiid lizards (*T. merianae* and *T. rufescens*) that occur in the Chaco region of central Argentina. Habitat heterogeneity is expected to increase the probability of coexistence among sister species (Tews et al., 2004). Therefore, we hypothesized divergence in landscape use in contact zones by both species, which would exploit high quality resources and take advantage of habitat heterogeneity. Within a regional context, animals that need either to maximize the availability of resources or to minimize interspecific interactions may select areas dominated by patches of a particular vegetation type (Jonshon et al., 2004).

We generated useful knowledge to guide conservation efforts including landscape-level process-oriented considerations, to contribute to avoid disruption of the evolutionary process and to ensure healthy biodiversity at all levels.

2. Methods

2.1 Study area

The study area was located in the province of Córdoba, central Argentina, which is an ideal natural scenario for the study of landscape-scale niche differentiation of *T. merianae* and *T. rufescens* because this area includes the southernmost contact zone between the species distributions. The province of Córdoba has a central-western mountain area with a maximum elevation of 2790 m a.s.l. surrounded by vast plains of 600-900 m a.s.l. The study area lies largely within the Gran Chaco, which is the largest dry forest in South America; vegetation in the region comprises a mosaic of xerophytic forests and scrubs (Zak & Cabido, 2002). The Gran Chaco is a highly threatened wooded region, strongly affected by extensive livestock raising, extractive forestry and poorly planned agricultural expansion (Zak et al., 2004, 2008). To the east, the study area also includes the Pampas region, which was originally composed of natural grasslands but which is currently severely degraded mainly due to the advance of crop farming (Cozzani et al., 2004).

2.2 Species data

We used a database that includes approximately 700 records of the presence of *T. rufescens* and *T. merianae* in central Argentina, which were gathered during field work conducted in the framework of a major project on lizard ecology developed by our research group at the Universidad Nacional de Córdoba, Argentina. Presence records were classified according to their locality of origin (69 localities of presence of *T. merianae* and 32 localities of presence of T. *rufescens*) (Fig. 1).

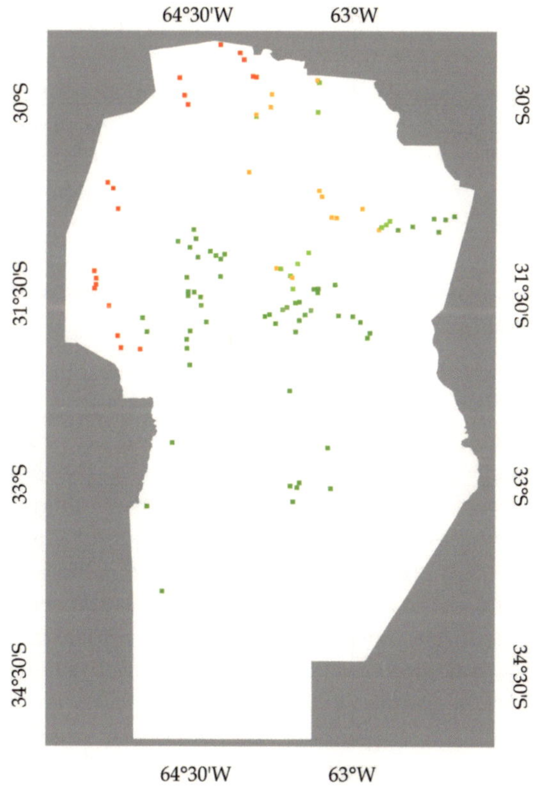

Fig. 1. Localities of presence of *T. merianae* (dark green: allopatry, light green: sympatry) and *T. rufescesns* (dark red: allopatry, light red: sympatry) in the province of Córdoba, central Argentina.

External polygons, considering presence of each species and the geographical characteristics of the study area were drawn. Localities were categorized as sympatric or allopatric depending on whether or not they were included within the intersection of the polygons or next to any heterospecific record of presence within a radius of 10 km (58 allopatric and 11 sympatric localities of presence of *T. merianae,* and 19 allopatric and 13 sympatric localities of presence of *T. rufescens*).

We defined the sample unit as a buffer area around the exact geographical coordinates of the locality of presence (Westphal et al., 2003). Sample units were circular plots of 2-km radius, which is equivalent to twice the area that contains the ecological range reported for other *Tupinambis* spp. (Mendoza & Noss, 2003; Winck, 2007). We intersected those areas with landscape cover features, i.e., we selected "mini-landscapes", for further characterization with landscape metrics (Westphal et al., 2003). To determine landscape availability we selected mini-landscapes at random within the distribution area delimited by the external polygon of the localities of presence for each species. We also quantified landscape availability in the contact zone considering the external polygon of the localities classified as sympatric.

2.3 Landscape analysis

The vegetation land-cover map of the province of Córdoba was created by the Multidisciplinary Institute of Plant Biology (IMBIV) of the National University of Córdoba and CONICET, Argentina (Zak, 2008). This map was obtained from the classification of Landsat 5 TM images together with phytosociological data. The researchers originally identified, described and mapped 19 land-cover types (Zak & Cabido, 2002). The classification of the Landsat imagery was based on the application of a maximum likelihood classifier using the sixth bands of the TM images and their Normalized Difference Vegetation Index (NDVI). Training sites were determined after analysis and field recognition of clusters defined by previous unsupervised classifications and the multivariate analysis of Braun-Blanquet (1950) phytosociological releves.

We grouped the original vegetation land-cover map according to the ecological function of the land-cover types for the bioecology of the study species. Among the environmental factors that might influence behaviors in Squamata, vegetation structure would be of great importance (Blouin-Demers & Weatherhead, 2001; Chiaraviglio & Bertona, 2007; Row & Blouin-Demers, 2006) because it provides alternative thermal environments for thermoregulation (Chiaraviglio, 2006) and might affect reproductive processes and life history traits (Cardozo & Chiaraviglio, 2008; Cardozo & Chiaraviglio, 2011). Therefore, according to the complexity of the vertical structure of the land covers, we determined three major vegetation classes: forest, shrublands and low vertical structures (LVS). Forest includes lowland forests and highland forests; shrublands includes lowland scrubs and highland scrubs; and LVS vegetation includes natural grasslands, halophytes, cordgrass, palustrine vegetation, cultural vegetation, saline zones, waterlogged soils, highland grasslands and bare soils.

To obtain consistent fragmentation metrics, we refined image classification by applying a moving window using the majority analysis (Baldi et al., 2006; Cardozo et al., 2007). We assessed the configuration of the three major vegetation classes in each mini-landscape by calculating the following fragmentation metrics: a. Class Metrics: proportional abundance in the landscape (PLAND) ; mean patch area (hectare) (AREA_MN); number of patches (NP); landscape shape index (LSI), which is a measure of the total edge of each class and increases as the patch type becomes more disaggregated; shape complexity (PARA_MN), which is calculated as mean perimeter area ratio; heterogeneity (IJI), which increases when the corresponding patch type becomes equally adjacent to all other patch types (i.e., maximally interspersed and juxtaposed to other patch types); proximity (PROX_MN), which increases as the neighborhood (defined as 2830 m to include the entire the area of the mini-landscape) is increasingly occupied by patches of the same type and as those patches become closer and more contiguous (or less fragmented) in distribution. b. Land Metrics: the total edge length in the landscape (LSI); contagion (CONTAG), which increases when all patch types are maximally aggregated and minimally interspersed (equal proportions of all pair-wise adjacencies); diversity (SHDI), which increases as the number of different patch types (i.e., patch richness, PR) increases and/or the proportional distribution of area among patch types becomes more equitable; contrast-weighted edge density (CWED) (metres per hectare), which increases as the amount of edge in the landscape increases and/or as the contrast in edges increases. We introduced the following edge contrast weights: forest-shrubs=0.25, forest-LVS=0.99, shrubs-LVS=0.75. All metrics were calculated using FragStats 3.3 (Cardozo & Chiaraviglio, 2008; McGarigal & Marks, 1995; Rutledge, 2003).

2.4 Analyses

We compared the landscape-scale habitat availability in the allopatric and sympatric localities areas of *T. merianae* and *T. rufescens* with Kruskall-Wallis test of landscape features of the random mini-landscapes. Then we compared the landscape-scale habitat use and selection between species by applying Wilcoxon test. Niche differentiation analysis was based on the comparison of landscape features of the sympatric localities between the two species. Intraspecific variations in habitat use were determined by comparing landscape features between allopatric and sympatric localities for each species by applying Wilcoxon test. We also analyzed the results in an information theoretical framework, which allowed us to examine various models including interactions among variables. We fit the generalized linear models (GLM) and employed the Akaike Information Criterion (AIC) to select the models (Westphal et al., 2003) that best identify the landscape features determining species distribution. The model with the lowest AIC was selected as the 'best' model (Mazerolle, 2006). These analyses were performed with R: A Language and Environment for Statistical Computing (2011).

3. Results

3.1 Landscape-scale habitat availability

The configuration of the available landscape presented a gradient from the distribution area of *T. merianae* to that of *T.rufescens* (SE-NW) (Table 1) of decreasing proportion (PLAND) and mean area (AREA_MN) of LVS patches, and increasing values of these metrics for the forest and shrubland areas. Along this gradient, toward the NW, LVS vegetation exhibited greater edge length because of increasing disaggregation (LSI), and increased patch shape complexity (PARA_MN). Shrublands became more interspersed (IJI), with more irregular edges (PARA_MN), than to the SE.

		Landscape availability *T. merianae* N=87	Landscape availability contact zone N=21	Landscape availability *T. rufescens* N=32	P value
Class Metrics					
PLAND	LVS	93.23	74.33	64.11	<0.0001
	Forest	4.26	25.64	26.09	0.0002
	Shrublands	2.51	0.03	9.80	>0.9999
AREA_MN	LVS	1347.24	849.46	684.06	<0.0001
	Forest	27.01	127.40	147.10	0.0002
	Shrublands	5.01	0.20	38.67	>0.9999
NP	LVS	1.47	3.19	4.28	>0.9999
	Forest	1.23	4.00	3.69	0.0042
	Shrublands	1.34	0.86	3.00	>0.9999
LSI	LVS	1.27	1.96	2.21	0.0004
	Forest	2.48	2.49	2.84	>0.9999
	Shrublands	3.02	1.30	2.58	>0.9999
PARA_MN	LVS	40.35	107.40	145.21	0.0005
	Forest	236.77	186.41	181.42	>0.9999
	Shrublands	262.30	635.03	382.32	0.0013
PROX_MN	LVS	21.99	28.67	54.19	>0.9999
	Forest	9.63	26.59	26.84	>0.9999
	Shrublands	16.64	0.01	21.28	>0.9999
IJI	LVS	77.77	10.71	48.06	>0.9999
	Forest	72.86	11.33	45.34	>0.9999
	Shrublands	69.73	97.25	86.62	0.0040
Land Metrics					
LSI		1.29	1.80	2.18	<0.0001
CWED		2.19	7.99	9.20	<0.0001
CONTAG		92.73	74.81	69.63	<0.0001
IJI		74.77	13.58	49.86	0.0005
SHDI		0.11	0.35	0.46	<0.0001

Table 1. Landscape availability in the distribution areas of *T. merianae* and *T. rufescens* in central Argentina.

The proximity (PROX_MN) among patches of the landscape cover types did not vary. Regarding mean landscape metrics, diversity (SHDI), landscape shape index (LSI) and contrast-weighted edge density (CWED) increased in the contact zone and in the distribution area of *T. rufescens*. By contrast, connectivity (CONTAG) decreased, showing more heterogeneous landscapes.

3.2 Landscape-scale habitat use and selection

Although occurring in landscapes with prevailing LVS vegetation distributed in few large patches, *T. merianae* selected landscapes with a greater proportion of forest and shrublands - about 20%- than the available average -7%- (forest :W=6542.00, P<0.0001; shrubland: W=6166.00, P=0.0008). *Tupinambis merianae* required forest patches of an average of 50 ha and shrub patches of 30 ha (Fig. 2 a).

Tupinambis rufescens also selected landscapes that are different from those available, with low proportion of LVS vegetation disaggregated in patches, and a high proportion of forest and shrubs, reaching 56%, which is slightly higher than landscape availability - 36%- (LVS: W=1210.00, P<0.0223; forest: W=931.00, P=0.1240; shrublands: W=951.00, P=0.2235). Landscapes selected by *T. rufescens* presented more forest patches than landscapes selected by *T. merianae*. *Tupinambis rufescens* required forest patches of approximately 224 ha and shrub patches of 188 ha (Fig. 2 b).

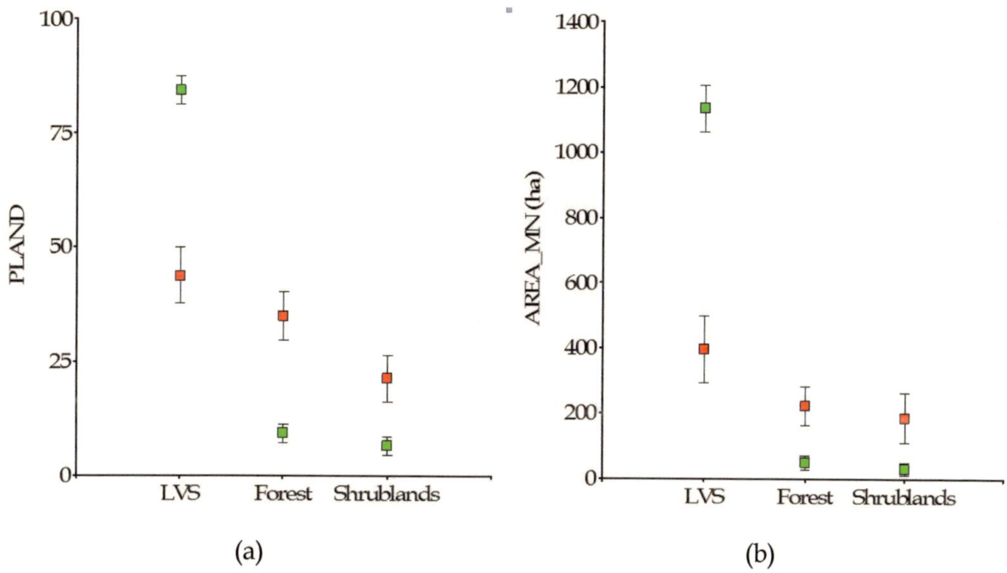

Fig. 2. Landscape use of *T. merianae* (green) and *T. rufescens* (red): (a) Percentage of landscape (PLAND); (b) Mean patch area (AREA_MN).

Species differed in the landscape-scale habitat use in terms of proportion of land cover types and mean area of the patches (LVS: PLAND, W=946.00, P<0.0001, AREA_MN, W=955.00, P<0.0001; forest, PLAND, W=2238.5, P<0.0001, AREA_MN, W=2212.5, P<0.0001; shrublands, PLAND, W=1995.00, p=0.0047, AREA_MN, W=2010.00, P=0.0032). Moreover, landscapes used by *T. rufescens* presented patches of LVS vegetation with more complex shape (PARA_MN) and more compact forest patches than landscapes used by *T.merianae*. *Tupinambis rufescens* used landscapes where LVS patches and forest patches exhibit greater proximity than landscapes used by *T. merianae*. The amount and contrast of edges (LSI and CWED) and diversity (SHDI) differed between landscapes used by both species (Table 2).

		T. merianae N=69	*T. rufescens* N=32	P value
Class Metrics				
NP	LVS	2.52	5.16	0.0001
	Forest	3.45	4.31	0.0389
	Shrublands	3.09	3.56	0.1161
LSI	LVS	1.72	2.50	0.0001
	Forest	2.68	2.68	0.6500
	Shrublands	2.96	2.73	0.6385
PARA_MN	LVS	76.87	201.44	<0.0001
	Forest	244.85	183.95	0.0132
	Shrublands	369.32	289.04	0.1980
PROX_MN	LVS	21.24	38.74	0.0003
	Forest	8.35	30.09	0.0021
	Shrublands	22.75	14.55	0.6023
IJI	LVS	58.22	56.44	0.6224
	Forest	53.35	47.75	0.6815
	Shrublands	72.92	73.85	>0.9999
Land Metrics				
LSI		1.77	2.31	0.0013
CWED		5.75	9.57	0.0033
CONTAG		82.60	62.56	0.0001
IJI		58.02	57.42	0.8055
SHDI		0.27	0.57	0.0002

Table 2. Landscape use of *T. merianae* and *T. rufescens* in central Argentina.

3.3 Landscape-scale habitat use and selection: Sympatry and allopatry

Tupinambis merianae both in allopatry and sympatry used similar landscapes in terms of proportions of land-cover types (LVS: W=307.50, P=0.1943; forest: W=451.00, P=0.2690; shrubs: W=360, P=0.6507). By contrast, *T. rufescens* in allopatry and sympatry used the landscape differentially. In sympatry, this species used landscapes with lower proportion of forest and shrubs, and higher proportion of LVS vegetation than in allopatry (LVS: W=289.00, P=0.0042; forest: W=161.00, P=0.0399; shrubs: W=151.50, P=0.0136). Landscape use did not differ between *T. rufescens* and *T. merianae* in sympatry, according to proportion of land cover types (LVS: W=145.50, P=0.6423; forest: W=134.00, P=0.8386; shrubs: W=136.50, P=0.9483) (Fig. 3a). Landscape-scale habitat use did not differ from landscape availability in either species (PLAND: LSV: H=55, P=0.7601; forest: H=1.67, P=0.4315; shrubs: H=0.10, P=0.9401).

Landscape-scale habitat use did not differ between *T. rufescens* and *T. merianae* in sympatry in terms of mean patch area (LVS: W=149.50, P=0.4859; forest: W=134.00, P=0.8386; shrubs: W=136.50, P=0.9483) or number of patches. *Tupinambis rufescens* in sympatry used landscapes with fewer patches of forest and shrubs than those used in allopatry (Table 3). In addition, shrubland patches were smaller and LVS vegetation patches were larger than in allopatry (LVS: W=284.00, P=0.0076; forest: W=177.00, P=0.1498; shrubs: W=155.00, P=0.0197) (Fig. 3b).

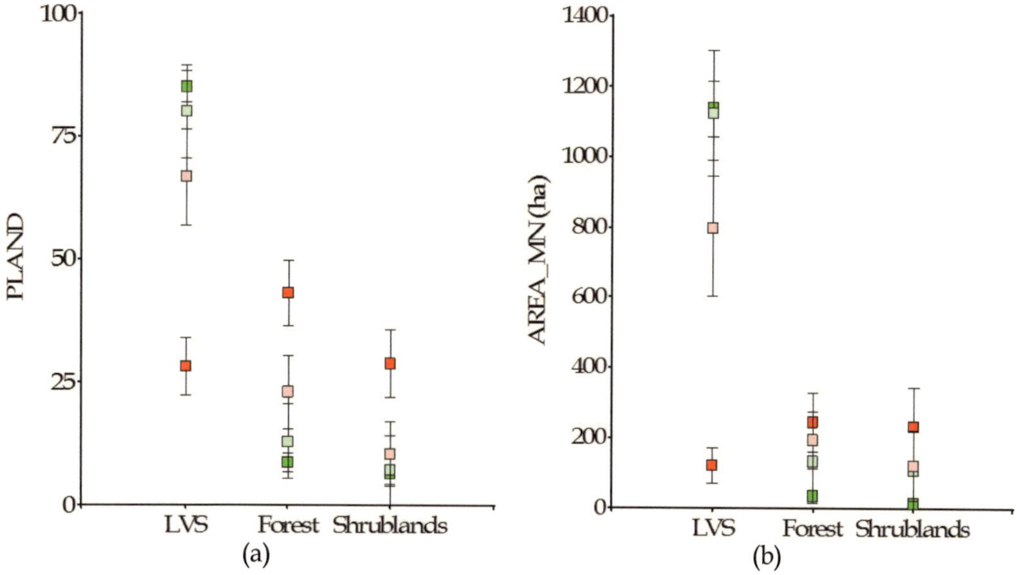

Fig. 3. Landscape use of *T. merianae* (green) and *T. rufescens* (red) in allopatry (dark) and sympatry (light). (a) Percentage of landscape (PLAND); (b) Mean patch area (AREA_MN).

Shape complexity (PARA_MN) of land cover types did not vary within or between species in sympatry (Table 3). *Tupinambis rufescens* in sympatry used landscapes with more dispersed forest patches (PROX_MN) than those used in allopatry. Landscapes used by *T.rufescens* in sympatry presented lower amount of edge (LSI) than those used in allopatry. *Tupinambis merianae* in sympatry used landscapes with shrubland patches more dispersed and with lower amount of edge than those used in allopatry.

		T. merianae allopatry N=58	*T. merianae* sympatry N=11	P value	*T. rufescens* allopatry N=19	*T. rufescens* sympatry N=13	P value	P value Interspecific sympatry
Class Metrics								
NP	LVS	2.59	2.18	0.8619	5.32	4.92	0.2704	0.1541
	Forest	3.40	3.73	0.2711	5.68	2.31	0.0020	0.3034
	Shrublands	3.55	0.64	0.4358	5.00	1.46	0.0084	0.9739
LSI	LVS	1.70	1.85	0.2344	2.76	2.14	0.0654	0.6843
	Forest	2.78	2.32	0.4050	2.78	2.45	0.3893	0.7440
	Shrublands	3.24	1.52	0.0190	2.93	2.09	0.1265	0.2809
PARA_MN	LVS	80.35	58.51	0.4166	229.22	162.98	0.0926	0.3529
	Forest	248.15	233.32	0.2870	202.17	145.50	0.1761	0.1777
	Shrublands	335.32	539.34	0.0613	287.42	294.24	0.8359	0.1473
PROX_MN	LVS	20.67	24.24	0.8061	48.75	24.87	0.0988	0.2314
	Forest	10.28	1.57	0.1304	40.33	8.49	0.0581	0.0972
	Shrublands	27.30	0.0016	0.0033	17.88	3.90	0.4035	0.1258

		T. merianae allopatry N=58	*T. merianae* sympatry N=11	P value	*T. rufescens* allopatry N=19	*T. rufescens* sympatry N=13	P value	P value Interspecific sympatry
IJI	LVS	66.51	16.77	0.0105	55.44	60.19	0.8415	0.2207
	Forest	57.40	33.08	0.3301	45.35	56.77	0.6892	0.8065
	Shrublands	71.81	78.52	0.1722	74.83	70.19	0.6886	0.5221
Land Metrics								
LSI		1.79	1.66	0.4460	2.64	1.83	0.0015	0.8845
CWED		5.65	6.27	0.3355	11.07	7.39	0.1028	0.9537
CONTAG		82.59	82.70	0.3614	57.21	70.39	0.1112	0.3236
IJI		65.71	19.58	0.0059	56.06	62.53	0.4839	0.2207
SHDI		0.28	0.23	0.4612	0.68	0.40	0.0148	0.2958

Table 3. Comparison of landscape use in allopatry and sympatry of *T. merianae* and *T. rufescens* in central Argentina.

The amount of edge of the land-cover types in the landscapes did not vary between species in sympatry. *Tupinambis rufescens* in sympatry used less diverse landscapes (SHDI) than in allopatry. *Tupinambis merianae* in sympatry used less interspersed landscapes (IJI) than in allopatry because the LVS vegetation was less juxtaposed than in allopatry. Land metrics and proximity of land-cover types did not differ between species in sympatry.

3.4 Model selection

According to the lowest AIC, the presence of *T. merianae* at landscape scale is determined by the proportion and the area of the patches of the shrublands, and the number of patches of forest (residual deviance: 80.55, AIC: 86.55), and the presence of *T. rufescens* by the mean area of the patches of shrublands and the proportion of forest in the landscape (residual deviance: 81.63, AIC: 87.63).

4. Discussion

The role of the landscape for reptiles has been largely discussed (Blouin-Demers & Weatherhead, 2001; Cardozo et al., 2007; Driscoll, 2004; Luiselli & Capizzi, 1997; Mac Nally & Brown, 2001; Marchand & Litvaitis, 2004; Stow & Sunnucks, 2004). However, the novelty of our approach lies in the importance of landscape conservation to the maintenance of ecological interactions between lizard sister species. Since landscape ecology analysis is useful to gain a better understanding of environmental suitability (Fouquet et al., 2010), the present work provides useful knowledge for the conservation of these species.

Identifying the habitat characteristics that regulate the ecological processes of reptiles is imperative to determine threats to the species (Urbina-Cardona et al., 2006). Our results showed that the configuration of the available landscape presented a gradient from the distribution area of *T. merianae* to that of *T.rufescens* (SE-NW) of decreasing proportion of mean area of LVS patches and increasing proportion of forest and shrublands area. Along this gradient, the landscape became more heterogeneous. These results show that landscape configuration is a main factor regulating the spatial distribution of ecologically similar species and has a central role in contact zones. Therefore, changes in distribution-related factors at landscape scale, such us habitat loss, might pose a threat for herpetofauna (Filippi & Luiselli, 2000).

Moreover, besides understanding how landscape patterns provide resource heterogeneity in the species distribution areas, it is vitally important to elucidate how these species use that heterogeneity i.e. , species might reveal diverse responses to the spatial variations in habitat resources (Cagle, 2008; Urbina-Cardona et al., 2006). We observed that species are selective on landscape patterns; for example, although occurring in landscapes with prevailing LVS vegetation distributed in few large patches, *T. merianae* selected landscapes with a greater proportion of forest and shrubs. *Tupinambis rufescens* also selected landscapes that are different from those available, with low proportion of LVS vegetation disaggregated in patches, and a high proportion of forest and shrubs. Considering the relevance of forest and shrublands to both species, the present results might guide conservation efforts including landscape-level process-oriented considerations.

Although morphological similarity among species induces niche similarity because behavioral and ecological activities of animals are associated with morphology (Losos, 1990; Pianka, 1986) we observed that these sister species differ in several features of the landscape-scale habitat use (e.g., shape complexity, patch proximity, length of edges and diversity). Taking into account the diversity of responses of the species, conservation strategies should be species-specific (Keogh et al., 2001). Our results indicate that *T. rufescens* and *T. merianae* have complex habitat requirements. Specialized habitat requirements make species more vulnerable to extinction (CITES, 2010; Santos et al., 2009; Webb et al., 2002). Since *T. rufescens* and *T. merianae* showed differences in landscape-scale habitat requirements, planning of conservation strategies should consider such interspecific heterogeneity.

Although the results obtained showed that species differed in the landscape-scale habitat use in their distribution areas, in the contact zone where species are in sympatry, and therefore have the same landscape-scale habitat availability, interespecific interactions would represent a significant pressure on the use of the resources. When we analyzed niche differentiation in terms of landscape-scale resources, we observed that species did not differ in the use of landscape resources in contact zones. Both species used landscapes with similar proportions of land-cover types, mean patch area and number of patches; further research is need to elucidate if the coexistence of these species could be explained by niche divergence at local scale. Furthermore, the results obtained enable us to get a better understanding of the strategies of the species in sympatric zones in terms of landscape-scale habitat use and selection. *Tupinambis merianae* both in allopatry and sympatry used similar landscapes. By contrast, *T. rufescens* in allopatry and sympatry used the landscape differentially, showing niche modification. In sympatry this species used landscapes with lower proportion of forest and shrubs, and higher proportion of LVS vegetation than in allopatry. Therefore, despite their ecological and morphological similarity, the species respond differently to spatial changes in landscape structure. Moreover, we remark the importance of landscape heterogeneity for the maintenance of species interactions in the contact zone.

5. Conclusion

Understanding the associations between landscape conservation status and distribution of sister species in contact zones might be useful to design conservation plans not only for individual species but also for ecological systems. Among the weaknesses of the regional conservation plans, poor information regarding behavioural ecology is one of the fundamental issues (The Nature Conservancy et al., 2005). To know whether species might be threatened by habitat change it is necessary to determine the relation between ecological processes and environmental patterns (Cardozo & Chiaraviglio, 2008). Our study provides useful knowledge about the important role of native forest and shrublands in allopatric and sympatric distribution areas of the lizard species. Furthermore, similar allopatry-simpatry systems might be occurring in this contact zone, which -as we mentioned above- coincides principally with the arid South American Gran Chaco. Considering that only 9% of the South American Gran Chaco is protected (The Nature Conservancy et al., 2005), we underline the need for efficient control of deforestation, protection of forest remnants and establishment of corridors. According to Beaudry et al. (2010) regional-scale conservation planning has to answer specific questions, such as the type of habitat that is needed and where it should be protected. The present work provides information that may be useful to guide conservation plans. Efforts to prevent habitat loss should involve preserving not only allopatric areas but also these critical heterogeneous sympatric areas where biological interactions might modify ecological processes of species (Brito et al., 2009; Santos et al., 2009).

6. Acknowledgments

We are grateful to the institutions that funded this study: Consejo Nacional de Investigaciones Científicas y Técnicas (CONICET), Fondo para la Investigación Científica y Tecnológica (FONCyT), MinCyT Córdoba -Préstamo BID-PID No. 013/2009-, Secretaría de Ciencia y Tecnología (SeCyT) and Universidad Nacional de Córdoba, Argentina. GC is scientist of the Consejo Nacional de Investigaciones Científicas y Técnicas (CONICET). SN is student of the Biological Sciences Doctorate of the Universidad Nacional de Córdoba and fellowship holder of the Consejo Nacional de Investigaciones Científicas y Técnicas (CONICET). MZ is Professor of the National University of Córdoba and senior scientist at the IMBIV. MCH is Professor and senior scientist of the National University of Córdoba.

7. References

Baldi, G., Guerschman, J.P. & Paruelo, J.M. (2006). Characterizing fragmentation in temperate South America grasslands. *Agr. Ecosyst. Environ*, Vol.116, pp. 197–208

Beaudry, F., Pidgeon, A.M., Radeloff, V.C., Howe, R.W., Mladenoff, D.J. & Bartelt, G.A. (2010). Modeling regional-scale habitat of forest birds when land management guidelines are needed but information is limited. *Biological Conservation*, Vol.143, pp. 1759-1769

Blouin-Demers, G. & Weatherhead, P.J. (2001). Habitat use by black rat snakes (Elaphe obsoleta obsoleta) in fragmented forests. *Ecology*, vol.82, pp. 2882–2896

Braun-Blanquet. J. (1950). *Sociologia Vegetal*. ACME, Buenos Aires

Bridle J.R., Baird S. & Butlin, R.K. (2001). Spatial structure and habitat variation in a grasshopper hybrid zone. *Evolution*, Vol.55, pp. 1832–1843

Brito, J.C., Acosta, A.L., Álvares, F. & Cuzin, F. (2009). Biogeography and conservation of taxa from remote regions: An application of ecological-niche based models and GIS to North-African Canids. *Biological Conservation*, Vol.142, pp. 3020-3029

Cagle, N. (2008). Snake species distributions and temperate grasslands: a case study from the American tallgrass prairie. *Biological Conservation*, Vol.141, pp. 744–755

Cardozo, G., Rivera, P.C., Lamfri, M., Scavuzzo, M., Gardenal, C. & Chiaraviglio, M. (2007). Effects of habitat loss on the genetic structure of the Argentine Boa Constrictor (*Boa constrictor occidentalis*) populations, In: *Biology of the Boa and Phytons*, Henderson, R., Powell, R. (Eds.), pp. (329–338), Eagle Mountain Publishing, Utah

Cardozo, G. & Chiaraviglio, M. (2008). Landscape changes influence the reproductive behavior of a key capital breeder snake (*Boa constrictor occidentalis*) in the Gran Chaco region. *Biological Conservation,* Vol.141, (2008), pp. 3050-3058.

Cardozo, G. & Chiaraviglio, M. (2011). Phenotypic plasticity of life history traits in relation to reproductive strategies in *Boa constrictor occidentalis. Evolutionary ecology,* (February 2011)

Carvalho, R.L., Antoniazzi, M.M., Jared, C., Silva, A.M.J., Santos, A.A. & Egami, M.I. (2006). Morphological, cytochemical, and ultrastructural observations on the blood cells of the reptile *Tupinambis merianae* (Squamata). *Comp Clin Pathol*, Vol.15, pp. 169-174

Castro, E.R. & Galetti, M. (2004). Frugivoria e dispersao de sementes pelo lagarto Teiu *Tupinambis merianae* (Reptilia: Teiidae). *Papéis Avulsos de Zoologia*, Vol.44, No.6, pp. 91-97

Cei, J.M. (1993). *Reptiles del noroeste, nordeste y este de la Argentina*. Mus. Reg. Sci. Nat. Monografía XIV, Torino.

Chiaraviglio, M. & Bertona, M. (2007). Reproduction and termoregulation as main factors influencing habitat choice in the Argentine Boa constrictor. In: *Biology of the boa and pythons,* Henderson R, Powell R (Eds), pp. (478–488), Eagle Mountain Publishing, Utah

Chuvieco, E. (2002). *Teledetección Ambiental: la observación de la Tierra desde el Espacio* (first ed.) Ariel, Barcelona

CITES- Convención sobre el Comercio Internacional de Especies Amenazadas de Fauna y Flora Silvestres (CITES). 2010. Criterios biológicos para la inclusión de especies. http://www.cites.org/esp/res/09/09-24R14.shtml#FN0

Colli, G.R., Peres, A.K. & Cunha, H.J. (1998). A new species of *Tupinambis* (Squamata: Teiidae) from central Brazil, with an analysis of morphological and genetic variation in the genus. *Herpetologica*, Vol.54, pp. 477-492

Cozzani, N.C., Sanchez, R. & Salba S. (2004). Nidificación de la loica pampena (Sturnella defilipii) en la provincia de Buenos Aires. Argentina. *Hornero,* Vol.19, No.2, pp. 47-52

Dammhahn, M. & Kappeler, P. (2008). Small-scale coexistence of two mouse lemur species (*Microcebus berthae* and *M. murinus*) within a homogeneous competitive environment. *Oecologia,* Vol.157, pp.473–483

Di Cola, V. & Chiaraviglio, M. (2010). Establishing species' environmental requirements to understand how the southernmost species of south american pitvipers (*Bothrops,* Viperidae) are distributed: a niche based modelling approach. *Austral Ecology,* Vol. 36, pp. 90-98

Driscoll, D.A. (2004). Extinction and outbreaks accompany fragmentation of a reptile community. *Ecol. Appl,* Vol.14, pp. 220–240

Filippi, E. & Luiselli, L. (2000). Status of the Italian snake fauna and assessment of conservation threats. Biological *Conservation*, Vol.93, pp. 219–225

Filippi, E. & Luiselli, L. (2006). Changes in community composition, habitats and abundance of snakes over 10 + years in a protected area in Italy: conservation implications. *Herpetol. J*, Vol.16, pp.29–36

Fouquet, A., Ficetola, G.F., Haigh, A. & Gemmell, N. (2010).Using ecological niche modelling to infer past, present and future environmental suitability for *Leiopelma hochstetteri*, an endangered New Zealand native frog. *Biological Conservation*, Vol.143, pp. 1375–1384

Goodman, R.M., Burton, F.J. & Echternacht, A.C. (2005). Habitat use of the endangered iguana *Cyclura lewisi* in a human-modified landscape on Grand Cayman. *Animal Conservation* (2005) Vol.8, pp. 397–405

Guisan, A. & Zimmermann, N.E. (2000). Predictive habitat distribution models in ecology. *Ecol. Model.*, Vol.135, pp. 147-186

Hatten, J. R. & Paradzick, C.E. (2003). A multiscaled model of southwestern willow flycatcher breeding habitat. *Journal of Wildlife Management*, Vol. 67, No. 4, (October 2003), pp. 774-815

Huey, R.B. (1974). Ecological Character Displacement in a Lizard. *Amer. Zool*, Vol.14, No.4, pp. 1127-1136

Huey, R.B. & Pianka, E.R. (1977). Patterns of niche overlap among broadly sympatric versus narrowly sympatric Kalahari lizards (Scincidae: *Mabuya*). *Ecology*, Vol.58, pp. 119–128

Ishwar, N.M., Chellam, R., Kumar, A. & Noo, B.R. (2003). The response of agamid lizards to rainforest fragmentation in the Southern Western Ghats, India. *Conserv. Soc.* Vol.1, pp. 79–86

Johnson, C.J., Boyce, M.S., Mulders, R., Gunn, A., Gau, R.J., Cluff, H.D. & Case, R.L. (2004). Quantifying patch distribution at multiple spatial scales: applications to wildlife-habitat models. *Landscape Ecology*, (2004), pp. 869–882

Keogh, J.S., Barker, D.G. & Shine, R. (2001). Heavily exploited but poorly known: systematics and biogeography of commercially harveted pythons (Phyton curtus group) in Southeast Asia. *Biological Journal of Linnean Society*, Vol.79, pp. 113-129

Law, R., Marrow, P. & Dieckman, U. (1997). On evolution under asymmetric competition. *Evol Ecol* Vol.11, pp. 485–501

Lopes, H.R. & Abe, A.S. (1999). Biología reproductiva e comportamento do teiú, *Tupinambis merianae*, em cativero (Reptilia, Teiidae). In: *Manejo y Conservación de Fauna Silvestre en América Latina*. Fang, T.G., Montenegro, O.L., Bodmer, R.E. (Eds.), Editorial Instituto e Ecología, pp. (259−272) La Paz, Bolivia

Losos J.B. (1990). The evolution of form and function: morphology and locomotor performance in West Indian *Anolis* lizards. *Evolution*, Vol. 44, pp. 1189–1203

Luiselli, L. & Capizzi, D. (1997). Influences of area, isolation and habitat features on distribution of snakes in Mediterranean fragmented woodlands. *Biodivers. Conserv.* Vol.6, pp. 1339–1351

Mac Nally, R. & Brown, G.W. (2001). Reptiles and habitat fragmentation in the box-ironbark forests of central Victoria, Australia: predictions, compositional change and faunal nestedness. *Oecologia*, Vol.128, pp. 116–125.

MacArthur, R. & Levins, R. (1967). The limiting similarity, convergence and divergence of coexisting species. *The American Naturalist*, Vol.101, No.921, pp. 377-385

Mc Garigal, K. & Marks, B. (1995). FRAGSTATS: spatial analysis program for quantifying landscape structure. USDA Forest Service General Tehnical Report PNW-GTR-351.

Marchand, M.N. & Litvaitis, J.A. (2004). Effects of habitat features and landscape composition on the population structure of a common aquatic turtle in a region undergoing rapid development. *Conserv. Biol*, Vol.18, pp. 758–767

Maynard Smith, J. & Parker, G.A. (1976) The logic of asymmetric contests. *Anim Behav.*, Vol.24, pp. 159-75

Mazerolle, M. (2006) Improving data analysis in herpetology: using Akaike's information criterion (AIC) to assess the strength of biological hypotheses. *Amphib Reptil*, Vol.27, pp. 169–180

Mendoza, F. & Noss. A. (2003). Radiotelemetría del peni (*Tupinambis rufescens*) en una zona del Chaco boliviano. In: *Manejo de fauna silvestre en Amazonía y Latinoamérica*. Selección de trabajos V Congreso Internacional. Rocío Polanco-Ochoa, ed. pp. (178-180), CITES, Fundación Natura, Bogotá

Pearce, J. & Ferrier, S. (2000). Evaluating the predictive performance of habitat models developed using logistic regression. *Ecol. Modell*, Vol.133, pp. 225-245

Peres-Neto P.R. (2006). Patterns in the co-occurrence of fish species in streams: the role of site suitability, morphology and phylogeny versus species interactions. *Oecologia*, Vol.140, pp. 352–360

Peters, J.A. & Donoso-Barros, R. (1986). *Catalogue of the Neotropical Squamata. Part II. Lizards an Amphisbaenians*. Rev. ed. Smithsonian Institution Press, Washington, DC.

Pianka E.R. (1986). *Ecology and natural history of desert lizards: analyses of the ecological niche and community structure*. Princeton University Press, Princeton, NJ.

R: A Language and Environment for Statistical Computing. (2011). R Development Core Team. R Foundation for Statistical Computing. Vienna, Austria. ISBN 3-900051-07-0. http://www.R-project.org

Roughgarden J. 1979. *Theory of population genetics and evolutionary ecology: an introduction*. Macmillan, New York.

Row, J.R. & Blouin-Demers, G. (2006). Thermal quality influences effectiveness of thermoregulation, habitat use, and behaviour in milk snakes. *Oecologia*, Vol.148, pp. 1–11

Rutledge, D. (2003). *Landscape Indices as Measures of the Effects ofFragmentation: can Pattern Reflect Process?* DOC Science International Series 98. Department of Conservation, Wellington

Santos, X., Brito, J.C., Caro, J., Abril, A.J., Lorenzo, M., Sillero, N. & Pleguezuelos, J. M. (2009). Habitat suitability, threats and conservation of isolated populations of the smooth snake (*Coronella austriaca*) in the southern Iberian Peninsula. Biological Conservation, Vol.142, pp. 344-352

Schaefer, J.F., Duvernell, D.D. & Kreiser, B.R. (2011). Ecological and genetic assessment of spatial structure among replicate contact zones between two topminnow species. *Evol Ecol* DOI 10.1007/s10682-011-9461-2

Stow, A.J. & Sunnucks, P. (2004). Inbreeding avoidance in Cunningham's skinks (*Egernia cunninghami*) in natural and fragmented habitat. *Mol. Ecol.*, Vol.13, pp. 443–447

Templeton, A.R., Robertson, R.J., Brisson, J., & Strasburg, J. (2001). Disrupting evolutionary processes: The effect of habitat fragmentation on collared lizards in the Missouri Ozarks. *PNAS*, Vol.98, No.10, (May 2001), pp. 5426–5432

Tews, J., Brose, U., Grimm, V., Tielborger, K., Wichmann, M.C., Schwager, M. & Jeltsch, F. (2004). Animal species diversity driven by habitat heterogeneity/diversity: the importance of keystone structures. *J Biogeogr*, Vol.31, pp. 79–92

The Nature Conservancy (2005), *Evaluación Ecorregional del Gran Chaco Americano/Gran Chaco Americano.* Fundación para el Desarrollo Sustentable del Chaco (De.S.del Chaco) y Wildife Conservation Society Bolivia (WCS). Fundación Vida Silvestre. Buenos Aires. Argentina

Urbina-Cardona, J.N., Olivares-Perez, M. & Reynoso, V.H. (2006). Herpetofauna diversity and microenvironment correlates across a pasture–edge–interior ecotone in tropical rainforest fragments in the Los Tuxtlas Biosphere Reserve of Veracruz, Mexico. *Biological Conservation*, Vol.132, pp. 61–75

Webb, J., Brook, B.W. & Shine, R. (2002). What makes a species vulnerable to extinction? Comparative life-history traits of two sympatric snakes. *Ecol. Res*, Vol.17, pp. 59–67

Westphal, M.I., Field, S.A., Tyre, A.J., Paton, D. & Possingham, H.P. (2003). Effects of Landscape Pattern on Bird Species Distribution in the Mt. Lofty Ranges, South Australia. *Landscape Ecology* Vol.18, (2003), pp. 413-426

Williams, J.D., Donadio, O.E. & Ré, I. (1993). Notas relativas a la dieta de *Tupinambis rufescens* (Reptilia: Sauria) del noroeste argentino. *Neotropica,* Vol.39, pp. 45-51

Winck, G.R. (2007). *Historia natural de Tupinambis merianae (Squamata, Teiidae) na estacao ecologica do Taim. Extremo sul do Brasil*. Santa Maria. RC., Brasil

Zak, M.R. & Cabido, M. (2002). Spatial patterns of the Chaco vegetation of central Argentina: integration of remote sensing and phytosociology. *Applied Vegetation Science,* Vol.5, pp. 213-226

Zak, M.R., Cabido, M. & Hodgson, J. (2004). Do subtropical seasonal forests in the Gran Chaco, Argentina, have a future? *Biological Conservation*, Vol.120, pp. 589–598.

Zak, M. & Cabido, M. (2008). What Drives Accelerated Land Cover Change in Central Argentina? Synergistic Consequences of Climatic, Socioeconomic, and Technological Factors. *Environmental Management,* Vol.42, pp. 181–189

Part 4

Conservation of Nature and Regional Development

8

Landscape Approach to Bio-Cultural Diversity Conservation in Rural Lebanon

Jala Makhzoumi, Salma Talhouk, Rami Zurayk and Riad Sadek
American University of Beirut
Lebanon

1. Introduction

Traditional rural landscapes in Lebanon as elsewhere in the eastern Mediterranean are as much a product of geographical setting and natural processes as they are of cultural modification and adaptations over time. A rich and diverse mosaic of woodland patches, degraded maquis scrubland, terraced perennial cropping of olives trees and vineyards, the rural landscape is characteristically a combination of 'natural' and 'cultural' ecosystems. Traditional rural landscapes combine agricultural, silvicultural and pastoral uses within an integrated management system. Multifunctional in use, sustainable environmentally, valued culturally, traditional rural landscapes are well adapted to poor the degraded environmental conditions in marginal terrain which are suitable for little else (Makhzoumi, 1997). Nevertheless, assessment and valuation and similarly development strategies fail to recognize their specificity. Attempts by the state in the 1950s to 'modernize' agriculture for the most part focused on monoculture farming and cash crops through blanket agricultural policies that do not recognize the value of traditional production systems and vernacular management practices. Nature conservation strategies have similarly sidestepped traditional rural landscapes. National strategies for nature conservation since the 1990s have prioritized on the protection of 'native' species, cedar forests and natural ecosystems, disregarding the potential of traditional rural landscapes as wildlife habitats, a repository of Lebanon's exceptional biodiversity. Research and development of rural regions are similarly narrow in approach, divided between the focus of scientists and engineers on environmental problems and that of social scientists on social and economic betterment. State management and administration replicates the disciplinary divide in national policies because ministries prioritize on one or another component of rural landscapes, for example agricultural production, environment, socio-economic betterment, with little coordination in planning and management. In combination, fragmentary planning and management fail to address the specificity of marginal rural settings as a unique mélange, part nature and part culture, tangible physicality and intangible socio-cultural.

Failure in planning and management are further aggravated by political marginalization that has left rural communities in Lebanon in need of social and economic development. Civil war (1975-1990), Israeli occupation in south Lebanon (1978-2000) and the war in July 2006 in turn depopulated much of the countryside, disrupted traditional rural lifestyles and undermined traditional rural economies. Forestlands were reduced from twenty percent of the land area in 1975 to 5 percent by the late 1990s. Degradation of traditional stone terraces that constitute the backbone of agriculture in marginal lands, the consequent soil erosion as well as destruction of Mediterranean rangeland constitutes a quarter of the total cost of environmental and natural resource destruction estimated by the World Bank at US$ 315 million in 1996 (Hamdan, 2002, p. 185).

Failure to appreciate the specificity of Mediterranean rural landscapes is not limited to Lebanon. Naveh (2008) critiques the 'ingrained tendency to fragmentize and take apart what is in reality whole and one' and in part the result of the rift between the 'biocentric' approach of scientists that focus on nature, ecosystem and environment and the 'anthropocentric' approach of social scientists, cultural and humanistic geography. The outcome has been a compartmentalized approach to rural development, a focus on 'nature' and the 'natural' or 'culture' and the 'cultural'. The imposition of north European planning to Mediterranean rural landscapes is equally to blame for these failures. Blanket policies for intensive agriculture production in marginal Lebanon, mirrors the EU Common Agricultural Policy (CAP). The damage caused by CAP is the outcome of CAP's inability to recognize the ecological and socio-economic conditions that prevail in fragile mountain landscapes in the Mediterranean that can only support traditional terraced cultivation (Naveh, 2008). Dismissive of the diversity and multifunctionality of traditional landscapes, large-scale intensive monocultures in marginal terrain eventually fail. In the process they destroy the sustainable traditional rural systems they came to replace and just as critically, they destroy the traditional practices and the values that are attached to them.

This chapter proposes a landscape approach to planning and management in rural marginal Lebanon. We shall argue that a landscape approach produces key advantages by integrating environmental, ecological and cultural values of marginal landscapes and as such addresses rural needs for health, decent living while protecting bio-cultural heritage. A community woodland project in Ebel-es-Saqi village in south Lebanon serves as a case study to demonstrate the landscape design approach. The underlying aim is to broaden the scope of landscape architecture, an emerging profession in Lebanon, beyond prevailing perception of the profession that is limited to urban environments and contemporary settings.

The chapter is divided into five sections. In the first, we shall define 'landscape' in the context of sustainable rural development in traditional marginal Mediterranean settings. The next section sets the background to the case study, Ebel-es-Saqi village, and the methodological framework of ecological landscape design. In sections three and four the methodological framework is applied to secure an expansive landscape reading of landscape Ebel-es-Saqi village and to write narratives for future development. In part four we review offshoot community initiatives that evolved from the landscape master plan. The paper concludes by revisiting the claims made in favor of a landscape design approach to protect bio-cultural diversity in rural marginal landscapes in Lebanon.

2. The landscape approach

Landscapes are the byproduct of human adaptations of natural settings for the purpose of securing shelter, food and/or for pleasure. 'Landscape' therefore implies tangible physicality (field, orchard, settlement, or region), the product, but also perceptions and cultural valuations attached to this physical setting in the act of production. By pairing 'product' and 'production', environment and people, landscape acquires a discursive elasticity that has encouraged use of the term by several disciplines, each focusing on one or another of the multiple meanings of the term (Makhzoumi, 2002). Scientists, for example, are concerned with the physicality of landscape which they use interchangeably with environment and ecosystem. Landscape ecologists, apply a holistic view of landscape as a "living system"composed of natural and managed components that evolve over time and that are contiguous in space from smallest mappable ecotone to the global ecosphere (Naveh and Lieberman, 1990). Social scientists on the other hand see landscape as a medium for interpreting traditions, constructing identities and unfolding cultural heritage. The 'humanized' conception of cultural geography, for example, has evolved from earlier objective ways of 'knowing about' landscapes and places, their material and tangible facts and features, towards deeper, 'empathetic ways' of understanding meanings, symbolic qualities and values imbedded in the socio-spatial dialectic (Adams et al., 2001). Landscape architects bridge the disciplinary spheres. They draw on the *instrumental* framework of scientists, the *interpretive* approach of social sciences which they combine with creative, lateral design thinking as they aspire to create places of significance and meaning. A landscape design reading therefore is integrative of the past while addressing futures, responsive to locale with awareness of larger spatial contiguities, sensitive and inclusive of shared cultural meanings and values (Figure 1).

Fig. 1. Schematic representation of the 'landscape' as the product of people-environment interaction, expansive spatially and temporally (source: Makhzoumi, 2010)

In this study, we shall argue that the expansive framework of landscape design offers several advantages when applied to traditional Mediterranean rural landscapes. First, breeching the natural and cultural sciences, approach and method in landscape architecture are integrative of the totality of the rural landscapes, 'natural' and 'cultural'. Landscape narratives for development as such address nature and biodiversity conservation just as they recognize the value of traditional rural landscapes as cultural heritage (Makhzoumi, 2012). Spatial in essence, the expansive integrative framework of landscape is successfully adopted in planning for nature conservation as well as serving as a medium for identity construction and heritage. The emerging discourse on the protection of cultural landscapes is increasingly being adopted as a framework for reaffirming local and/or regional identity and heritage. For example, the Council of Europe 2000 declaration states that "landscape contributes to the formation of local culture and it is a basic component of the European natural and cultural heritage, contributing to human well being and consolidation of the European" (http://www.coe.int/t/dg4/cultureheritage/Conventions/Landscape/default_en.asp).

A second advantage lies in the contextualized narrative proposed by landscape design. Responsive to place and inclusive of local community needs and aspirations while aiming for economic and social betterment, a landscape design approach is therefore bottom-up rather than generic and top-heavy. Landscape design and planning is likely to counter indiscriminate application of imported development and conservation policies that disregard the specificities of Mediterranean contexts, whether ecological or cultural (Makhzoumi and Pungetti, 1999; 2008). The integrative and interdisciplinary approach of landscape design is more likely to foster multifunctional scenarios for development. Applied to traditional rural landscapes, an interdisciplinary landscape design approach can lead to sound planning strategies.

3. The Ebel-es-Saqi case study

Ebel-es-Saqi is a village of 3,448 inhabitants, 70 km drive from the coastal city of Sidon. The village woodland was totally destroyed in the last decades of Ottoman rule (circa 1900). In the 1960s-1970s, the Ministry of Agriculture undertook a campaign to reforest degraded village woodlands[1]. Ebel-es-Saqi was one such village, its common land (Arabic himas), reforested with pine, cypress and eucalyptus trees. Budgetary limitations accounts for the selection of inferior, fast growing species rather than oak, hawthorn and cedar trees that were native to the region and that composed the original Ebel-es-Saqi woodland. Only the gentle slopes of the north-eastern aspect were reforested (12.6 hectares, one third the total area of Ebel-es-Saqi common land). The rocky, south-eastern slopes were difficult to plant and remained bare. More significantly, authority over the newly reforested woodland was removed from the village to the Ministry of Agriculture, albeit temporarily pending establishment of the woodland. The enforcement of top-heavy forestry protection laws prohibited the village community from use and management. Having lost stewardship, the local community eventually lost interest altogether in woodland.

Ebel-es-Saqi woodland project was one of several post-occupation recovery initiatives in south Lebanon following Israeli withdrawal in 2000. The woodland had survived the ravages of war through protection of the United Nations International Force in Lebanon (UNIFIL) that established its headquarters in the village. Meanwhile, the mixed pine and cypress trees, 30+ years old, had become an impressive landscape in a region that through war and occupation had lost most of its woodlands (Figure 2). UNIFIL alerted United Nations Economic and Social Commission in West Asia (UN-ESCWA) to the beauty of Ebel-

[1]Woodlands were re-established in sixty six villages covering the four regional governorates: 10 villages in Mount Lebanon; 21 villages in the Bekaa; 21 villages in the North; and 20 villages in the South (Personal communication with Mr Fadi Asmar, Ministry of Agriculture).

es-Saqi woodland urging its protection. The challenge of linking woodland protection and community betterment was not clear. This chapter's lead author was consulted on possible approaches and asked to define a project scope and objectives and was thereafter commissioned with the preparation of a landscape master plan. Commissioning a landscape architect was itself significant, since perception of the profession in Lebanon is more readily associated with urban commercial settings than with rural development. The latter was due to the fact that Ebel-es-Saqi woodland landscape was too small and humanized to be valued as 'nature' and too 'natural', to be commissioned to an architect.

Fig. 2. Ebel-es-Saqi village woodland is a landmark in a forestless landscape.

The challenge facing the landscape design was how to integrate physical and human recovery and in addition how to integrate environmental and socio-economic developmental objectives. Initial visits to the site and early negotiations with UN-ESCWA confirmed that a landscape approach and methodology would be adopted to define the project objectives. These were threefold: to protect the village woodland as a wildlife habitat; benefit local community economically from the woodland; and promote local awareness of environment and natural resources (Makhzoumi, 2003).

The methodological framework of ecological landscape design (Makhzoumi, 2000; Makhzoumi and Pungetti, 1999) was applied to secure a holistic reading of woodland and village landscape and similarly to write expansive narratives for community-inclusive protection. The interdisciplinary framework of ecological landscape design affected assembly of the project team, which included faculty and students from the American University of Beirut. The team included a conservation biologist, ecosystem scientist, ecologist, GIS expert, agriculturalist, architect and a landscape architect.[2] Comprehensive surveys were undertaken of the woodland and village landscape, which included geomorphology and land cover, on-site surveys of woodland flora and fauna, photographic documentation of views and vistas from within the woodland, interviews and focus groups with the local community. The research extended to include an archival search of historic village maps from the French Mandate (circa 1930s), which were instrumental in recalling Arabic vernacular placenames that had been transcribed for the entire village on the archival map. One such name was 'hima', referring to the Ebel-es-Saqi common land. Ironically, Arabic placenames do not appear on more recent cadastral maps in Lebanon. The findings of the interdisciplinary survey are herein presented, the layers woven into a landscape reading of place and people (Makhzoumi, 2003).

4. Reading Ebel-es-Saqi landscape

Ebel-es-Saqi village's location and landscape features are typical of Lebanese mountain terrain (Freyha, 1957; Murr, 1987). The village's origin is biblical having been established in the proximity of the spring of Ebel from which the village derives its name (Arabic, 'Ebel of waters')(Feghali, 2002). The village straddles one of two hilly peaks, 684 meters above sea level (m.a.s.l.), enabling defense and a commanding view of the surroundings. The hillside southeast of the village is terraced and cultivated with olive trees all the way to the Hasbany River Valley. Village woodland occupies the northeastern aspect of the second hill at 700 m.a.s.l. The geology at Ebel-es-Saqi is predominantly of sedimentary and calcareous rocks with minor sandstone and basalt formations. Red soils dominate 70% of the village cadastral area with discontinuous grey soils that form the substrata beneath the Hasbani River southeast of the village. Village cadastral area is 750 ha of which 44% constitutes the hima, 34% olive tree cropping, 14% arable land and 8% built-up settlement area (Figure 3). The hima is predominantly of grassland and degraded garigue with advanced maquis scrubland along the Hasbany River verges. Ebel-es-Saqi woodland, the forested part of the village common land, constitutes a mere 5% but has a considerable visual impact. Olive tree cropping is on private land (Figure 4). Specimen olive trees were estimated at several hundred years, claimed by the local community as their Roman heritage. Each household possesses an olive orchard as olives are central to the local diet and often the only means of livelihood. Declining marketability invariably undermines profitability and affects the value of the traditional rural landscapes. Field crops, mainly wheat and barley, traditionally grown on gentle slopes to the northwest of the village are in decline. Built-up settled area is compact, consisting of traditional stone houses with pitched, terracotta roofs. Residents returning to the village after 2000 have restored the traditional village houses but also favor 'contemporary' building styles and materials.

---

[2]The Master Plan Team consisted of conservation biology, Salma Talhouk, ecosystem and environmental resource management, Rami Zurayk, GIS expert, Dani Leshia , ecologist, Riadh Sadek, agriculturist, Khaled Sleem, graduate students Ranya Nasrallah and Rhea Selwan, junior architect, Fatima Qaissi and the project leader, architect and landscape architect, Jala Makhzoumi.

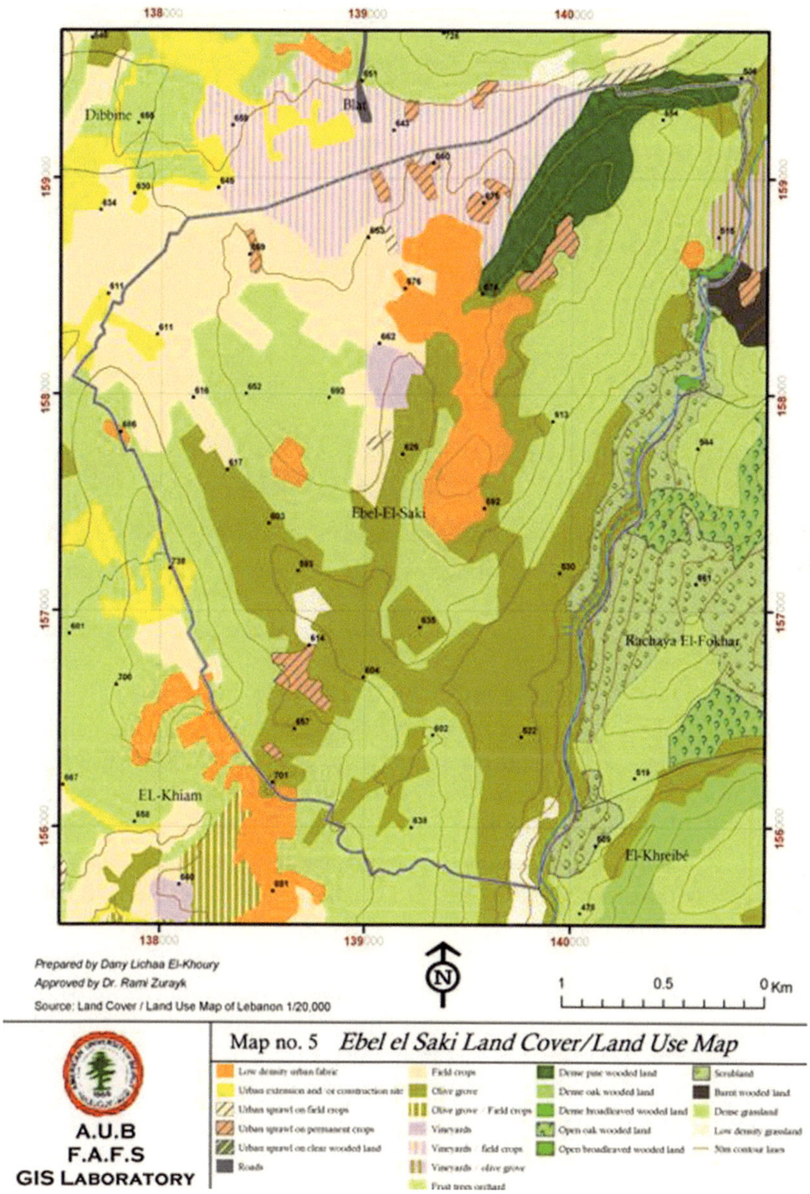

Fig. 3. Land cover, Land use map for Ebel-es-Saqi (Makhzoumi, 2003)

Fig. 4. Olive cropping southeast of the village extends all the way to the Hasbany River Valley.

Literature on floral diversity in Ebel-es-Saqi and the surrounding region reflect the potential of traditional rural landscapes as wildlife habitat for Lebanon's exceptionally high biodiversity (Ministry of Agriculture and UNEP, 1996). Earlier floral surveys for Ebel-es-Saqi, Marjayyoun, Hasbani River and Hasbaya by Paul Mouterde (1965) list a wide variety of floral species. Many species and varieties were specific to Ebel Esaqi, for example Calendula palaestina, Centaurea crocodylium, and Scabiosa palaestina var. palaestina. On-site collection and identification of the plant species extended over three full days in November 2002. A total of 41 plant species were collected and 38 identified. Thirty of the plants collected were perennial, the remaining annuals, bulbs, crocuses and succulents. Five tree species were identified in the woodland: three species of pines, Pinus brutia, Pinus halepensis and Pinus pinea; one species of oak and a gum tree, Eucalyptus spp. The tree density within the woodland varied considerably from 37 trees/1000m^2 to 64 trees/1000m^2. Faunal diversity, based on site visits but also from observation of what could be inferred with some certainty from surveys in the region, include smaller mammals, mole rats, *Spalax leucodon ehrenbergi*, field voles, *Microtus guentheri*, tortoise *Testudo graeca* and two types of lizards, *Lacerta laevis* and the hardun *Laudakia stellio* and one snake species, the Montpelleir snake, *Malpolon monspessulanus*. By far the most abundant wildlife is avifaunal. One eagle was observed on site, more in the surrounding areas. Ebel-es-Saqi residents are avid hunters. They spoke of the diversity of birds that visit the forest throughout the year.

Beyond village woodland, ecological significance as indeed landscape character lies in the entire village landscape, the treeless *hima,* open scrubland descending to the Hasbani River, abandoned terraces and those managed, cultivated with olives. The responsiveness of each landscape component to physical setting ensures that the whole, the rural landscape mosaic, is sustainable environmentally, while landscape heterogeneity and connectivity provide a diverse and continuous wildlife habitat.

Earlier in this paper we argued that the concept of 'landscape' includes not only tangible physicality but equally intangible, socio-cultural perceptions and valuation of the physical. A cultural reading of Ebel-es-Saqi landscape was secured to complement the reading of physical landscape. A diversity of methods was adopted to secure a cultural reading of Ebel-es-Saqi and establish in-depth understanding of local community expectations of the woodland and the village landscape. Initial interviews with the local community members and focus groups (Makhzoumi, 2003), revealed puzzlement as to why the woodland was favored. Local community members argued that the investment should have targeted olive agriculture that was of far greater benefit to village. The community repeatedly referred to the status of Ebel-es-Saqi as a model Lebanese village in the 1960s. The bait al fallah (Arabic, 'house of the farmer'), a folklore museum for traditional agricultural implements, was repeatedly mentioned as proof of the village's elevated status before its destruction during the civil war (1975-1990).

Nevertheless, the village community was grateful that their woodland was now the source of international attention with funding allocated for its development.

A survey of aesthetic preferences in Ebel-es-Saqi confirms the findings of earlier interviews (Selwan,2004). The survey indicates that olive agriculture, the main source of livelihood, came in second place in terms of landscape aesthetic preference, while the village landscape's distinctiveness and local identity, more readily associated with the Ebel Spring, came in first place[3]. The choice is understandable considering that water is the key to agriculture in a semiarid region but more obviously because of the origin of the village's name. More significantly, the treeless part of the hima, the rocky outcrops, ranked ahead of the woodland itself, which reflects local valuation of wildlife resources that are abundant in degraded scrubland and constitute an important part of the local important part of the local diet. The accumulated scores assigned to each landscape component indicate that traditional communities distinguish between aesthetic preference of landscapes, for example Ebel Spring and village woodland, and valuation of landscape in terms of usefulness, for example olive agriculture and open scrubland (ibid). Regardless of age, affiliation or education, the people of Ebel-es-Saqi value all components of the village landscape including traditional management routines, social practices of harvesting wild plants and hunting and agriculture related festivities such as olive harvesting. The analysis of semi-structured interviews reveals that far from a passive container, the discourse emerging from Ebel-es-Saqi reflects landscape as an enabling medium through which local identities are defined and redefined, in response to socio-economic and political influences, both local and regional (Makhzoumi, 2009; 2004). Perceptions of village identity too were changing. Older inhabitants and farmers associated identity with agriculture (particularly of olives), arguing

[3]Selwan uses photographs of key landscape features in Ebel-es-Saqi (Ebel Spring, olive groves, agricultural fields, rocky outcrop of the Hima and woodland), asking respondents to rank photographs according to their aesthetic preference with a ranking scale from 1 to 5. Background characteristics of respondents (gender, age, level of education, income and employment in agriculture) and 'fringe' questions on biodiversity evaluation were also included.

that value and beauty reside in landscape usefulness. Young community members on the other hand associated village identity and distinctiveness with the woodland, reflecting greater awareness of environment and nature conservation (ibid) (Table 1). The totalizing, holistic local perception of the village landscape, experiential rather than strictly visual, inclusive of past while looking to the future, was as important in shaping the landscape master plan as the team's reading of the physical setting.

Traditional Conceptions	Emerging Conceptions
Agriculture related (village olive orchards)	Environment related (village woodland)
Valuation associated with agricultural productivity and usefulness (agricultural, woodland and pastoral livelihoods)	Valuation associated with nature and environment (biodiversity conservation and nature tourism related livelihoods)
Predominantly interviewees age <30-35	Predominantly interviewees age >30
Olive orchards and Ebel spring prominent in the discourse on identity/heritage	Village woodland and war memorial gardens dominate the discourse on identity/heritage
Inherited landscape values passed on within family, extended family and village community	Landscape values influenced by global trends, international projects and environmental education at schools

Table 1. The conception of landscape, place, identity and heritage in post-Israeli occupation Ebel-es-Saqi Village evolves in response to economic, political and environmental influences (source: Makhzoumi, 2009).

5. Writing Ebel-es-Saqi landscape

The holistic reading of Ebel-es-Saqi secured by the project team broadened the aim of the landscape master plan spatially beyond the woodland to embrace the entire *hima* and programmatically to address cultural preferences, intangible community needs and the valued rural heritage. The first step towards developing the landscape master plan was to re-aggregate the deconstructed, layered reading into Ecological Landscape Associations (ELAs), broad landscape character zones that are at once a framework for conceptualizing landscape and an operational tool for writing futures narratives (Makhzoumi, 2000). Based on geomorphology, vegetation cover, traditional uses and management, three broad ecological landscape character zones were identified for (Figure 5): Zone I: Pine Woodland is visually dominant, but poor ecologically because no native tree species were included in reforestation. Replacement with native oak species and associated shrubs recommended were seen as a means to reclaim the natural heritage that survives in village domestic gardens and in the collective memory of the local community; Zone II: Rocky Outcrop, the open landscape and rugged terrain of this zone makes it the richest in terms of species diversity. With the exception of dwarfed, remnant oak trees, the landscape is rich in flowering bulbs and annuals (Figure 6). The master plan calls for recognition and protection of this zone and for reduced public access to encourage biodiversity responsive uses, for example honeybee-keeping that the Ebel-es-Saqi is well known for; Zone III: The Hasbani Valley Ecotone is significant ecologically as the meeting of two ecosystems, open scrubland of the Hima and the Hasbani riparian ecosystem. It is of ecological value because the river corridor links the hima with the outlying landscape.

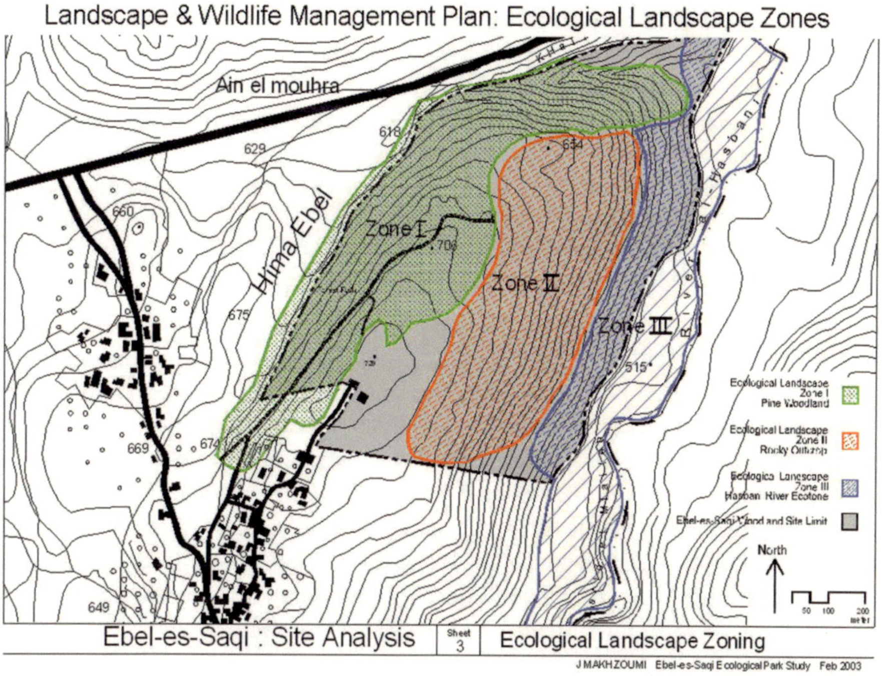

Fig. 5. Ecological landscape character zones were identified for the hima Ebel-es-Saqi (Makhzoumi, 2003).

Fig. 6. Zone II: Rocky Outcrop, an open landscape, rugged terrain rich in floral diversity.

These three zones provide a sound basis for conservation and sustainable management while constituting the building blocks for the landscape master plan concept (Figure 7). In parallel the concept of 'multifunctional landscapes' was adopted in developing the landscape master plan. Traditional Mediterranean rural landscapes are multifunctional, integrating multiple uses on the same land area thus ensuring efficient and sustainable use of land, natural and human resources. As a contemporary concept in landscape planning (Brandt and Vejre, 2004; Tress et al, 2001), multifunctional landscapes ensure the integration of a project's multiple objectives in two ways. The first conceptualizes the woodland as a venue for alternative livelihoods from nature-related tourism and equally as a place to promote rural culture. The master plan proposes the 'Ebel Market' at the entrance of the ecological park to promote the village's historic and traditional rural heritage and to market local produce such as olives, honey and dried mountain herbs. The second strategy provides for flexible recreational and community activities for the local community and the surrounding villages that include informal spaces, panoramic outlooks, promenades and sport courts. The diversity of formal and informal spaces aims to re-engage the local community with the protected site as well as generating interest in environment and nature conservation. Also a proposed 'Science Centre' with a small lecture hall and references on biodiversity would serve as the focus for environmental awareness for schools, local communities of the surrounding villages and beyond (Figure 8).

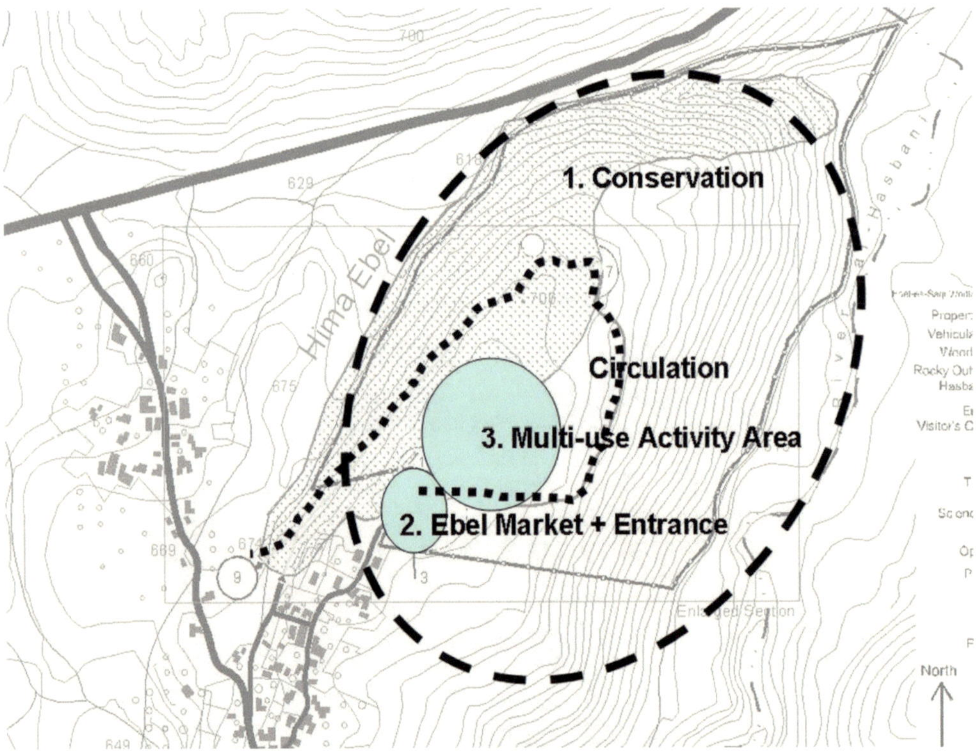

Fig. 7. Schematic concept for the Ebel-es-Saqi woodland landscape master plan (Makhzoumi, 2003)

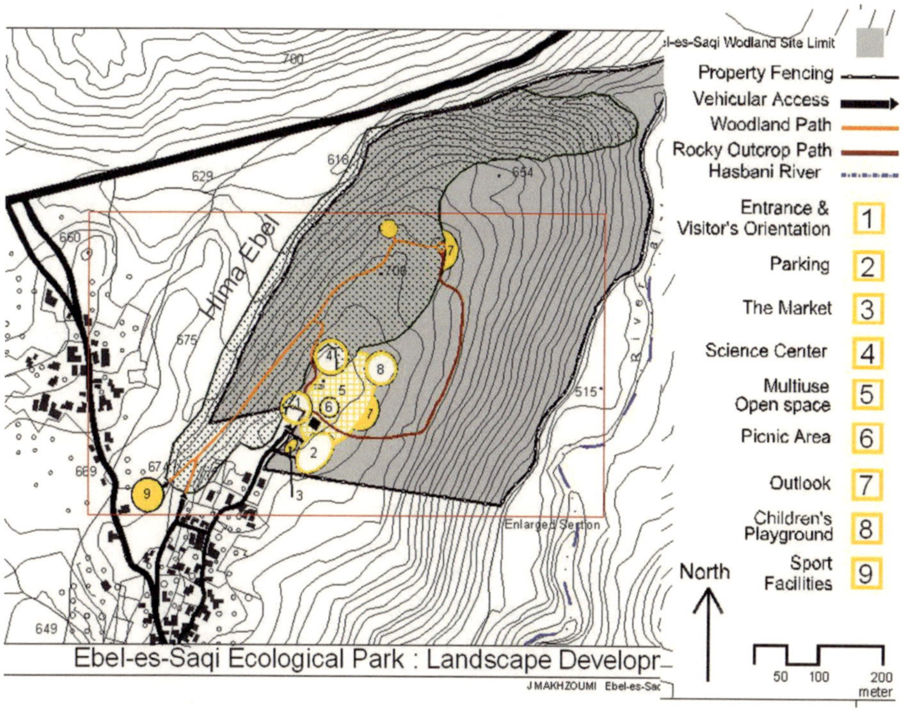

Fig. 8. The Ebel-es-Saqi woodland landscape master plan (Makhzoumi, 2003)

To summarize, the far from completed landscape master plan provides a flexible framework that recognizes the village's history and combines it with its present needs while promoting awareness of the environment and natural resources. The efficacy of the master plan as a sound basis for conservation and development and as an enabling framework for local community enjoyment and action was demonstrated in theyears that followed[4].

6. Ebel-es-Saqi woodland landscape master plan: Offshoot initiative

6.1 Reclaiming the Hima

The community- and culture-inclusive approach of landscape promoted dialogue between the design team and the village inhabitants and renewed local interest in the woodland, i.e. as a potential for economic development, and reaffirmed local pride in their village heritage. And although few in the village comprehended the technical specificities of the landscape design, the master plan was nevertheless instrumental in empowering and mobilizing them to seek funding for further development. The first line of action was to reclaim authority over the hima, which was a key for the success of the project for two reasons. First, from its inception the project was community inclusive, seeing the project as a means rather than an end.

[4]The landscape master plan was completed in 2003. An international campaign was made by UN-ESCWA to introduce the Ebel-es-Saqi project together with three others projects it had initiated to secure funding for implementation. The Ebel-es-Saqi woodland was selected by Mercy Corps (http://www.mercycorps.org/), which was already funding other projects in south Lebanon. Thereafter, Mercy Corps secured funding for implementation of the master plan.

Protecting the woodland was to support local community claims to stewardship of the woodland. Second, international funding for implementation had as a prerequisite local community ownership. International agencies are increasingly hesitant to fund projects through the state and/or the ministries.

Towards this aim, the project team with representatives from Ebel-es-Saqi municipality with the support of UN-ESCWA approached the Ministry of Agriculture to discuss legislative and procedural steps to reinstate local authority of the hima. Negotiations were settled in favor of the village with completion of the master plan in the spring of 2003. An official decree by the Ministry of Agriculture conceded that authority of the woodland reverts to the village. Encouraged by their success, the village community looked to reclaim its cultural heritage, and to negotiate with Mercy Corps the prospects of reviving the bait al fallah.

6.2 Recognizing the village rural heritage

Bait al Fallah was a folkloric museum that represented historic rural lifestyle and had been the pride of the Ebel-es-Saqi village. The museum was housed in one of the village stone houses, itself an example of vernacular building traditions, and exhibited a collection of traditional agricultural implements. The Bait al Fallah must have been significant architecturally for it to be included in Friedrich Ragette's (1980) survey of rural architecture in Lebanon before its destruction during the civil war (Figure 9). Although Ebel-es-Saqi's status as a model village and the museum are not apparently linked, both figured prominently in interviews with the community, as if destruction of the museum put an end to the elevated village status. The Bait al Fallah was as such an important part of village shared identity and rural heritage.

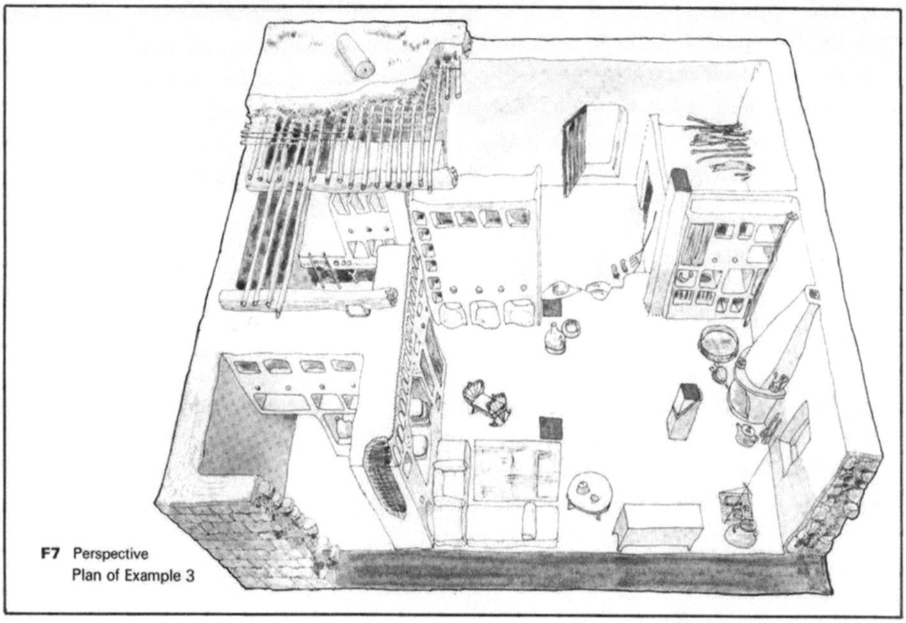

Fig. 9. Three dimensional construction of the *Bait Al Fallah*, folkloric museum of Ebel-es-Saqi as surveyed by Frederic Ragette (1980)

In 2005 Mercy Corps agreed to fund reconstruction of the *Bait al Fallah* folkloric museum. As heritage architecture, the Ebel Market proposed by the master plan was the appropriate site for relocation. A local architecture firm was commissioned to prepare the plans and implementation details based on Ragette's drawings[5]. Rebuilt, the Bait al Fallah is today a prominent showpiece for the local community, one of the few folkloric museums that embody Lebanon's traditional Mediterranean rural culture (Figure 10).

Fig. 10. View of the Bait Al Fallah (foreground), the Visitor Centre (background) and the upper edge of the woodland (left)

6.3 Broadening the agenda for biodiversity

The ecological basis for developing the landscape design and the expansive reading of the village landscape were instrumental in reconfiguring the nature conservation agenda to include the diversity of the rural mosaic, the entire landscape of Ebel-es-Saqi. Having secured funding for implementation of the Ebel-es-Saqi landscape master plan, Mercy Corps in turn commissioned the project to the Society for the Protection of Nature in Lebanon (SPNL). As a comprehensive framework and tangible plan, the landscape design fulfilled SPNL's focus on protecting the country's avifaunal wealth. Lebanon's geographic location on the path of winter-summer bird migration routes, its extensive plant cover and diversity of natural habitat makes it an important habitat for migrating birds. Birds travel along parallel routes to the forest, taking advantage of the wind currents favored by the slopes of Mount Hermon and the extensive shelter provided by the woodland (SPNL 2004). Through consultation with Birdlife International, Ebel-es-Saqi woodland was declared a bird migration "hot spot" and recognized as an Important Bird Area (IBA). Beyond the woodland, SPNL listed the entire rural landscape mosaic as defined by the master plan as a potential bird habitat (ibid) (Table 2). Associating the woodland with Birdlife International in turn bolstered the landscape master plan's proposal for the woodland to provide livelihoods from nature-related tourism, albeit bird watching (Figure 11). SPNL proceeded to build local capacity through workshops to train local

[5]Architect Hana Alamudin was commissioned to prepare the design and implementation drawings for the *Bait Al Fallah*.

guides, plan tour packages for visitors to spend more time in the village to increase socio-economic benefits from nature-related tourism. Capacity building and the awareness campaigns were all the more significant considering that excessive and indiscriminant hunting is endemic throughout the region.

Zone I	Zone II	Zone III	Zone IV	Zone V	Zone VI
Pine forest	**Scrubland, Outcrop**	**Hasbani River Ecotone Valley**	**Hasbani River**	**Crop Fields**	**Olive Grove Buffer Zone**
Levant Sparrow hawk	Short-toed Eagle	Night Heron	Teal	White Stork	Song Thrush
Barn Owl	Lesser Kestrel	Grey Heron	Water Rail	Crane	Bulbul
Tawny Owl	Swift	Sparrow hawk	Little crake	Black Kite	Graceful Warbler
Little Owl	Skylark	Corncrake	Moorhen	Montagu's	Olivacious
Woodcock	Swallow	Water Pipit	Coot	Harrier	Warbler
Turtle Dove	Hoopoe	Wren	Kingfisher	Crested Lark	Upcher's Warbler
Wood lark	Wheatear spp.	Blue Throat		Short-toed Lark	Olive Tree
Redstart	Woodchat shrike	BlackbirdSedge		Bee-eater	Warbler
Blackbird	Rock Sparrow	Warbler		Tawny Pipit	Orphean Warbler
Great Tit	Rock Bunting	Sardinian Warbler		Linnet	
Masked Shrike	Cinereous Bunting	Syrian Serin		Black-headed Bunting	
Chaffinch					
Greenfinch					
Goldfinch					

Table 2. Bird species associated with the zones designated by the Ebel-es-Saqi landscape master plan (source: SPNL, 2004)

Fig. 11. Ebel-es-Saqi IBA Poster (source: SPNL, 2004)

6.4 Recognizing the rural landscape heritage

Another contribution of the landscape master plan, albeit less tangible, was re-introduction of vernacular concepts of nature conservation in Lebanese villages. The concept embodies community-based nature conservation, practices that ensure rights and obligations to protect communal grazing land, except under severe climatic conditions, such as drought, when the ban would be breached. Interpretation of the hima is landscape/context specific, differing from one region to another and equally between different Lebanese villages (Selwan, 2004). In essence, hima is an embodiment of community-based conservation. Far from being a relic vernacular, the concept of hima continues to function as an operational framework for sustainable land use and efficient management practices, a living cultural heritage (Makhzoumi, 2009). Revival of the concept is significant not only because it is integral to Lebanon's rural heritage but equally because community protection serves as an alternative to the prevailing, top-heavy state-declared and managed nature conservation.

Ebel-es-Saqi hima boldly marked on old cadastral maps uncovered early on by the landscape master plan team was further developed by SPNL that adopted the traditional reference, 'Hima Ebel', to refer to the IBA. The absence of prescribed conservation measures associated with the designation of an IBA encouraged adoption of hima by SPNL as a framework for managing the Ebel-es-Saqi woodland project.

Implicit in the adoption of the term 'Hima' is recognition of the value of traditional practices and vernacular wisdom and in promoting practical and socially equitable management of ecosystem goods and services. The concept of hima has since been recognized by IUCN as an effective alternative to formal protected areas and as representing site specific, community inclusive nature conservation (http://www.iucn.org/where/asia/index.cfm?uNewsID=255).

7. Conclusion

This paper has argued that a landscape architecture reading of marginal rural settings is inclusive of physical and tangible environment and intangible cultural practices, perceptions and aspirations. The Ebel-es-Saqi case study demonstrates an approach, method and outcome that starts with a dynamic, holistic reading of the village landscape and follows through by constructing an open-ended, multifunctional framework to accommodate ongoing development, to empower and to enable the village community to reclaim stewardship of the woodland. The ecological landscape design methodology proposed here alerts designers that landscapes evolve over time and that they are contiguous across the spatial hierarchy encouraging them to look beyond tangible physical landscape to intangible human dimensions, local conceptions and valuation of the village landscape as shared heritage and identity. As a result, the landscape master plan, far from being an 'end-product' serves as an enabling place and community responsive framework for development. The interdisciplinary composition of the project team was a key to informing the landscape approach and guiding a sustainable writing of the woodland landscape. Pooling together their respective expertise, the authors of this paper concede to the complementarity of their academic and professional expertise as a necessary platform for sustainable rural development.

Underlying the specifics of the case study is the call to recognize biological and cultural the value of marginal rural landscapes in the eastern Mediterranean. From a conventional, agricultural development perspective, these degraded landscapes are not 'productive' and thus not of much value. From the standpoint of national nature conservation policies they are equally peripheral because they are not 'natural', but managed, lived-in. Valuation of traditional rural landscapes are equally lost between environmental scientists that focus on measures for preventing soil erosion and mitigating pollution of water resources and architects that prioritize the historical, vernacular built heritage in rural towns and villages. But nature in the Mediterranean as discussed at the start of this chapter has been managed traditionally to form a 'surrogate/constructed nature', part nature-part culture to borrow Tilley's term, the value lies in the totality. The value of the landscape approach lies partly in that it embodies the 'totality'. Tangible and intangible, physical and human, resources and processes, past and present are imbued in its form, pattern and cultural valuation of rural landscapes. Similarly, the discursive elasticity of 'landscape' as straddling the social sciences, sciences and design disciplines broadens outlook, scope and method of inquiry in landscape architecture, enabling an integration of the various disciplinary concerns. Expansive temporally to include not only the present but equally pasts and futures, spatially, to look beyond a designated site, the landscape approach balances social, economic, ecological and environmental developmental objectives.

Accepting that marginal rural landscapes cannot be protected in their entirety, they need to be recognized at a national scale within an integrated planning approach that incorporates socio-cultural, economic and environmental development with landscape as the conceptual and operational medium. Such integrated landscape planning will promote nature conservation, agricultural production, tourism and cultural heritage protection which lie at the heart of traditional rural landscapes in marginal Lebanon. Strategic interventions such as the Ebel-es-Saqi project become a means for the recognition and revival of rural traditions albeit within contemporary socio-economic and political contexts.

8. References

Adams, P. Hoelscher, S. & Till, K. (Ed(s).) (2001).*Textures of Place. Exploring Humanist Geographies,* University of Minnesota Press, Minneapolis.

Brandt, J. and Vejre, H. (Eds.)(2004) Multifunctional Landscapes. Volume I. Theory, Values and History (Southampton, WIT Press).

Freyha, A. (1957). *The Lebanese Village,* Al Ahliya Press, Beirut (Arabic).

Feghali, K., (2002). *Lubnan fi Mawsu'a,* International Publishers, Beirut (Arabic).

Hamdan, F. (2002). The ecological crisis in Lebanon, In: *Lebanon's Second Republic. Prospects for the 21st century,* Ellis, K. (Ed.), pp. 175-187 University Press of Florida, Gainsville.

Makhzoumi, J. (2012) "Is rural heritage relevant in an urbanizing Mashreq? Exploring the discourse of landscape heritage in Lebanon". In I Maffi and R Daher (eds) Practices of Patrimonialization in the Arab World: Positioning the Material Past in Contemporary Societies. I.B. Tauris, London (in progress)

Makhzoumi, J. (2010) "Marginal Landscapes, Marginalized Rural Communities: Sustainable Postwar Recovery in Southern Lebanon". In H Al Harithy (ed) Lessons in Postwar Reconstruction: Case Studies from Lebanon in the Aftermath of the 2006 War. Routledge, London, pp. 127-157.Makhzoumi, J. (2009). Unfolding landscape in a Lebanese village: rural heritage in a globalizing world, *International Journal of Heritage Studies,* Spring 2009 (forthcoming).

Makhzoumi, J. (2004). The contested landscape of Ebel-es-Saqi: traditional rural landscapes in a post traditional World, *Proceedings of Conference for of the International Association for the Study of Traditional Environments,* Sharja, UAE, December 14-16, 2004.

Makhzoumi, J. (2003). *Ebel-es-Saqi Ecological Park: Site Assessment and Landscape Development Proposal,* Project Report, The United Nations Economic and Social Commission for Western Asia, Beirut (unpublished).

Makhzoumi, J. (2002). Landscape in the Middle East: an Inquiry, *Landscape Research 27, 3,* pp. 213-228.

Makhzoumi, J. (2000). Landscape ecology as a foundation for landscape architecture: applications in Malta, *Landscape and Urban Planning, 50, 1 & 3,* pp. 167-177.

Makhzoumi, J. (1997). The changing role of rural landscapes: olive and carob multi-use tree plantations in the semiarid Mediterranean, *Landscape and Urban Planning, Volume 37,* pp. 115-122.

Makhzoumi, J. & Pungetti, G. (2008). Landscape Strategies, in: *Mediterranean Island Landscapes. Natural and Cultural Approaches*,Vogiatzakis, I., Pungetti, G.and Mannion, A (Eds.) pp. 351-375, Springer-Verlag, Heidelberg.

Makhzoumi, J. & Pungetti, G. (1999) *Ecological Landscape Design and Planning: The Mediterranean context*, EF & N Spon, London.

Ministry of Agriculture & United Nations Environmental Programm, (1996) *Biological Diversity of Lebanon, Comprehensive Report,* Common Business Press, Beirut.

Mouterde, P. (1965). *Nouvelle Flore du Liban et de la Syrie, Volume I, II and III,* Dar El Machreq, Beirut.

Murr, G. (1987) *The Lebanese Village: An Old Culture in a New Era,* Librairie Du Liban, Beirut.

Naveh, Z. (2008). Forward, in: *Mediterranean Island Landscapes. Natural and Cultural Approaches,* Vogiatzakis, I., Pungetti, G. and Mannion, A.(Eds.), pp. ix-xi Springer-Verlag, Heidelberg.

Naveh, Z. & Lieberman, A. (1990) *Landscape ecology* (First edition 1984, Springer-Verlag, New York.

Ragette, F. (1980). *Architecture in Lebanon. The Lebanese House During the 18th and 19th Centuries,* Caravan Books, New York.

Selwan, R. (2004). *Aesthetic landscape preferences and nature conservation: A case study in rural South Lebanon,* Unpublished MS Thesis, American University of Beirut, Beirut.

Tress, B., Tress, G., van der Valk, A. and Fry, G. (Eds)(2003) Interdisciplinary and Transdisciplinary Landscape Studies: Potentials and Limitations, (Wageninger, Delta Series 2).

9

The Nature Conservation of Baikal Region: Special Natural Protected Areas System in Three Environmental Models

Tatyana P. Kalikhman
Institute of Geography of Siberian Branch of Russian Academy of Sciences, Irkutsk Russia

1. Introduction

Lake Baikal is one of the most unique places on Earth. This is the oldest and deepest lake in the world. It contains 20 % of the world's fresh water and is home to more than 2,000 endemic species of plants and animals. Therefore, current conservation of the waters of Lake Baikal and the surrounding region, is affecting the state of its ecosystem. In 1996, the Baikal area and its immediate surroundings were included in the list of World Natural Heritage Sites by UNESCO. In 1999, a special federal law "On protection of Lake Baikal» was passed. This is Russia's only law regarding the management of the lake and the Baikal region.

The main way to preserve, maintain and restore the biotic and landscape diversity is through territorial nature conservation. The territorial nature conservation activities are all of the efforts to protect areas of different categories, status and regimes of protection. In accordance with federal law "On specially protected natural areas», the main categories of protected areas are zapovednik (nature reserves or strictly protected areas), natural park (regionally declared), national park (federally declared), and zakaznik (refuges established at the federal and regional levels). There are also other, less significant (in terms of conservation) categories of protected areas.

The two most widely used approaches to conservation of protected areas are: the basin approach and administrative approach. The basin approach is driven by the goal of preserving the biotic and/or landscape diversity of the territory using natural boundaries of lake basin, for instance, a watershed as limits for management. For example, the special management structure (Lake Chaplain Steering Committee) was creating for protected of the transboundary Lake Champlain basin (The Lake, 1994). An administrative approach stresses the uniformity of institutional establishments and economic conditions within the territory, in particular for protected areas, such as those in Germany which are subject to separate federal lands (Bishop et al., 2002).

In the Baikal region, the basin approach covers the entire drainage basin of Lake Baikal which contains units of four administrative entities within the Russian Federation: the Buryatia Republic, the Zabaikalsky Kray, the Irkutsk Oblast' and the Tyva Republic. A significant part of the basin of Lake Baikal is in Mongolia and is beyond Russia's control.

The basin approach takes into account the condition of ecosystems and natural conditions within the region in order to plan activities related to protection of the area.

The administrative approach seeks to standardize management of protected areas within the separate territories pertaining to Lake Baikal. The established institutional features of the Irkutsk Oblast', Buryatia Republic, and Zabaikalsky Kray are factored into protected area planning, including those related to land-use problems, the relationships of protected area regulations at federal and regional levels, in order to achieve unified management of protected areas within the administrative unit.

A new approach is proposed in this chapter. Called the integral, it is intended to overcome the shortcomings of the basin and the administrative approaches. It is known that the boundaries of natural areas and areas formed by the administrative and territorial division may not coincide. In the Baikal region an example of this mismatch is the Baikal Natural Territory (BNT). Determination of the BNT, which was defined in the aforementioned federal law, is the key to guiding its protection. There was therefore a need to identify and develop a new integral approach to overcome administrative boundaries in a single plan or to achieve harmonious administration of the surrounding territory, not just the area limited by the watershed of Lake Baikal. This chapter will discuss each of the three approaches for comparison and analysis.

2. Basin approach to the study of ecosystems of Baikal watershed

Natural circumstances isolated the basin of Lake Baikal as a region possessing high biotic and landscape diversity. There are the unique ecosystems, a large number of rare endemic species of flora and fauna, and numerous endangered species. In general, the ecology of the Lake Baikal basin is relatively well understood. A nearly continuous mountain chain of the periphery of the basin contributes to local endemism of the flora and fauna. On the other hand, in the basin contains the intersections of areas representing different geographical zones. There is a complex pattern of floristic-faunistic and ecosystem-typological interaction, which extends to adjacent biogeographic areas and across the northern part of the Asian continent.

The biotic and landscape diversity of Lake Baikal is determined by the latitude-zonal, provincial and elevation-zonal differentiation. Ecosystems form 3 main types of environments: tundra, taiga and steppe. Biomes of the major mountain systems have significantly greater diversity compared to adjacent plains. Typical of the Baikal basin, the overlap latitudinal-zonal and elevation-zonal patterns leads to taiga and forest-steppe ecosystems that are largely mountainous and steppes that have highland and lowland variants. Forest-steppe within the basin, in most cases, forms an almost continuous band of zonally elongated areas, southward of the taiga mountain systems. In general, forest-steppe communities differ in the maximum structural and biotic (adaptive types and forms of life) diversity (Bannikova, 1998; Gunin et al., 1998). Fauna in the steppe is apparent, but there is low diversity, and therefore the sustainability of these ecosystems in the region is low (Lavrenko et al., 1991).

The extent of the basin of Lake Baikal enables provincial biogeographic differentiation. There are differences in the flora of eastern and western parts of the basin. The main forest species are two different types of larch (*Larix sibirica* and *Larix dahurica*) and they share "Taiga Forest on the Southern Siberian" and "Baikalo-Dzhugdzhursky Taiga Forest" areas (Atlas of Transbaikalia, 1967). There is apparently no similar differentiation of fauna from west to east in the taiga zone. More significantly the forest-steppe zone and the river valleys beyond its borders contain forest-meadow species.

The ecosystem diversity of the basin of Lake Baikal is almost 3/4 of the continent's north without the subtropics. There are general geographic regularities of this phenomenon. The first is the placement of the basin in the middle of zonal spectrum of the continent, a dense arrangement of the zonal bands of high gradients increase aridity, the presence of high mountain systems with a full range of landscapes and ecosystems of elevation zones for the corresponding latitude and longitude intervals (Gunin et al, 1998 .)

The largest and most unique ecosystem in the basin is the Lake Baikal ecosystem. In addition to its ancient history and geological and geographical characteristics, Lake Baikal is unique in the amount of diversity and endemism of living plants and animals found there. More than 2,600 species have been cataloged and 84 % of them are endemic. Of particular interest are freshwater sponges, invertebrates amphipods, and the endemic freshwater seal, which is only mammal that lives in Lake Baikal (Present and Future, 1996). The relatively large ecosystem of Lake Baikal basin can be classified as larch forest-steppe (Bannikova, 1998), meadow tansy steppe (Lavrenko et al., 1991), sandy-pebbly desert with almost no ephemera as "extreme types of desert vegetation" (Grubov, 1963).

Analysis of the ecosystem of Lake Baikal basin reveals uneven distribution in space and varying degrees of disturbance of ecosystems. Very few disturbed ecosystems are found in the high mountains (Khangai, Baikalsky, Barguzinsky, Ikatsky Ranges of the North-Baikal and Hentey-Chikoysky Highlands) or the midlands (Hentey, mountain ranges of southeastern Transbaikalia). Small populations over large areas usually do not pose a threat to natural systems. Some of these territories are part of the protected area (Huvsgul and Zabaikalsky national parks, Baikalsky, Sokhondinsky, Dzherginsky zapovedniks). The middle and lower elevations of the Lake Baikal basin are characterized by mild to moderate degrees of ecosystem disturbance. In the southeastern part of the valley, the plains and hummocky areas, disturbance is moderate and even severe. Local disturbance in the largest lowland riparian and lacustrine ecosystems and particularly in the areas of water collecting in Mongolia can be extreme.

Creating year	Number	Square, ha	Part of Baikal basin square in Russia, %
before 1917	3	about 150 000	0.48
1960	5	143 300	0.45
1970	9	614 300	1.95
1980	31	2 033 700	6.46
1990	36	3 038 000	9.64
2000	31	4 748 300	15.07
2002	34	3 531 621	11.21
2005	29	3 293 613	10.45
2010	30	3 295 807	10.46

Table 1. Changing the number and square of protected areas of the Russian part of the Lake Baikal basin

The process of creating new protected areas in the Russian part of the basin intensified during the "perestroika" period, but stagnated over the last decade (Savenkova, 2001; Kalikhman, 2007). In the 1980's, the zapovednik Baikalo-Lensky (1986), Pribaikalsky and Zabaikalsky national parks (1986) were created. In 1981, the zakaznik Pribaikalsky in Buryatia was established. By 1990, the network of protected natural territories in the Russian section of the Baikal basin included 4 zapovedniks, 2 national parks, 24 zakazniks and about 120 registered natural monuments. The total area of protected natural areas is more than 3 million hectares, or 9 % of the Russian part of the basin. In the 1990's zapovednik Dzherginsky and Tunkinsky National Park were created (Table 1).

Mongolia's part of the Lake Baikal basin contains 3 strictly protected areas (Bogdhan Uul (biosphere reserve), Khan Khentii, Hordol Sardag), 7 national parks (Noen Khangai, Terelj, Hangayn Nuruu, Huvsgul, Horgen, Khustain Nuruu and Tarvagatay Nuruu), 3 nature reserves (Batkhaan, Nagalkhaan; Hogno Khaan), 3 monuments (Bulgan Uul, Tulga Uul Togoo Uranus, and Husiyn Naiman Nuur) (Savenkova, Erdenetsetseg, 2000, 2002; Special Protected, 2000, Atlas of Mongolia, 2009). Table 2 lists the establishment and growth trends of protected areas in Mongolia.

Creating year	Number	Square, ha	Part of Mongolia square, %
1778	1	41 600	0.03
1957	3	66 400	0.04
1965	9	236 200	0.15
1976	10	5 547 900	3.52
1977	11	5 613 800	3.56
1991	19	8 793 100	5.58
1992	21	8 825 300	6.00
1993	26	12 629 800	8.01
1996	31	16 452 000	10.00
1998	42	18 251 586	11.67
2000	48	20 530 588	13.10
2005	50	21 370 602	13.64
2010	61	21 832 321	13.94

Table 2. Increase in the number and square of protected areas in Mongolia

The distribution of protected areas in the basin of Lake Baikal uneven (Figure 1). Irkutsk Oblast's small part of the basin is almost completely covered by reserve land (Pribaikalsky National Park, zapovednik Baikalo-Lensky, and two zakasniks). This is an almost continuous strip of protected area along the northwest shore of the Lake Baikal. In the Buryatia Republic, the largest protected areas are close to Lake Baikal, while the remaining are small zakasniks. Protected areas of the Zabaikalsky Kray are small territories in square, but mainly protected the environments of the regions of rivers source.

Mongolian has more recently rapidly increased the number of units of different kinds of protected areas. In the central part of Lake Baikal basin in Mongolia, there is very little protected territory. There are only three minor areas: the Bogdhan Uul strictly protected area, and the Khorgo and Khustain Nuruu national parks. In 2003, Tuzhiyn Nars National Park was established in this part of the basin, but its effectiveness is still unknown.

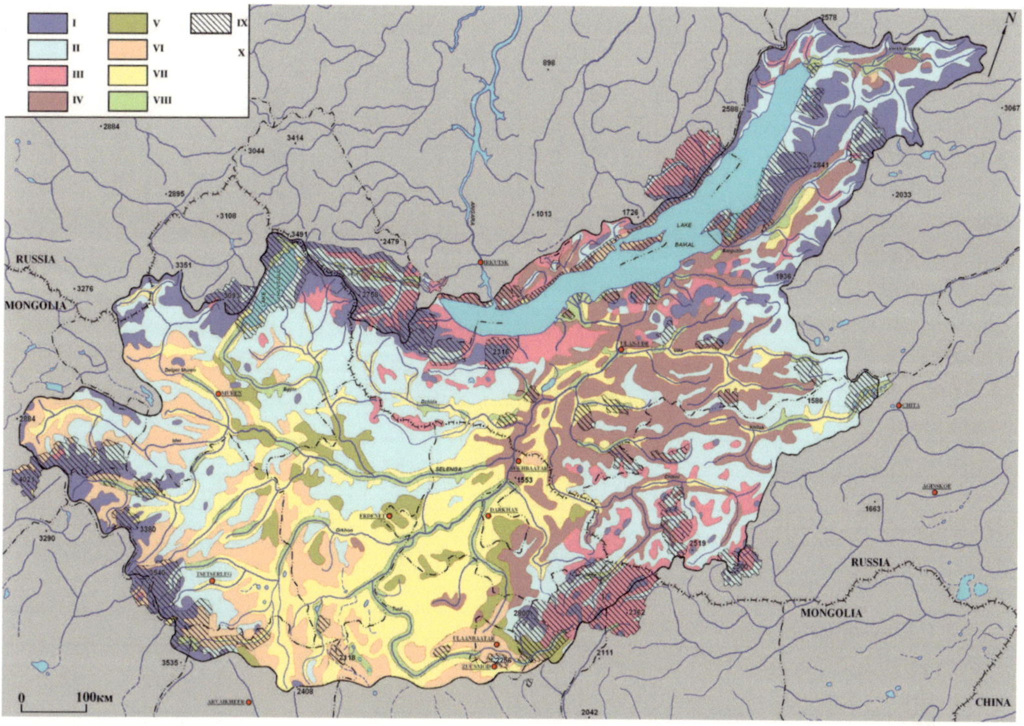

Ecosystem groups: I - high mountains wilderness and glades, II - mountains forests with larch (*Larix sibirica, Larix dahurica gmelinii*), III - mountains forests with cembra pine (*Pinus sibirica*) and fir (*Abies sibirica*), IV - forest with pine (*Pinus silvestris*), V - forest-steppe, VI - middle high mountains steppe, VII - plain or valley steppe, VIII - rivers glades; *Objects*: IX - special natural protected areas.

Fig. 1. Ecosystem groups and special natural protected areas in Lake Baikal basin

The uneven distribution of protected areas within the Lake Baikal basin has led to a relatively incomplete coverage of protection for different types of ecosystems (Table 3). It is evident that the most valuable in terms of biodiversity is forest-steppe. The steppe is poorly represented in the valleys of the Selenge, Orkhon, and Hilok rivers. Alpine belt ecosystems are protected only around the Lake Baikal and on the periphery of Mongolian side of the basin, as well as in adjacent Tunka Valley (Tunkinsky National Park). Typical and unique biomes are protected along the shore of Lake Baikal, except along the northern and southern lakeshores.

Thus, of the 375 different types of ecosystems identified from several sources (Belov et al, 1972; Mikheev, Ryashin, 1977; Yunnatov, Dashnyam, 1979; Ecosistems, 1995; Savenkova, 2002), only 127 (33.9 %) have been legislatively approved for conservation.

Most of protected ecosystems are in the middle- and low- elevation forests, or high-altitude glacial-nival and tundra settings. This is due to the preferential location of protected areas in the high and middle parts of the basin: Baikalsky, Barguzinsky, Baikalo-Lensky, Sokhondinsky and Dzherginsky zapovedniks and Hordol Sar'dag, Otgon Tenger, Bogdkhan Uul strictly protected areas; Hangayn Nuruu, Tarvagatay, Terelj, Huvsgul, Tunkinsky, Zabaikalsky, Pribaikalsky national parks; Angirsky, Atsinsky, Burkalsky, Ivano-Arakhleisky, Pribaikalsky, Snezhinsky, Uzkolugsky, Ulyunsky, Frolikhinsky zakasniks.

Ecosistems in height-zonal differentiation / Variety of ecosistems in latitude-zonal status	mountain			plain			hydrogenic			
	High-mountain	Middle-mountain	Low-mountain	Small-mountains	Elevated	Lower	Mounting-river's	Plain-river's	Lake's and lake-river's	Bay-lake's
Zone's										
glacial-nival	1	–	–	–	–	–	–	–	–	–
height-mountain desert (tundra)	6	4	–	1	2	–	–	–	–	–
mountain-forest	1	11	10	2	4	–	–	–	–	–
forest-steppe	–	2	1	1	1	1	–	–	–	–
meadow-steppe	2	1	4	2	2	–	–	–	–	–
steppe	2	3	6	3	4	2	–	–	–	–
dry-steppe	–	3	–	–	2	1	–	–	–	–
desertification-steppe	–	1	3	–	1	1	–	–	–	–
Out-of-zone's										
hydromorphic, mountain	–	–	1	2	–	1	5	–	–	–
hydromorphic, plain	–	–	–	–	–	–	–	4	5	4
aquatic: include Lake Baikal	–	–	2	–	–	1	–	–	3 / –	–
TOTAL:	17	27	27	12	16	7	5	4	8	4

Table 3. The number of different types of ecosystems of Lake Baikal basin, stored within the boundaries of protected areas

Left out of conservation efforts are these types of ecosystems: aquatic, including Lake Baikal itself (the only exeptions are Chivyrkuisky Bay in the Zabaikalsky National Park and the three-kilometer strip along the Barguzinsky zapovednik); forest-steppe; desert-steppe; steppe on gently undulating plateaus; hilly ridges and depressions with steppe and lacustrine communities in hydromorphic in Mongolia; and low elevation plains (including saline environments where lake-levels fluctuate).

The traditional basin approach in studying the structure of protected areas in the Baikal region can adequately reflect the effectiveness of the protected areas system in terms of coverage of the biotic and landscape diversity. But this approach ignores other important environmental features, such as political institutions and economics.

From the viewpoint of basin approach, non-uniform placement of the main categories of protected areas within the basin of Lake Baikal reveal shortcomings of the existing system of territorial environmental protection. In addition, because of the basin approach, protected areas at the periphery of the basin of Lake Baikal sometimes intersect the boundary of the basin. Therefore, for a complete picture must violate the principle of the basin and to include areas outside the basin in order to protect the basin.

3. Administrative approach and consideration of the institutional features

The administrative approach to territorial nature protection may differ significantly between regions with similar natural and socio-economic conditions. This is due to regional differences in nature conservation legislation, federal control over local politics, and the leadership of a region's head with respect to nature protection. Oftentimes, the typical and most common regional ecosystems and landscapes are overlooked. In contrast to the basin approach, the emphasis is on protecting unique and rare communities, not the unique, but commonplace. That is why in each region has created its own "Red Book", a list of rare species of animals and plants in a specific territory. However, a single institutional framework for a protected area (through legislation, administration, and economic conditions), enables consistent and coordinated efforts for nature conservation.

Two examples of the administrative approach to conservation can demonstrate the specific weaknesses this approach, as well as reveal possibilities for overcoming them through the creation of transboundary protected areas.

3.1 Comparison of protected areas of the Irkutsk Oblast' and Krasnoyarsky Kray

In a system of protected areas of the Irkutsk Oblast' (without the Ust-Orda Buryat autonomous district) and the Krasnoyarsky Kray (without Dolgan-Nenets and Evenk autonomous districts) one can see the following similarities: relatively large regions containing areas of pristine nature; latitudinal similarities of natural conditions wherein southern part are Sayan mountain taiga, a central taiga-covered plain (south-taiga pine forests of the Leno-Angarsky Plateau and the Yenisei Ridge) with alternating steppe and forest regions (steppe valley of the Angara and Olkhon, Achinsk and Minusinsk steppe) and northern areas of taiga in permafrost (larch forests of northern Middle Siberian Plateau in the Nizhnyaya and Podkamennaya Tunguska rivers territory); common history of development activities in the valleys of large rivers: the Kansk-Achinsk industrial area in the Krasnoyarsky Kray and the Irkutsk-Cheremhovsky industrial area in the Irkutsk Oblast' both containing open coal mines, timber production, and hydropower development; and the primary forest production areas in Russia.

Contrasts include differences in the distribution of protected areas and the area occupied by them. In the Irkutsk Oblast' the total protected area is 2048.1 thousand hectares, or 2.7% of the administrative region. In Krasnoyarsky Kray, the protected areas are uniformly distributed and comprise 3616.4 thousand hectares, or 5.1% of the area. Lake Baikal is in the Irkutsk Oblast'. The lake is one of the largest in the world and has status of World Natural Heritage site. Thus its protection is goal of nature protection of the Irkutsk Oblast', but most of the natural areas in the region but beyond Lake Baikal are regarded as less important to the guidance of development or conservation.

Krasnoyarsky Kray adopted a regional law "On specially protected natural areas in the Krasnoyarsky Kray" immediately after the adoption of the March 1995 federal law "On specially protected natural areas". The regional law specified the following new categories of protected areas at the regional and local levels: state natural micro-reserves, state natural mikrozakazniks, protected wetlands, biological stations, green areas, protected water bodies, riparian zones, urban forests and urban parks. In the Irkutsk Oblast′ enacted a regional law on protected areas in 2007, but did not provide any detailed regional actions, it simply adopted the provisions of the federal law.

Krasnoyarsky Kray's "Scheme development and distribution of protected areas" prioritized zakazniks as the main biodiversity preservation (mostly of individual species) mechanism in the region. In the Irkutsk Oblast′, the natural park was determined to be primary protected areas units, which in addition to meeting the general goals of conservation of landscape and biotic diversity, are designed to help develop recreational resources, creating a basis for the development of ecological tourism in the region and reflect the modern world trends toward tourist access and to natural areas.

Comparative analysis of the protected areas system in the Krasnoyarsky Kray and Irkutsk Oblast′ allows one make specific recommendations for measures to improve the performance of protected areas and to enable network planning (Kalikhman, Sokolov, 2005). In the Krasnoyarsky Kray, more rigorous implementation of the plan has recently motivated the reduction in number of zakazniks to preserve the beaver after sharp increase in population and observed evidence of overpopulation. Krasnoyarsky Kray protected areas are mainly intended to preservation of wildlife, but are also important for the conservation of plant communities and landscapes to support recreational resources. Irkutsk Oblast′ will need to consider creation of new protected areas and to determine the mechanisms of their organization. It is extremely important to provide ways to reserve land for future protected areas as well as to balance the relationship of development to nature conservation in both the Lake Baikal basin and the rest of the region.

3.2 National Park within the administrative boundaries

Among Russia's Baikal protected areas the most radical form of administrative approach is in Tunkinsky National Park (TNP). Part of the park is including in the Baikal Natural Territory and is the only one in Russia organized within the administrative boundaries of the eponymous district of the Buryatia Republic. There is no evidence in the 20-years existence of TNP that there had been active protected nature within the administrative boundaries. It is clear that creation of the TNP within the administrative boundaries of the Tunkinsky district created so-called institutional contradictions or institutional overlap. These made the implementation of federal law "On specially protected natural areas" difficult, as the law declared that "National parks are unique to federal property" (Article 12, Clause 5). But TNP could not be entirely federal, because within its borders were villages, farms, private land, and resorts. The boundaries were established during Soviet times, and since then settlements were given new powers as "municipalities" and former collective and state farms became "agricultural land" which were included in the land market in line with the updated Land (2001) and Town Planning (2004) codes. TNP, based on the requirements of the federal protected-areas law, was supposed to be completely devoid of possible economic development and non-ecological land uses.

The most acceptable solution to the problem of competition between two land-users within common land borders may be the divide the land between district land and the park's land. A national park should include the lands that are most valuable for protection of ecological and landscape diversity, and for recreational use. This process should be mandatory withdrawal from a national park intensively used agricultural lands and territories of settlements for the effective and legitimate economic development of the district.

The logic of combining the administrative boundaries of the district and a park does not allow an optimal way to undertake nature conservation adjacent to the boundaries of the park. It is extremely important and valuable for preservation of biodiversity in areas adjacent to the Okinsky and Zakamensky districts, specifically on the northern slope of the Tunkinsky range, part of Kitoysky range and on the southern slope of the Hangarulsky ridge. These areas are also important for the effective conservation of rare species like the first migratory species to be protected: snow leopard and reindeer.

The proposed version of TNP would include areas for the conservation of rare animals species as well as mountain taiga, mountain landscapes and small areas of steppe on the northern slope of the Tunkinsky range, part of the Kitoysky range, and the Bolshoy Sayan range, which divides Russian territory from Mongolia to the west of Lake Huvsgul. This new area located north of the existing boundaries of TNP expands recreational opportunities due inclusion of popular tourist destinations: the highest point of the Vostochny Sayan mountains at Munku-Sardyk; the source of the Belyi Irkut River, Lake Ilchir, "The Valley of a hundred sources" on Shumak river at the confluence of Pravy Shumak River (108 radon, thermal and mineral sources); the valley and mountains of Arhut which bends around the northeastern part of the Tunkinsky range. The revised TNP would be more effective for environmental protection and tourism in the park, because it allows expansion of the environmental "nucleus" and recreational opportunities for visitors in the areas adjacent to the Okinsky and Zakamensky districts and removes the contradictions of the radical administrative approach to conservation (Kalikhman 2007).

4. The integral approach within the boundaries of the BNT

The boundaries of natural areas and administrative-territorial boundaries do not always coincide. We have considered two approaches to conservation in protected areas. The basin approach solves biodiversity and landscape preservation based on boundaries coincident with watershed boundaries. The administrative approach establishes uniformity of economic and administrative activities within the protected areas. Ways to overcome the limitations of both the basin and administrative approaches are: to establish transboundary protected areas and to create a complex nature conservation plan.

4.1 Transboundary protected areas

The first attempts to overcome the shortcomings of the administrative approach to the territorial nature protection are projects to organize transboundary protected areas (TBPA). A TBPA is two or more protected areas, located on both sides of a border, and having common, or at least similar, legal bases and managed through a coordinated. The principal requirements for the creation of new TBPA are the following criteria: 1. There should be high (global) significance of territory in terms of conservation of biodiversity and ecosystems. Often this is linked to the preservation of rare species, including migratory species, distribution area which is located on the territory of neighboring states (Convention, Bonn, 23.6.1979; Agreement, Netherlands, 06/10/1996); 2. There should be good preservation as defined by common practice in similar areas; and 3. There must be similar protected-areas laws and the potential for consistent decision-making for conservation in adjacent territorial units. A favorable factor is the preexistence of special natural protected areas within the territories of future TBPA (Kalikhman at al., 2005).

Transboundary protected areas allow: the avoidance territorial conflicts in nature conservation which are the main problem associated with the administrative approach and a lack of consideration of the natural boundaries of natural communities; the adopting of common or similar legal frameworks; and the organization of a single or similar management approaches in the protected areas.

The unity of the natural conditions suggests the potential creation of four TBPA between Russia and Mongolia: "From Huvsgul to Lake Baikal", "Selenga" and "Hentey-Chikoy Highlands", as well as cross-border zapovednik at the source of the Delger-Muren River at the border between the Tuva Republic and Huvsgul Aimak. It should also be noted that the TBPA "Hentey Chikoy Highlands'" could be separated into two units: "Hentey Chikoy Highlands" and "Sokhondo" (or " Source of the Amur River").

4.2 Complex plan of nature conservation

Another way to overcome the shortcomings of the basin and the administrative approach is to create complex regional plans for nature conservation and natural resources use. One of the most complete and comprehensive instruments for the design of a system of nature protection is a "Territorial Complex Plan of Nature Conservation of Lake Baikal" (TerCPNC Baikal), established by the act of the former USSR on 13.04.1987: "On measures to ensure the protection and rational use natural resources in the basin of Lake Baikal in 1987-1995" (Territorial, 1990). TerCPNC Baikal focused on the need for long-term conservation of the ecosystem of Lake Baikal. In addition, the plan required the pursuit of optimal solutions for socio-economic development problems and the improvement of production efficiency. TerCPNC Baikal took into account the "Standards of acceptable impacts on the ecosystem of Lake Baikal and its watershed" which had been a little earlier. The choices were determined by estimating the cost of conservation, which was comparable to the economies of the region's manufacturing industries. In even the most optimal alternative, about a third of the revenues from production in the region were to be aimed at preserving the environment. It became clear that there a lack of economic instruments for financing an effective environmental policy.

The first steps of employing TerCPNC Baikal were to analyze the degree of sensitivity of the natural ecosystems of the region to human impacts and to estimate the contribution of components of the self-regulation processes of in the complex ecosystems of the Lake Baikal and its basin. It is possible to obtain the necessary understanding of environmental regimes and the allocation of ecological zones. Carried out within the framework of the TerCPNC Baikal, ecological zoning generated three zones of regulation within the boundaries of each mode of natural resources and economic activities. Such zoning was later used to create a zoning of the BNT. Integral approach is using TerCPNC Baikal's ideas.

4.3 The concept and principles of the BNT

The first and only federal law that pertains to natural object is "On protection of Lake Baikal" (1999). The law requires protection of BNT. Such a task should be comprehensive and cover all aspects of contemporary nature conservation. Despite the framework and the declarative nature of the law, it created the opportunity to surmount the limitations and contradictions caused by the administrative boundaries or natural boundaries of the watershed of Lake Baikal for environmental management.

The BNT region was defined by the federal law: "Baikal natural territory is a territory which includes Lake Baikal, the water protection zone, adjacent to Lake Baikal and its watershed area within the territory of the Russian Federation, protected areas adjacent to Lake Baikal, and adjacent to the Lake Baikal area up to 200 kilometers to the west and north-west of it" (Chapter 1, Article 2, p.1). The enactment of the Lake Baikal law was a requirement for inclusion under World Heritage Convention of UNESCO. The convention requires that there is a single legal and management approach for the efficient operation and proper conservation of WNHS. Lake Baikal was added to the convention in 1996.

Acceptance of a BNT as defined by the federal law is the key to management of development activities relative to conservation of nature in the area. Obviously, the BNT is outside of development zones, and this allows basin or administrative approaches to be used for the analysis of environmental protection needs. Therefore for the BNT has promoted the development of an integral approach, which allows the managers to overcome the administrative segmentation of planning by creating a common or similar administration of the territory, and one not limited by natural boundaries of Lake Baikal basin.

Advantages of using an integrated approach to BNT is displayed in a territorial nature-protection model that includes ecological (its essence is represented by considering the basin approach), institutional and economic components. In this system, the wording of the institutional model of the conservation of BNT must precede the formulation of an economic model of environmental management.

5. BPT in the institutional model of nature conservation

The Baikal law contains a number undefined and ambiguous terms or discrepancies, of which three are most significant. The first is that the definition of BNT cannot be associated with the boundaries of WNHS "Lake Baikal". It is clear that WNHS "Lake Baikal" and BNT are different territories, though the primary objective of the law was to regulate management of the natural heritage of the region. The second discrepancy is the limited list of territories encompassed by the BNT. It is unclear whether this list includes all areas that must be in BNT, whether additions to the list are permitted, or whether it was intended only to highlight the dimensions of BNT. This ambiguity of the definition of BNT delaying the implementation of the law. A discussion of the relationship between boundaries of the central ecological zone of BNT and WNHS "Lake Baikal" lasted six years and generated no fewer than five possible conclusions. The third discrepancy is the mention of the "watershed area within the territory of the Russian Federation". The Russian's Lake Baikal basin consists of two parts, the least well-known of which is a small area located near the source of the Delger-Muren River in Tuva Republic. This river system flows into the Selenga which flows from Mongolia into Lake Baikal (Savenkova, 2001, 2002). This second area is geographically unrelated to the federal law governing the BNT. All three of these inconsistencies reflect problems with the *first element of the institutional model*. Correctly fixing these problems will ensure harmonization of existing and future regulations at the start and throughout the process of implementation of the law to protect Lake Baikal.

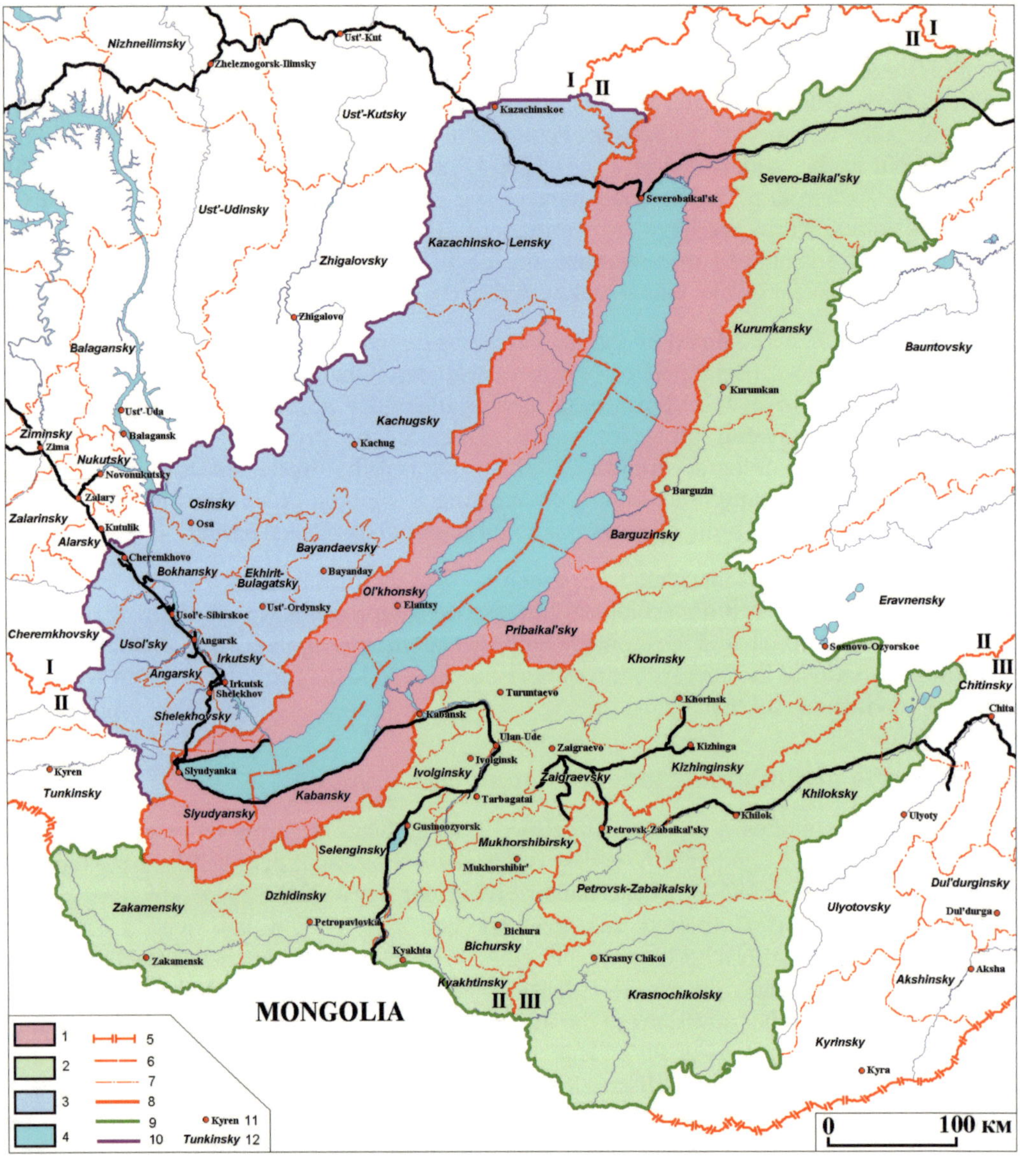

Functional ecological zones of BNT: 1 – central, 2 – buffer, 3 – of the atmosphere impact, 4 – Lake Baikal – the part of the central ecological zone; *Borders of*: 5 – states, 6 – regions (administrative units), 7 – districts (local administrative units, municipality), 8 – central ecological zone, 9 – buffer ecological zone, 10 - ecological zone of atmosphere impact; *Administrative names*: 11 – the centre of district (rayon), 12 – district (rayon); I - Irkutsk Oblast', II – Buryatia Republic, III – Zabaikalsky Kray.

Fig. 2. Functional zoning Baikal Natural Territory

Figure 2 shows the latest version of the location of the central ecological zone of BNT, which coincides with WNHS "Lake Baikal". The figure demonstrates a buffer zone to the southeast of the WNHS and an atmospheric influence ecological zone of to the north-west. Mapping of the zones was undertaken by the Institute of Geography of Siberian Brunch of Russian Academy of Science (RAS) in May 2006 and was approved by the Russian Government in November 2006.

Territorial planning WNHS "Lake Baikal" was determine by the activities within the protected areas within its boundaries. Within WNHS "Lake Baikal" there are 3 zapovedniks, 3 national parks, 6 zakazniks and 2 recreational areas. Over 70 % of the shoreline is contained in protected areas. In addition to the WNHS "Lake Baikal" the following categories of land: "land settlement", "forest lands", "agricultural land", "land of state reserve" etc. All of these categories are distinguished as well by different levels of lands ownership: federal, regional and local.

Therefore, with regard to the institutional model of BNT and WNHS "Lake Baikal" and a new system of municipalities, consider the contradictions inherent in the relatively recent law imposed upon the boundaries and management of protected areas, settlements and agricultural enterprises that were established during the Soviet era. This conflicting circumstance reflects the *second element of the institutional model*, which creates the need for development of solutions to eliminate the evolutionary or sequential inconsistencies in the law and in conservation management.

5.1 The system of special protected natural areas

The system of special protected natural areas within the BNT at the beginning of 2006 included 5 zapovedniks, 3 national parks, 23 zakazniks and two recreational areas. Figure 3 (and table 4) shows the location of the main categories of protected areas within the BNT at the beginning of 2002 (Kalikhman, 2008 a). After 2002, the government began to reduce the total area to committed to protected areas. The reasons for reducing the number and area of protected areas are several, but are primarily associated with industrial logging. There is noteworthy absence of aquatic protected areas in the BNT on the Lake Baikal, and only a small part of the water surface of Lake Baikal is included in the existing coastal protected areas. It can be assumed if Lake Baikal's surface had been protected before the construction of the Irkutsk hydroelectric station and dam, the impact of rising the level of Lake Baikal (1958-1962) as a reservoir probably would have prevented the undertaking of the project, which has had the most significant modern impact on the ecosystems of the lake, especially those affecting fish and coastal habitats. Table 5 shows the proportion of protected areas by ecological zones in the BNT and WNHS "Lake Baikal".

Number protected area	Name of protected area	Notes
	Zapovedniks	
1	Baikalo-Lensky	federal
2	Baikalsky	federal
3	Barguzinsky	federal
4	Dzherginsky	federal
5	Sokhondinsky	federal, part including in BNT

Number protected area	Name of protected area	Notes
	National parks	
6	Pribaikalsky	federal
7	Zabaikalsky	federal
8	Tunkinsky	federal, part including in BNT
	Zakazniks	
9	Altacheisky	federal
10	Angirsky	regional
11	Atsinsky	regional
12	Atsul'sky	regional, liquidating 2002
13	Borgoisky	regional
14	Burkalsky	federal
15	Butungarsky	regional
16	Verkhne-Angarsky	regional
17	Ivano-Arakhleisky	regional
18	Irkutny	regional
19	Kabansky	federal
20	Kochergatsky	regional
21	Kurtunsky	regional, liquidating 2003
22	Kizhinginsky	regional
23	Krasny Yar	federal
24	Magdansky	regional
25	Mokheisky	regional, liquidating 2004
26	Pribaikalsky	regional
27	Ptichy (Sushinsky Kaltus)	local, liquidating 2002
28	Snezhinsky	regional
29	Stepnodvoretsky	regional, liquidating 2004
30	Tagleisky	regional, liquidating 2004
31	Tugnuisky	regional
32	Tukolon'	regional
33	Uzkolugsky	regional
34	Ulyunsky	regional
35	Frolikhinsky	federal
36	Khudaksky	regional
37	Shirokaya Pad'	local, liquidating 2002
38	Enkhaluksky	regional
	Recreational sites	
39	Baikalsky Priboi-Kultushnaya	local
40	Lemasovo	local

Table 4. Basic categories of Special Natural Protected Areas list in Baikal Natural Territories borders

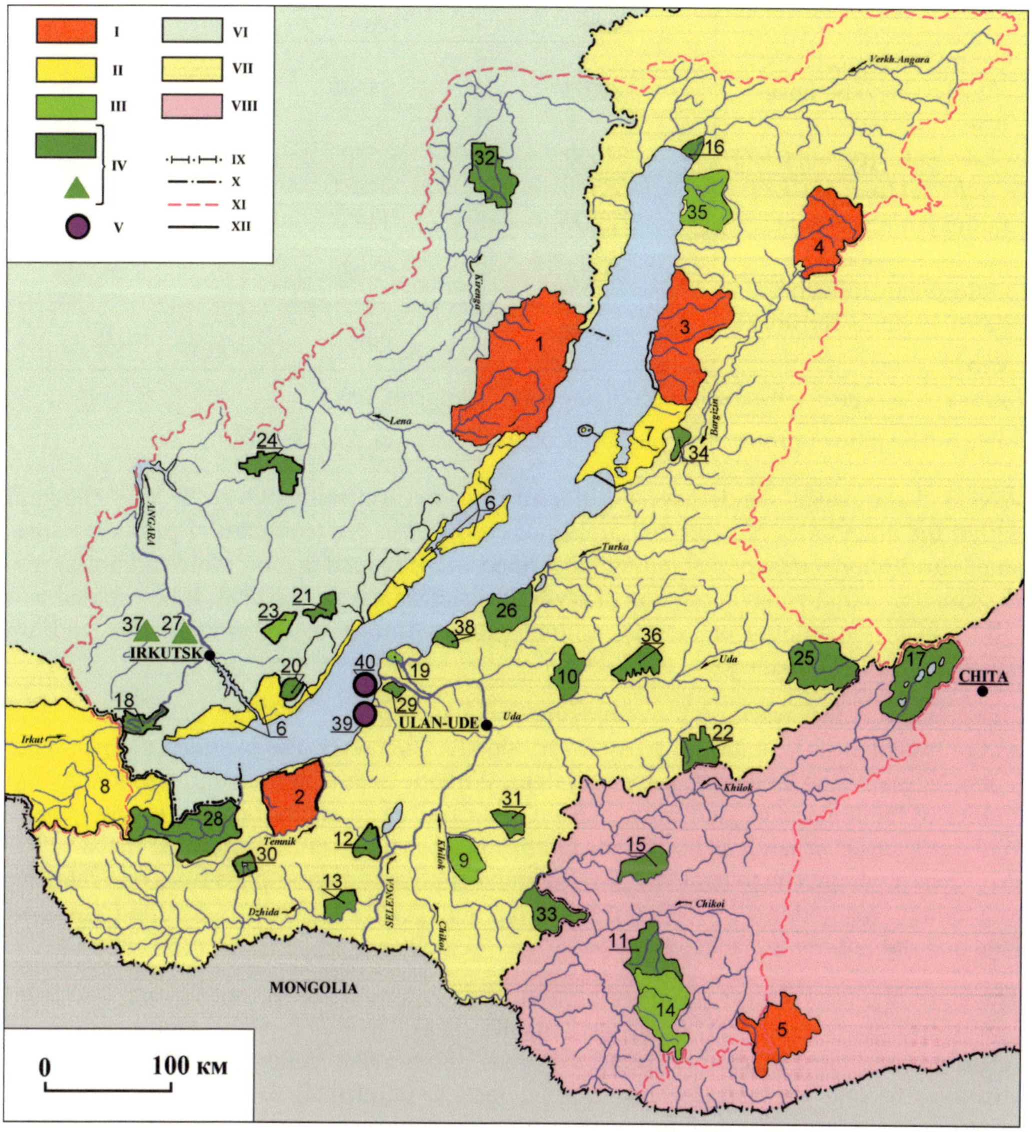

Footnote: The protected areas numeration are complying with a numeration in the table 4. *Protected areas category:*: I - zapovedniks (strictly protected areas), II - national parks, III - zakazniks (refuges) of federal meaning, IV - zakazniks (refuges) of regional and local meaning, V - recreational sites; *Administrative units (subjects of Russian Federation)*: VI - Irkutskaya Oblast', VII - Buryatia Republic, VIII - Zabaikalsky Kray; *Borders of*: IX - state's, X - regions (administrative units), XI - Baikal Natural Territory, XII - protected areas.

Fig. 3. Special Protected Natural Areas of Baikal Natural Territory (the status 2002 - a year had largest area and number of protected areas)

Name of territory	Square of BNT ecological zones, km2	Square of protected areas, km2	Ratio of protected areas, %
Central ecological zone (CEZ) or WNHS "Lake Baikal", include:	*89071	24801	27,84
- *Lake Baikal* (part of CEZ)	31500	520	1,65
- *mainland* (part of CEZ)	57571	24281	32,18
Buffer ecological zone	213875	11457	5,36
Ecological zone of atmospheric influence	83212	2380	2,86
Baikal Natural Territory (BNT)	386158	38638	10,01

Foot note: * The square is the result of specifying data (State report, 2006)

Table 5. The ratio of protected areas and ecological zones of BNT in 2004

WNHS "Lake Baikal" lands have a different category and status. Over 20 % of the land within the boundaries of protected areas not classified as "special natural protected areas land". Institutional misunderstanding (enshrined in the federal law on protected areas) was the situation with by the liquidation of several zakazniks within the BNT. It has not led to a reduction in the amount of "special natural protected areas land" as zakasniks land are usually related to the category "forest lands".

The problem is that work on the surveying and registration of WNHS "Lake Baikal" land and protected areas land has progressed very slowly. Such work also includes a transition to a new legislative on land management system. Without such work, territorial planning and studies of the impacts of land transfer from one category to another cannot by completed. Without transfer of lands it is difficult to effectively manage either protected areas or the WNHS as a whole. The federal law "On the transfer of land or land plots from one category to another" was updated in 2005, but it did not simplify the procedures for the transfer of land and did not create a better process.

Therefore, the *third element of the institutional model* is a mechanism transferring land from one category (agricultural land, settlements land, land of state land's stock or "goszemzapas", etc.) to another (protected areas, recreational facilities, etc.), as well as for changing the status of land (federal, regional, local or municipal). The main obstacle to the transfer of land is a "defective" mechanism. The translation process is referred to in Art. 8 of the Land Code. The new version of the federal law on the transfer of land states that such transfer is permitted only in exceptional cases. Previously, "exceptional cases" meant only those occurring during the creation of protected areas. Now the allocation of land conservation, historical, cultural, recreational and other values is particularly valuable. The new version of the law is a formality and needs to make normative. An effective legal mechanism for transfer of land needs to be created.

In accordance with the Urban Planning Code of Russia shall be subject to special regulation of urban development "... in cases where, without introducing special rules for use of the territory ... it is impossible or difficult" (Article 6). In WNHS "Lake Baikal" objects of special urban planning regulations (Article 4) may be traditional territories of indigenous peoples and the settlements within the boundaries of protected areas. On the northeast shore of Lake Baikal territories traditionally inhabited by indigenous peoples are defined by the natural boundary of the Shegnanda River Evenk clan "Revival" and the area of settlement of the Evenk in the village of Kholodnaya and its surroundings.

There are 46 settlements in WNHS "Lake Baikal" within protected areas boundaries, and they may be the subject of many urban planning regulations. Only the the village Davsha of Barguzinsky zapovednik is directly involved in the activities of protected area. In addition, the boundaries of settlements in protected areas are not always clearly defined or confirmed by the Committee on Land Resources of the municipal administration. Obviously, the inclusion of settlements within the boundaries of protected areas is the result of poorly informed and poorly thought-out solutions for the organization of Pribaikalsky National Park (in 1986) and Tunkinsky National Park (in 1991). The boundaries of national parks and their functional zoning projects have been identified only in the framework of an earlier forest arrangement of the Forest Department. In the future boundaries of Pribaikalsky National Park and Tunkinsky National Park must be approved by the Russian government.

The presence of two or more types of land-users at selected sites of protected areas leads to conflicts between local communities and of protected-areas administrators. Section 3.2 has already been mentioned as producing similar problems in Tunkinsky National Park as it was organized within the administrative area (in WNHS "Lake Baikal" and BNT is 0.1 part of the park square). In this area, so-called "development zones" of settlements are not yet included in the Urban Planning Code. Therefore, the proposal to establish boundaries should encompass not only issues pertaining to settlements and guided by urban planning regulations, but also refinement and approval of the boundaries of national parks and its functional zones within territories, and providing land surveys.

Consequently, the *fourth element of the institutional model* is the mechanism for implementing land surveys and state registration of lands in accordance with the planning legislation.

5.2 Planning of new protected areas

Establishment of new protected areas in WNHS "Lake Baikal" extends protection of nature to conserve, maintain and restore biotic and landscape diversity. Among the planned protected areas within WNHS "Lake Baikal" are "natural parks" (24 out of 29 planned protected areas) (Savenkova 2001, 2002). Natural parks are most common in countries such as USA and Germany. In the state California state natural parks system are 185 units, the first was created in 1902. Only 8 national parks more larger size: Yosemite, Sequoia, King Canyon, Channel Islands, Death Valley, Joshua Tree, Redwood, and Lassen Volcanic. Common square of these categories of protected areas are comparable (Guide 2004, 2006; Ostertag R&G, 1998). In Germany, national parks are called, in fact, natural parks, as all these protected areas are regional and are subject to the ministries of environment of individual federal lands of the country, not a federal ministry. In all there are 13 such parks in 9 of the 16 federal lands (Bishop at al., 2000).

The natural parks within WNHS "Lake Baikal" could become important components of the spatial organization of conservation, restoration and maintenance of the biodiversity and landscape diversity, as well as the development of recreational and tourist activities. In the Baikal region there are no natural parks, despite the many proposals that have been made. Difficulties in their creation are related to institutional conditions mentioned above. It goes without saying that this category of protected areas "natural parks" can withdraw land from existing traditional economic uses. Natural parks can have their own administration in contrast to the zakazniks and natural monuments. At the same time, smaller parks in the area are more compact and manageable compared to the national parks. Natural parks can serve as buffer between the high status of protected areas, national parks and zapovedniks, and can be established in resource development areas. It is also important that the parks can be serve the local population as well, and thus reduce the recreational load on zapovedniks and national parks. Consequently, the *fifth element of the institutional model* is to create natural parks, a new category of protected areas in region.

Thus, the proposed institutional model allows evaluation of the effectiveness of protected areas to conserve, maintain and restore natural systems. The use of five main elements of the model permit the development of a protected-areas system and the necessary institutional changes in the sequence of the nature protection, including: land surveying and public registration of land in all categories and types of ownership within the BNT; the establishment of borders of settlements and their "development zones"; resolution of conflicts between users of nature resources in the disputed areas of BNT; the creating a real mechanism for transferring land from one category to another for sustainable land use planning BNT; the approval by the Russian government of the boundaries of Pribaikalsky and Tunkinsky national parks with respect to the necessary use of land for agriculture and settlement; definition of recreational areas around Lake Baikal reservation and land conversion to the category of "recreational land"; and creation of a natural parks system within the BNP, which would be a new category of protected areas in Baikal region.

6. BNT in the economic model of nature conservation

The primary objective of activity of protected areas is to conserve biotic and landscape diversity. This objective is achieved in the process of solving relevant problems provided the expenses are adequate and economically justified. In the case of protection of nature, the economic aspect of issues to be tackled is not always amenable to straightforward and unambiguous assessment. Action to reduce biotic and landscape diversity should undergo feasibility study. Global environmental concerns are transformed in transition to regional and local level. Regional and local economic interests cannot afford to make large expenditures on behalf of nature conservation.

Such logic is evident in the BNT. The formation of BNT was constructed into account on the existing structure of nature resources using environmental and economic interests of individual actors in the region. The law "On protection of Lake Baikal" zoning BNT on the central ecological zone, buffer ecological zone, and ecological zone of atmospheric influence. The names of zones give an indication of the polarization of the interests of nature resources.

Within the BNT is valid only simple model and estimate the costs of biodiversity conservation. These estimates are based so-called "Baikal factor" can justify the receipt of federal grants for economic development and social development of the Buryat Republic and compensate for economic losses (Kalikhman, 2008 b). In environmental economics such problems have long been resolved within the concept of externalities (external effects), as well as factors "external (externalities) of costs" to society and future generations (Coas, 1993). But they are not applicable in Russia with a dominant resource economics.

History of the creation of protected areas within the BNT began after two reductions in the number of protected areas in 1951 and 1961. In the 1970s Baikalsky zapovednik (1969) and Sokhondinsky zapovednik (1974), Burkalsky (1978) and Kabanskiy (1967 local, since 1974 federal) federal zakazniks were created. And beginning in the mid-1980s, Baikalo-Lensky (1986) and Dzherginsky (1992) zapovedniks, Pribaikalsky (1986), Zabaikalsky (1986) and Tunkinsky (1991) national parks, and Altacheysky (1966 local, 1984 federal), Frolihinsky (1976, 1988) and Krasny Yar (1994, 2000) federal zakazniks were established.

At the same time the academic community ushered in a new global paradigm of sustainable development that now dominates the principles of environmental protection. Reflection of global trends on the national system of territorial protection of nature is presented in Table 6.

Areas of development of activity	Traditional approaches	Principles of sustainable development
Strategy of utilization of natural territories	Exclusion of the maximum possible area from economic utilization	Functional differentiation and spatial optimization of the areas of nature management
Strategy of management of natural territories	Ideological declarative, voluntarism and utilitarianism	Current normative legal base with legalized pattern of land use
Economic bases	Requirements for large expenses on protection and scantiness of budgetary financing	Combination of budget and off-budget sources of funding
Nature and Man	Minimization of human presence in nature	Technological support of human access to nature

Table 6. General trend towards the development of protected areas

6.1 The overall economic assessment

The main categories of protected areas, such as zapovedniks, national parks and zakazniks are state environment organizations and funded from the federal budget. Such legislation establishes the status of institutions as a mechanism of complete or partial withdrawal of these territories from economic use.

Consequently, the *first element of the economic model* can be considered to be the use status of protected areas by state budget organizations for solving environmental problems. The dominant of the economic model of BNT protected areas is federal budget funding, which is usually associated with the effectiveness of protected areas.

Over the past five years, funding has more than doubled. According to the Ministry of Natural Resources, only 66 % of the estimated annual funding requirements for the current contents of the state natural reserves and national parks a being met.

Table 7 shows a hierarchy of tasks for the main categories of protected areas under the Federal Law "On Specially Protected Natural Areas". The problem of protection of natural areas to preserve biodiversity and maintain the natural environment and facilities is a priority for the main categories of protected areas. Achieving this task must be reliable and adequate funding. Therefore, the existing of 66 % provision of protected areas and continued growth the budget of the protected areas can be considered satisfactory level of protection even when recognizing that there is a lack of funding.

Tasks	Zapovedniks	National Parks	Zakazniks
Protection of natural areas	1	1	1
Protection of historical and cultural sites	-	2	-
Research activity	2	5	-
Implementation of Environmental Monitoring	3	6	3
Environmental education	4	3	-
Participation in Environmental Assessment	5	-	-
Assisting in the training of scientists	6	-	-
Adjustable Tourism and Leisure	-	4	4
Restoration of natural and cultural complexes	-	7	2

Table 7. The priority tasks of the main categories of protected areas

Financial support for scientific activity can be considered to be at normal levels. The own research programs are supplemented by cooperative projects with staff members of academic institutions and universities in Irkutsk and Ulan-Ude as well as from other Russian and foreign research centers. Such cooperation is instrumental in enhancing the publishing and informational activities through the use of the resources provided by project participants. The success of ecological education is facilitated by the fact that the protected areas administrations are located in cities and towns, so that school and university students can be recruited as volunteers to participate in museum, exhibition and excursion activities.

In accordance with the recently firmly established system of budget and off-budget financing of protected areas 20 % of the own funds are added to the 66 % of the federal budget component as well as 6 % are provided by local budgets, 6 % by grants from international environmental foundations, and 2 % are received from sponsors (Ministry, 2006).

The most marked influence on the activity of protected areas within the BNT was exerted by the Global Environment Fund (GEF) during 2000-2004 as well as its project titled "Biodiversity conservation" (Russia, Baikal component). Under these programs the Pribaikalsky National Park, for example, obtained grants in the following amounts: 550.2 thou Rbls. (2001), 74.7 thou Rbls. (2002), and 126.1 thou Rbls. (2003). During the same period the Barguzinsky, Baikalo-Lensky, Baikalsky and Dzherginsky zapovedniks obtained under GEF grants more impressive funds: from 3 to 10 mil. Rbls.

6.2 The economy of landuse on BNT

In recent years there has been land registration conducted by the government. To assess their own economic viability, this procedure must be applied to all protected areas.

Protected areas on BNT are land users, and the estimates of value are based on assessments of forest and land resources. Such analyses are usually carried out once every 10 years, and if necessary more often. Table 8 shows the assessment for the main categories of protected areas. In Baikalo-Lensky zapovednik last forest inventory was carried out more then 35 years ago, over 11 years before the reserve was created. For the other protected areas, last forest inventory was carried out 10-25 years ago.

Name of protected areas	Years of	
	forest arrangement	**land arrangement**
	Zapovedniks	
Baikalo-Lensky	1975 (before creation of zapovednik)	–*
Baikalsky	1980-81	–
Barguzinsky	1980-81	–
Dzherginsky	1981	–
Sokhondinsky	1991	2004
	National parks	
Zabaikalsky	1991	2003
Pribaykalsky	1992	–
Tunkinsky	1995	–
	Zakazniks	
Altacheisky	1989-1990	–
Burkalsky	2000	–
Frolikhinsky	1999-2000	–
Atsinsky	2000	–
Ivano-Arakhleisky	1996	–

**Foot note*: «–» do not to made

Table 8. Carrying out the latest forest arrangement and land arrangement activities in protected areas.

Currently, the procedure of land use analysis is complemented by the work of land surveying and registration of all protected areas. The cost of surveying the land protected areas has not yet been determined. For the owners of private land, land prices based on free market value depends upon the areas in which the land is situated and unit area value. The main work in protected areas is to estimate the cost of surveying the boundaries. Ownership of the land adjacent to protected areas, outside the boundaries, is another important consideration. Federal lands dominated the BNT. It is therefore logical to transfer the costs associated with surveying to the federal budget. Thus, the *second element of the economic model* is the federal financing of the cost of both surveying and land and forest management in protected areas.

Within the BNP, the economy of land use in protected areas varies. The boundaries of the state nature zapovedniks approved by the Russian government transferred the lands to the category of "protected areas land". They have a federal status and have no significant settlements. Therefore there are no conflict with the local population. Pribaikalsky and Tunkinsky national parks are not approved under the Russian government borders. Their boundaries are consistent only with the former federal forestry service, for which they were designated. Nature has a different value, including those not derived from economic activity, and this leads to conflict. Federal zakazniks had long been in abeyance from 2004 to 2008-2009. Now, these reserves are divisions of zapovedniks or national parks. Their funding is part of the budget for zapovedniks and national parks. Regional zakazniks are found on lands of the federal forest fund, but are subordinated to the regional authorities and funded from regional budgets. The situation is different in different regions, however: in the Irkutsk Oblast', they practically unmanaged and are not financed, in Buryatia Republic and Zabaikalsky Kray the situation is more favorable.

6.3 Recreation at BNT

Recreational activity is not among the priorities of the protected areas, as shown in Table 7. Budgetary growth of protected areas is used to perform the basic environmental functions. Instead of differentiating the territory of protected areas based on permitted and prohibited activities, the Department of State policy in the field of environmental protection suggest transfering zapovedniks to the category of national parks. Moreover, the proposed development of the protected areas system in the direction of recreational resources and the creation of tourism infrastructure is supported. These simple ideas to increase economic activity and funding of environmental management of protected areas have been discussed above. In the irrational (non-economic) ideology of biodiversity conservation there is always a contradiction between its economic assessment on global and regional levels (Rumina, Karachevtsev, 2005).

The first attempt to study the development of recreation on the lake Baikal was undertaken in 1994 as a project commissioned by the World Bank's "Master Plan for ecotourism in the region of Lake Baikal" (The Master Plan, 1995). The basic concept of this plan are to: not exceed the maximum permissible load level of socio-economic, cultural, historical, ecological relationships in the region, including the unique communities of flora and fauna and cultural heritage; maximize opportunities and economic benefits to the local population; and maximize preservation of natural areas, national parks and reserves in the area of Lake Baikal and its waters by increasing the effectiveness of environmental stimuluses.

Since the advent of ecotourism development reports regularly promote the focus of recreational activity in protected areas. It is known that ecotourism is one of the most successful industries in the world. The formula of ecotourism in protected areas is to reduce the separation of permanent and temporary places of stay of visitors. Permanent residence refers to permanent occupation and temporary refers to the brief visits (Kalikhman at al., 2005; Shirokov at al., 2002). On BNT the recreational activities is determined by the demands for visits to Lake Baikal and the ability of protected areas to satisfy this demand. It should be noted that the proposals of quality services to stay in protected areas is extremely limited.

It is clear that extra-budgetary economy of protected areas depends directly on the flexible and operational records of demand for the visit and their competitiveness in comparison to the services for other recreational activities. Thus, *the third element of the economic model* is the development of recreational activities in protected areas. Such activities can satisfy the increasing demand for visits to Lake Baikal and can be used to increase funding of conservation.

7. Conclusion

Specially protected natural areas of the Baikal region have typically used either the basin or administrative approach. In this chapter we propose an integral approach that allowing us to overcome shortcomings of the basin and administrative approaches. There is a need to develop an integral approach and its linked to the emergence of the law "On protection of Lake Baikal." The emergence of the law determining BNT indicates a new environmental paradigm, which is based on: the possibility of transcending the limitations and contradictions caused by the obligatory account or administrative boundaries or the boundaries of the Lake Baikal basin in environmental management; the scope of nature conservation in terms of preservation of biotic and landscape diversity the key areas on BNP with a common regulatory and legal framework similar management; and the need to transfer the emphasis from the traditional declaration of environmental regulations in the modes of regulations on the conservation of the natural environment in accordance with the purpose of ecological zones BNP. The functional model of the territorial nature protection on BNP formalizes the transition to the new environmental paradigm. Corresponding to this paradigm an integral approach to nature protection is based on the institutional and economic models, as well as a system of recreation in protected areas within the BNT and WNHS "Lake Baikal".

8. Acknowledgment

This study has been possible with financial support of the Russian Geographical Society, Grant N 10/06/2011.

9. References

Atlas of Mongolia (2009). - Ulaanbaatar. - 248 pages.

Atlas of Transbaikalia (Buryat ASSR & Chitinskaya Oblast') (1967). - Moscow, Irkutsk: SDGK. - 176 pages.

Bannikova I.A. (1990). The structure and the function of the forest-steppe cover as a exponent natural ecosystem sustainable // Methodological aspects of the environment state valuation. - Puschino. - P.65-67.

Bardahanova T.B., Mikheyev A.S., Puntsukova S.D., Radnaev B.L. (2001). Methodology for determining the environmental costs in the region. - Ulan-Ude: Buryat Scientist Centre SB RAS. - 129 pages.

Belov A.V., Garaschenko A.V., Krotova V.M., Lapshina E.I., Peshkova G.A., Ryashin V.A., Frolova M.V. (1972). Vegetation of the Eastern Siberia south. Scale 1: 1500000. - Noscow: SDGK.

Bishop K., Green M., Phillips A. (2000) Models of national parks. Moscow: Biodiversity Conservation Center, 2000. - 216 pages.

Bobylev S.N., Khodjayev A.Sh. (2003). Environmental Economics. - Moscow: Moscow State University. - 567 pages.

Coas Ronald (1993) The problem of social costs // The firm, the market and right. - Moscow: "Ltd". - 192 pages.

Guide to the National Parks of the United States (2006). - Washington, DC: The Book Division National Geographic Society. - 480 p.

Guide to the State Parks of the United States (2004). - Washington, DC: The Book Division National Geographic Society. - 384 p.

Ecosystems of Mongolia: The map scale 1:1000000 (1995). Moscow – 15 pages.

Gunin P.D., Vostokova E.A., Matyushkin E.N. (1998). Protection of ecosystems in Inner Asia (1998). - Moscow: Nauka. – 221 pages.

Grubov V.I. (1963). The Central Asia botanic-geography zoning // Plants of Central Asia. – Moscow, Leningrad: USSR AS. – P. 10-69.

Kalikhman A.D., Kalikhman T.P., Khidekel V.V. (2005). Trails in natural areas of Lake Baikal. - Irkutsk: "Ottisk" - 114 pages.

Kalikhman T.P. (2007). Tunkinsky National Park, people or nature // Wildness protection, 2007 - Vol. 12 - P. 41-66.

Kalikhman T.P. (2008 a) Baikal natural territory in the institutional model of nature conservation // Geography and Natural Resources. - N 3. - p. 65-74.

Kalikhman T.P. (2008 b). The Baikal natural territory in economic model of nature conservation // Geography and Natural Resources, 2008. – N 4. – P. 330-337.

Kalikhman T.P., Sokolov V.A. (2005). The creation and the development of special protected natural areas in Irkutsk Oblast' and Krasnoyarskyy Kray // Forest rating and forest arrange. – N 2 (35). – P. 45-57.

Lavrenko E.M., Karamysheva Z.V., Nikulina R.I. (1991). Steppe of Eurasia. – Leningrad: Nauka. – 145 pages.

Mikheev V.S., Ryashin V.A. (1977). Landscapes of the Eastern Siberia south. Scale 1: 1500000. – Noscow: SDGK.

Ministry of Natural Resources (2006), http://www.mnr.gov.ru

Ostertag R., Ostertag G. (1998). California State Parks. A complete recreation guide. - Seattle, WA: The Mountaineers. - 384 pages.

Rumina E.V., Karachevtsev I.L. (2005). Specificity of biodiversity conservation in the spatial and temporal aspects // Environmental Economics. - № 1. - P. 112-118.

Savenkova T.P. (2001). Protected areas of Lake Baikal basin. - Irkutsk: Institute of geography of the SB RAS. – 185 pages.

Savenkova T.P. (2002). Protected areas of Lake Baikal basin. Atlas. - Irkutsk: "Ottisk", 2002. - 96.

Savenkova T.P., Erdenetsetseg D. (2000). The protected natural areas system development in Mongolian part of Lake Baikal basin // Geography and Natural Resources. – N 2. – P. 131-138.

Savenkova T.P., Erdenetsetseg D. (2002). Spesial protected natural areas in Baikal Natural Territory // Questions of Geography (Mongolian state university). - N 2. – P. 45-53.

Shirokov G.I., Kalikhman A.D., Komissarova N.V., Savenkova T.P. (2002) Eco-tourism: Lake Baikal. Baikal region. - Irkutsk: "Ottisk". - 192 pages.

Special protected areas of Mongolia (2000). - Ulaanbaatar. -105 pages.

State Report on Lake Baikal and measures for its protection in 2005 (2006). - Irkutsk. - 410 pages.

"Territorial Complex Plan of Nature Conservation of Lake Baikal" (TerCPNC Baikal): basic position (1990). – Moscow: Giprogor (in 2 parts). – 303 & 403 pages.

The Lake Champlain Basin Atlas. – Burlington, VT: Northern Cartographic, 1994 – 71 p.

The Master Plan for ecotourism in the region of Lake Baikal (1995). Report for submission to the Government of Russia, under the coordination of the World Bank. - 248 pages.

Yunnatov A.A., Dashnyam B. (1979). The map of vegetation in Mongolian people republic. Scale 1: 1500000. – Moscow: SDGK.

10

Trail Impact Management Related to Vegetation Response in Termessos National Park, in Turkish Mediterranean

Meryem Atik, Selçuk Sayan, Osman Karagüzel and Emrah Yıldırım
University of Akdeniz Faculty of Agriculture Department of Landscape Architecture, Antalya
Turkey

1. Introduction

Recreation is a pleasant activity that people realize as function of enjoyment, relaxation and refreshment. Protected natural areas around the world are established and managed to accommodate visitor access and recreation while protecting natural and cultural resources from degradation (Wimpey & Marion, 2011). At the same time, recreation settings alone often include natural areas with special ecological, biological or cultural significance. Therefore recreational impacts on natural areas have become a fundamental concern.

Regarding the disturbances that natural and semi-natural areas are subjected to, hiking trails are largely inevitable wherever recreation use occurs (Cole & Spildie, 1998). Admiting that the enjoyment of scenery is often a positive, desired recreational pursuit in national parks, it can also cause environmental disturbance, particularly in scenic areas (Edington & Edington, 1986). At the same time scenic quality is normally a primary consideration in the development of recreation areas. Therefore care must be taken to ensure that the sources of recreation within a natural environment must be carefully considered to prevent scenic degradation along trails.

Marion & Leung (2001) posited that protection of the natural structure and vegetation along trails is important for establishing sustainable land development in protected areas. Protection of the diversity of flora and fauna found in national parks is one primary basis for nature conservation and is a primary attraction for visitors. But in order to prioritize 'protection' in natural areas, ecological knowledge about the impact of trails can benefit both nature conservation and recreational use.

As a path through a forest or other natural site (Kirkpatric, 1980), trails are the most important element of recreational activities. As linear corridors, trails provide space for walking, viewing scenery, viewing wildlife, studying cultural and for educational visits exploring nature and history. Bell (2001) reported that trails can provide a means of access into the outdoors and nature, stimulating exercise in attractive surroundings and a variety of scenery, where Greer (2000) remarked that trail corridors help people rejuvenate themselves through fitness activities and contact with their environment.

In order to provide safe access to protected areas and to enable facilities for recreational hiking, biking, sightseeing or wildlife observation, a well-designed and managed trail system is crucial. Lynn & Brown (2003) reported that trail-users felt that the naturalness of the site contributed most to their overall recreation experince. On the other hand trail extension and widening had a large effect on all experience indicators, except naturalness.

Trail impact is any undesirable visitor-related change of the natural environment caused by the use of trails. Trail impacts include a variety of problems. Even low levels of trampling disturbance on trails reduce ground vegetation height, cover, and biomass, and may alter species composition by eliminating fragile species (Aust et al., 2005). As a critical issue for protected area management, trail impact has been identified through several different perspectives. Early studies of trail impacts focused on impact severity and environmental factors that affecting trail degradation (Leung & Marion, 2000). Similarly, Farrel & Marion (2002) emphasized trail impacts related to visitation. Nepal (2003) examined trail conditions, Nepal & Nepal (2004) and Turton (2005) summerised visitor impacts on trails, whereas Hall & Kuss (1989) and Bhuju & Ohsawa (1998) examined vegetation responses to trail impacts.

Common trail impacts include vegetation loss, compositional changes in the vegetation, soil compaction, exposure of plant roots (Marion 1994; Marion & Leung 2001; Farrel & Marion 2002; Marion et al., 2006). Nepal & Way (2007) reported that there were significant differences between control and trail-side quadrats in relative vegetation cover and overall species richness. Visitor-based deterioration has also been documented as increased trail incision, muddiness and widening (Cole, 1978; Farrel & Marion, 2002). Overall, the greatest impact of trail construction and maintenance was opening up tree and shrub canopies (Cole, 2004).

Excessive trail-related impacts to vegetation, soil or wildlife could represent an unacceptable departure from natural conditions and processes (Marion & Lueng, 2001). In other words, Cole (1981) remarked that current attemps to minimise changes due to recreation impacts were often inadequate due to lack of ecological information. Similarly Bayfield (1979) underscored that despite a large number of nature trails in Britain, there has been little investigation of their use and effectiveness.

Likewise there is very little information on recreation impacts in Turkey despite the fact that it has a high potential for nature conservation in over 1000 areas, 45 of which are national parks (Atik et al., 2005; Çevre ve Orman Bakanlığı 2010; Çevre ve Orman Bakanlığı, 2011; Resmi Gazete 2008). However, protection of vegetation along the trails is still important for securing sustainable recreation use in protected areas (Marion & Leung, 2001) considering increasing visitation to national parks.

The trail condition depends on trail user-related factors, slope, altitude, and vegetation type. Recreation impact varies depending upon the original environment and different types of ecosystems (such grassland, forest or desert) in which recreation takes place (Liddle 1997, in Nepal 2003). The chapter describes a study of the impact of trail use on vegetation (Mediterranean sclerophyll forest) in Termessos National Park and to develop a code of conduct for trail management. Termessos National Park was selected because hiking is the most intensive human use in this park and because experiencing trail impact may be unacceptable to visitors.

2. Material and method

2.1 Termossos National Park

Termessos National Park was chosen as the research area due to its natural and archeological features, easy visitor access, its status as the most visited protected area in Antalya, and the presence of more than one trail for comparison (Figure 1). Covering 6702 hectares, it is located in the northwestern Antalya and was established as a national park in 1970 (Anonymous, 1970). Found by tribes known as Solymians around 2000 B.C., Termessos was an important city of the ancient Psidian province of Anatolia. Temples, a gymnasium, agora, theatre, odium and infrastructure elements (walls, streets and cisterns) are important cultural assests in the city embedded in the Mediterranean vegetation characterized by sclerophyllous trees and shrubs that are known as macchia. There is high species diversity in this region. Therefore Termessos National Park atrracts people who have interests in both culture and nature (Figure 2). An average of 24.000 people (including both individual visitors and group tours) annually visit the park.

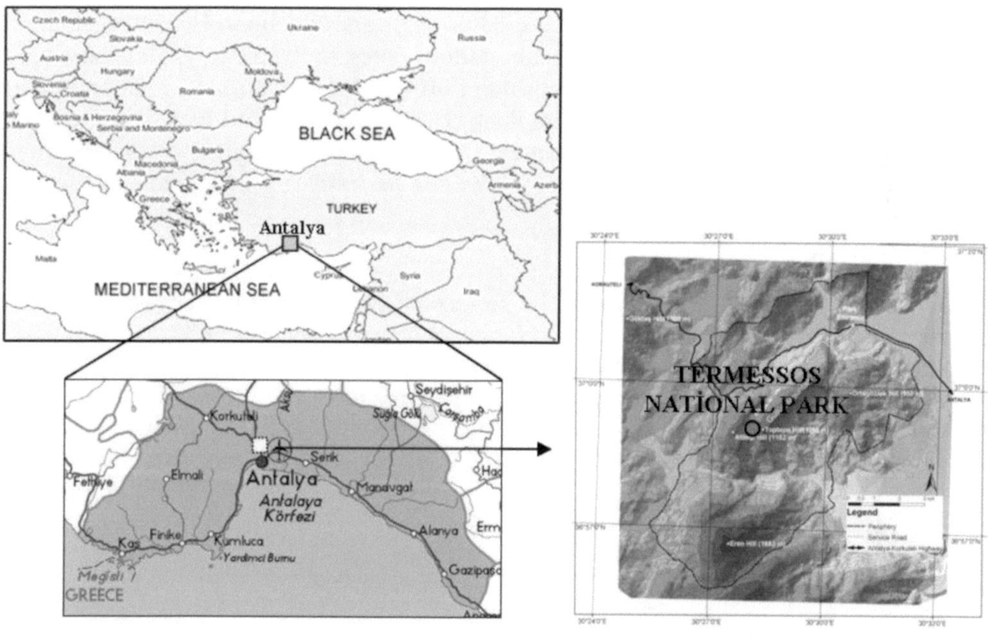

Fig. 1. Location of Termessos National Park

Characterised by very hot and dry summers and rainy, temperate winters, the area occasionally receives snow and frost above the 800-meter elevation level. There are local micro-climates extending into cool to mild conditions that enable greater species diversity than a typical Mediterranean climate.

Unlike other ancient coastal settlements around Antalya, natural characteristics of Termessos were well preserved. Thus rich flora and fauna and numerous natural habitats between 250 and 1265 m elevation are present in the park. The natural diversity of flora, avifauna, lichens, mammals and insects have been studied (Alçıtepe, 1998; Sert, 2000; Yalım & Çıplak, 2002;

Tufan et al., 2005; De Marinis & Masseti 2009). Tufan et al. (2005) noted that the wide variation in its topography composed of high mountains, valleys and deep canyons provide habitats for a rich diversity of lichen. They identified 152 species that are the indicator of air quality.

Fig. 2. Views from natural and historical characteristics of Termessos National Park

Alçıtepe (1998) identified 686 plant species; 80 of them are endemics. Dense vegetation in the national park supports mountain goats, fallow deers and golden eagles. Sert (2000) identified 113 bird species from 32 families and affirmed that natural vegetation was the integral component of bird habitats where thick cover of macchia and includes several tree species: *Pistacia terebinthus, Quercus coccifera, Quercus infectoria, Myrtus communis, Smilax aspera, Laurus nobilis, Daphne sericea, Arbatus andrachne, Juniperus oxycedrus* and *J.excelsa.*

2.2 Data analysis

Walking paths have long been present in Termessos National Park. Located in different sections of the park, five trails were chosen for examination of their impacts on vegetation. A sampling point was selected randomly on each of the five trails. At each point, indicators were measured in six plots lined up to traverse the trails: three plots were selected within the 10m straddling the central axis of the trail and three more plots were selected between the 15th and 20th m from the center on either side of the trail. The latter were used as controls (Figure 3).

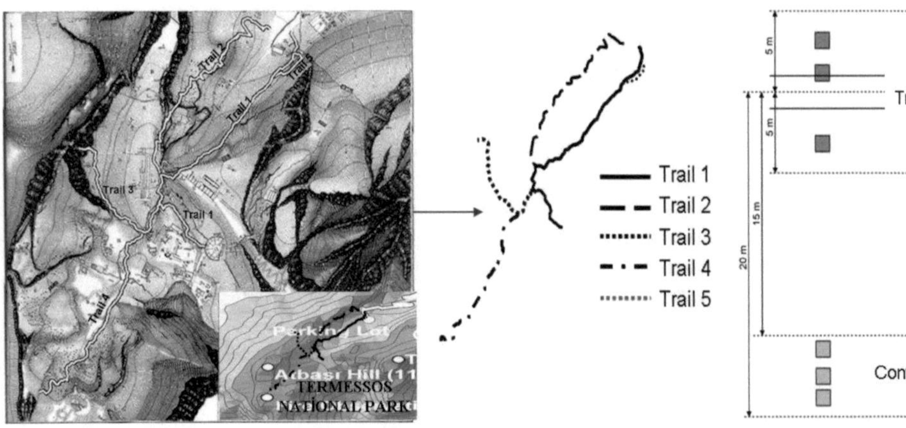

Fig. 3. Location of analysed trail system and sampling design in Termessos National Park

Vegetation sampling was designed according to Kutiel et al. (1999), Leung & Marion (2000) and Whinam & Chilcott (2003). Species richness (#) or the average number of species in each plot, vegetation height (cm) or average height of plants (cm) found in each plot and vegetation cover (%) based on a visual assessment of plant condition in the ground cover in each plot were collected at each plot. Data processing and analyses were conducted with SPSS 9.0 software. Data related to the trail impacts were analysed by ANOVA (McPherson, 1990).

3.Trail Impact on vegetation in Termessos National Park

As the main access into natural environment, trail conditions, and trail impacts may be influenced by use-related factors (Farrel & Marion, 2002). In this study, the 'use' factor was assumed to be the main source of vegetation change on the five main trails in Termessos National Park. Each trail sampling point prossessed different aspect, elevation and slope characteristics (Table 1). Vegetation characteristics varied from macchia to mixed forest to alpine.

Trail characteristics	*Trail 1*	*Trail 2*	*Trail 3*	*Trail 4*	*Trail 5*
Altitude (meters; start-end)	854-998	886-996	982-1060	1003-1107	893-916
Aspect	West	South-east	Noth-East	South	West
Trail length (meters)	952	772	482	676	158
Average slope (%)	15	14	16	15	15
Vegetation	Macchia, Mixed forest	Mixed forest	Mixed forest	Machhia, alpine	Mixed forest

Table 1. Site characteristics of the trails

Data collected for species richness, plant height and vegetation cover revealed a statistically significiant relationship to trail 'use' (Table 2). With regard to species richness, Trail 1, Trail 2 and Trail 4 showed similar variances, while Trail 3 was unlike the others. In terms of plant height, no significant differences were found between trails; however Trail 1 had the lowest apparent impacts from 'use'. Though Trail 2 exhibited higher vegetation cover (50,7%), there were no obvious differences in vegetation cover among the other trails. In terms of trail impact management, the effect of 'use' on trails on selected impact indicators proved that the recreational use significantly affected the vegetation richness (P≤0,05), plant height (P≤0,05) and vegetation cover (P≤0,01) along trails.

Trails	Species Richness (#)	Plant Height (cm)	Vegetation Cover (%)
Trail 1	4,83 a	26,3 c	26,8 b
Trail 2	4,76 a	33,3 ab	50,7 a
Trail 3	3,47 b	34,0 a	36,2 b
Trail 4	4,76 a	28,2 bc	35,7 b
Trail 5	4,29 ab	28,6 bc	33,5 b
Significance	P≤0,019	P≤0,034	P≤0,004

^{z}: Within each characteristic (column) mean separations by Duncan's multiple range test at 0.05 level.

Table 2. Effect of trail 'use' on species richness, plant heigh and vegetation cover

3.1 Response in species richness

Species richness is regarded as one of the thresholds for desirability in many recreation studies. Means of species richness on the trails are presented in Table 3. Species richness on Trail 1, with a mean of 4,58, is higher than found on the other trails. Analysis of data indicated that there was no significant difference between the richness measures of Trail 1, Trail 2, Trail 4 and Trail 5. Only Trail 3 was found to have a significantly different average number of species.

Analysis revealed that each trail's environmental characteristics (P≤0,001) and use impact on those trails (P≤0,01) independently and significantly affected species richness, however there was no significant interactive effect between trail and use factors. In contrast to predicted recreational impacts, the average numbers of species on trail plots were higher than in control plots (for Trails 1, 2, 3 and 4). Similar outcomes were confirmed by Bright (1986) and Bayfield and Brookes (1979) in evergreen vegetation environments. From an ecological perspective this was apparently due to disturbance allowing increased light, water and nutrients in open trail sections compared to those factors in closed tree canopies of the control plots.

TRAILS	**USE**		**Means of Trail**
	Use	Control	
Trail 1	4,83	4,42	4,58a
Trail 2	4,76	4,16	4,46a
Trail 3	3,47	2,76	3,11b
Trail 4	4,76	3,78	4,27a
Trail 5	4,29	4,33	4,31a
Means of use:	4,40a	3,89b	
Significance			
Trail (T):	***		
Use (U):	**		
T XU:	NS		

In each column and line, means that are significantly different at the 5% level followed by different letters according to Duncan's multiple range test
NS , **, *** : Not significiant and significiant at the level of 0,01 and 0,001 respectively

Table 3. Trail impact on species richness (#)

3.2 Response in plant height

Plant height often responds to the physical trail impacts. Sun and Liddle (1993) concluded that plant height appears to be strongly associated with resistance to trampling. Our results showed no statistical differences between trails in terms of plant height (Table 4). But mean plant height on Trail 5 has the highest value with 36,09 cm. There is evidence that plant height was significantly affected by use (U) (P≤0,05) as heights were sigificantly greater in control plots where mean plant height was 38,08 cm higher than in plots along the trails. The environmental context of the trails (T) alone the interaction between the environment and use (T X U) were not significantly related to plant height.

TRAILS	USE		Means of Trail
	Use	Control	
Trail 1	26,32	37,85	32,08a
Trail 2	33,36	38,15	35,76a
Trail 3	34,07	35,66	34,87a
Trail 4	28,25	35,26	31,75a
Trail 5	28,68	43,50	36,09a
Means of use:	30,13b	38,08a	
Significiance			
Trail (T):	NS		
Use (U):	*		
T XU:	NS		

In each column and line, means that are significantly different at the 0,5% level followed by different letters according to Duncan's multiple range test
NS , * : Not significiant and significiant at the level of 0,5%, respectively

Table 4. Trail impact on plant height (cm)

3.3 Response in vegetation cover

Hammit & Cole (1998) stated that the effect of vegetation cover on environmental durability is complex. As well as being impacted by trail use, vegetation cover can also be impacted by site characteristics. A significant variation between trails was observed for vegetation cover (Table 5). The highest mean vegetation cover was found on southeast-facing Trail 2 (49%), where sun intensity and duration might encourage plant growth. Trail 3 and Trail 4 showed similar coverage, and Trail 1 and Trail 5 had the lowest mean coverage amounts. Neither the environmental factors (T), nor the use (U), nor the interaction between two (T X U) were found to be significantly related to mean vegetation cover.

TRAILS	USE		Means of Trail
	Use	Control	
Trail 1	26,878	34,211	30,544c
Trail 2	50,767	48,600	49,683a
Trail 3	36,211	48,222	42,217b
Trail 4	35,778	43,689	39,733b
Trail 5	33,589	31,311	32,450c
Means of use:	36,644a	41,207a	
Significiance			
Trail (T):	NS		
Use (U):	NS		
T XU:	NS		

In each column and line, means that are significantly different at the 5% level followed by different letters according to Duncan's multiple range test
NS : Not significiant

Table 5. Trail impact on vegetation cover (%)

4. Trail impact management

Protected areas offer high quality nature-based recreational spaces for people. Andres-Abellan et al. (2006) pointed out that an increasing visitation to natural areas derives from people's desire to enjoy nature. But it is commonly agreed that even most careful recreational use could cause damage to the natural environment and thus reduces one's experience of nature.

Vegetation change can provide remarkable evidence of the negative impacts of land use on natural environments and can be used as an effective tool in conservation planning and visitor management in national parks. Most previous studies have pointed out that recreational use in protected areas is inevitable. Due to undesirable consequences of visitor use, trail management initiatives are needed.

Trails are generally regarded as an integral to parks and recreation areas, as they provide access to roadless areas and offer recreational opportunities (Aust et al. 2005). According to Wimpey & Marion (2010) a system of formal trails is a fundamental and essential component of the infrastructure of protected natural areas as it facilitates visitor access and supports sustainable recreational opportunities and experiences. However in order to effectively plan recreation, trail conditions and trail impacts must be thoroughly examined. Altan (1990) suggested that ecological principles and ecological relations must be taken into account during the planning phase to guide the human-nature relationship. Planning, on the other hand, requires holistic analysis of all impacts (Sinden, 1976).

Through educational and regulatory actions, managers can influence or control all use-related factors (Aust et al., 2005). Opening new trails would imply diminished natural space and further environmental damage. Dispersed trail use in protected areas can be an alternative for some areas, but park administrations outh to critically consider whether the resulting changes would be acceptible. Other possible solutions are to regulate visitor cycle, to rotate use of trails during specific periods, and to disperse hikers. Trail management measures need to be based on the results of recreation impact studies and thorough the knowledge of trails's characteristics.

The significance of nature conservation and the values of landscapes are closely related, so there is a need for effective management of recreational resources (Tzatzanis et al., 2003). Despite the long presence of numerous hiking trails in Termessos National Park since its establishment in 1970, there has been little investigation of the state of its natural environment. The trail system of Termessos National Park was created to enable views of the cultural heritage scenery, of the interesting topography, and of the diverse nature generally. Here, we have examined five trails in the park to assess vegetation responses to the use of these trails, taking into consideration each trail's site characteristics. We now offer some recommendations for site and visitor management on each trail.

Trail 1

From a managerial perspective, assessment of the vegetation response to trail use along Trail 1 showed that plants were most affected by physical impacts as average height was higher in control sites. On the other hand species richness was high as well due to plentiful sun light along the trail edges enabling diversity despite the high amount of visitor use. According to Cayuela et al. (2008) water, nutrient-level amelioration, and light deprivation in the Mediterranean forests in particular might affect species diversity. This also suggests that the natural vegetation is more likely to withstand visitors' use in the environmental conditions of Termessos National Park.

The elevation of Trail 1 climbs from 854 to 998 meters over a distance of 952 meters. The transect crossed vegetation containing mainly macchia and mixed forest (Figure 4). Beginning from a parking area and climbing to theatre, the first half of the trail crosses limestone which slopes steeply in both lateral directions. The most characteristic species on exposed trail slopes that experience surface erosion was *Astragalus hamosus*, which gives way to *Acantholimon acerosum* in control plots. The trail's width varied from 50 to 150 cm. The second half of Trail 1 was nearly flat as it crossed trough settlement ruins and vegetation (both on the trail and in the control plots) predominantly comprised of *Quercus coccifera, Ruscus aculeatus, Osyris alba, Daphne gnidioides.* The most typical plant was *Anagyris foetida* which was only found in control plots, where good shade encouraged *Allium cepa, A. orientale* and *Ornithogalum onites.*

With respect to trail use, Trail 1 functions as a gateway for most visitors to the park. However high trail use is clearly evident through the diminished vegetation cover and reduced plant heights along the trail. Visitor cycling along Trail 1, particularly in spring when vegetation is growing most rapidly, may be an effective management option. Aside from this, surface damage done by wild pigs in the park, should also be considered as it is most common in spring and autumn. Some other management tools may be needed to be developed to manage these impacts along Trail 1.

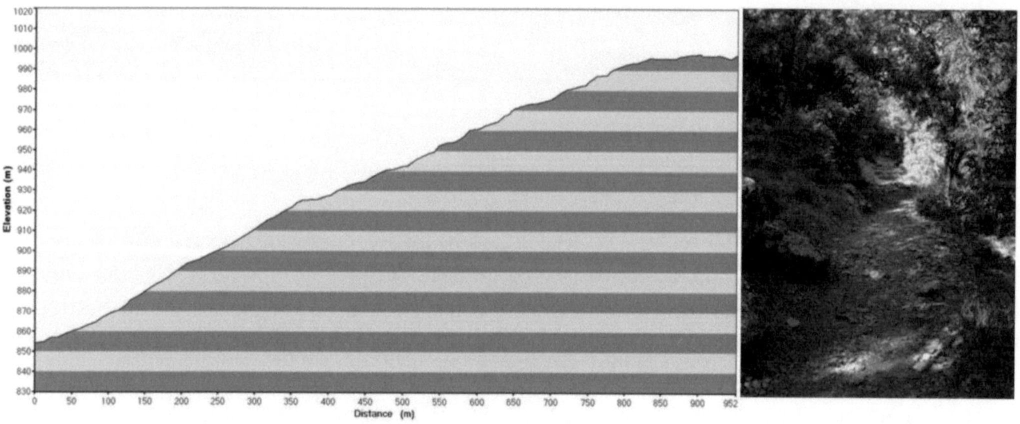

Fig. 4. Elevation profile and a view from Trail 1

Trail 2

Plant morphology is one possible determinant of resistant and resilience to the impacts (Whinam & Chilcott, 2003). Soil compaction due to high visitor use along trails leads to diminishing vegetation cover and deteriorating appearance, but on the other hand it creates a durable surface for walkers and diminishes the propensity to create ancillary trails.

With the second highest rate of visitor use after Trail 1, Trail 2 is mainly a return path or exit to the parking area. This trail descends (or ascends depending on the direction) from 996 meters to 886 meters over a 772-meter distance (Figure 5). Thick vegetation along the Trail 2 limits visitor's impacts and mitigates the erosion risk along its steep slope. *Quercus coccifera, Quercus infectoria* and *Quercus cerris* form a tall tree canopy while common species of *Ruscus aculeatus, Oysris alba, Asparagus acutifolius, Smilax aspera, Jasminum fruticans, Rhamnus oleoides* comprise the understory cover along Trail 2. Only a small number of *Cupressus sempervirens, Juniperus excelsa* and *J. oxycedrus* were found in control plots, where abundance of *Nectaroscordum sicilum* and *Iris albicans* indicated of the abundance of shade and moisture.

Thick vegetation and the abundance of ruins along Trail 2 kept the physical trail impact to a minimum. Because species richness, which was found to be the highest among the trail, vegetation height and vegetation cover were all relatively high, typical trail use could be allowed to continue along Trail 2 with no management change except for a careful monitoring of visitor numbers.

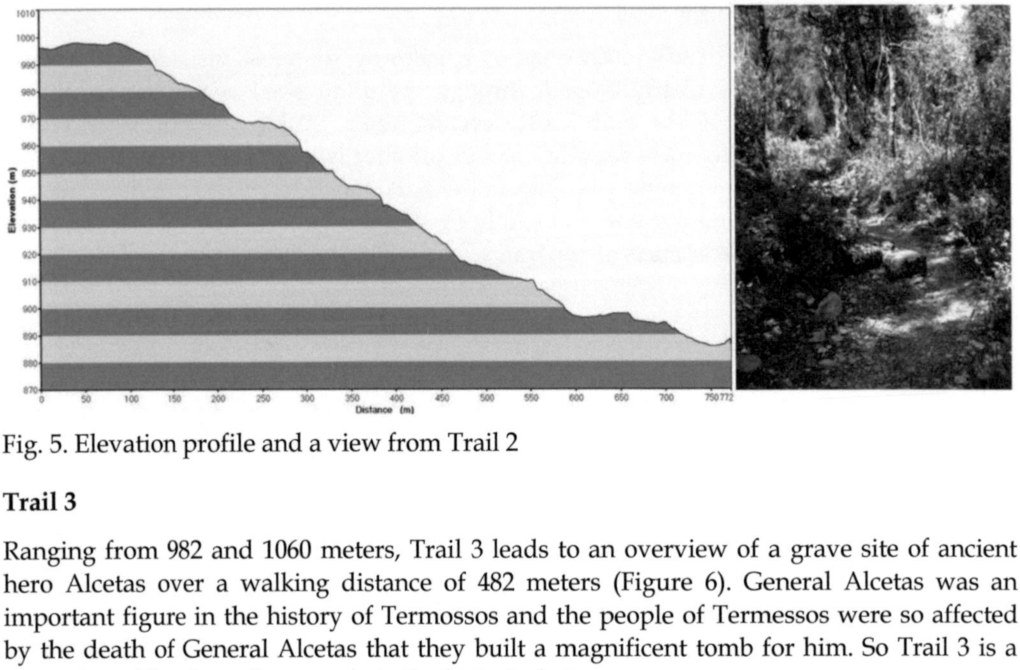

Fig. 5. Elevation profile and a view from Trail 2

Trail 3

Ranging from 982 and 1060 meters, Trail 3 leads to an overview of a grave site of ancient hero Alcetas over a walking distance of 482 meters (Figure 6). General Alcetas was an important figure in the history of Termossos and the people of Termessos were so affected by the death of General Alcetas that they built a magnificent tomb for him. So Trail 3 is a natural corridor that takes people to the historical site.

Trail 3 can be characterised as having low to moderate visitor use. People knowingly take this trail to access the monument. Northeast-facing Trail 3 is characterised by thick mixed forest of tall trees and bushes where *Pinus brutia, Quercus coccifera, Q. infectoria, Rhamnus oleoides, Styrax officinalis, Colutea melanocalyx* compose a tall overstory. *Carex divulsa, Althea cannabina, Bryonia cretica, Doronicum vulgaris* typically cover the forest floor along the trail.

As Hammit and Cole (1998) emphasised, trails' impact on vegetation above tree line is often severe. Specied richness was higher in trail plots than in control plots because there is typically insufficient sun and space for plant growth. Compared to other trails in the park, species richness was lowest along Trail 3. But mean plant height and vegetation cover were higher in control plots and were considerably higher on the trails.

Vegetation response in Trail 3 appeared to be due to the thick vegetation and typically limiting trail characteristic of high slope with 16% and closed tree canopy. Abundance of light on the trail sides encourages species diversity; however physical trail impact was evident with low plant height and vegetation cover in Trail 3.

Because of its vegetation response, the most effected element along Trail 3 was species richness which was linked to the physical site characteristics rather than to trail use. Therefore the tolerable vegetation response in terms of plant heigh and vegetation cover would allow continued moderate levels of use of Trail 3 in combination with monitoring of the changes in vegetation.

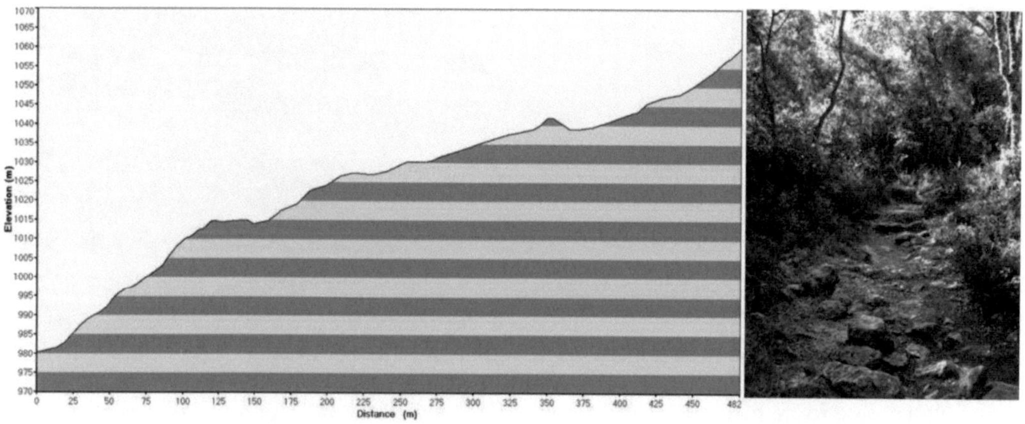

Fig. 6. Elevation profile and a view from Trail 3

Trail 4

Mountain areas are physical environments that very sensitive to change. They are characterised by short growing seasons and a combination of factors that makes subsequent regeneration of damaged vegetation more difficult (Holden, 2000). Located between 1003 and 1107 meters, Trail 4 presents the upper forest line of a mixed forest (Figure 7). Providing a great view of the entirety of the national park and of the Antalya environs particularly, the highest point on Trail 4 is also used as a fire observation centre by the park administration.

Trail 4 faces south and is exposed to high and persistent winds, particularly at the 1107-meter elevation. Characterised by macchia at the lower elevations and alpine flora at the higher elevations, vegetation on Trail 4 becomes dwarfed and stocky by exposure to the constant winds where *Juniperus excelsa* and *J. oxycedrus* morphologically dwindle on a 15% slope. Particularly brisk day and night temperatures promote herbaceous species of *Catapodium loliaceum, C. rigidum, Bromus sterilis, Melica minuta, Poa trivialis, Melica minuta, Brizia humulis, Trifolium arvense, T. resupinatum, T. campestre, T. angustifolium, Cruciata taurica, Crucianella latifolia, Biscutella resupinatum, Geranium purpureum, Buplerium heldreichii* and *Stachys cretica* along the sides of the trail. *Ostrya carpinifoia, Pistacia lentiscus, Pistacia terebinthus, Juniperus excelsa, Rhamnus pyrellus* are woody tree species that are typically found in the control plots.

At tree line, negative trail impacts were observed on Trail 4. Soil erosion and surface exposure exaggerated by the site characteristics of steep slopes and high elevation.

Although species richness was high along the open trail, the impact was clearly indicated by low plant height and vegetation cover values in trail plots. Both plant height and vegetation cover was remarkably higher in control plots. Therefore, visitor use must be reduced on Trail 4, either by orientating visitors towards a more direct descend or by closing the trail during certain seasons to be determined by further ecological research.

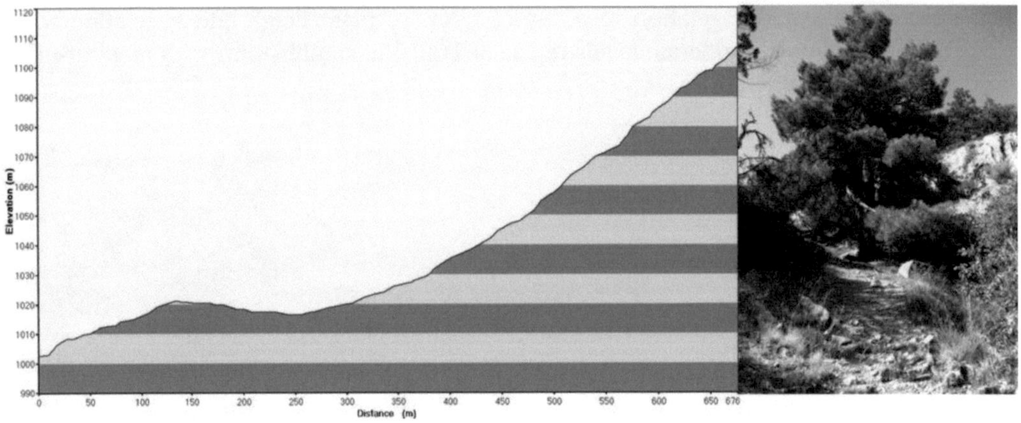

Fig. 7. Elevation profile and a view from Trail 4

Trail 5

As an alternative route to Trail 1, Trail 5 has very low visitor use. The trail's elevation ranges from 893 to 916 meters in a short distance of 158 meters (Figure 8). Located in closed-canopy, mixed-forest vegetation, species richness (particularly of perennial herbs) was enhanced by filtering in light from the sides of the open trail. Typical species were *Clinopodium vulgare, Erodium moshatum, Anthiriscus nemorosa, Cerastium glomeratum, Carex divulsa, Althea cannabiana* along the trail. A dense cover of *Quercus coccifera, Q. infectoria, Q. cerris, Styrax officinalis, Ostrya carpinifolia, Jasminum fruticans, Colutea melanocalyx, Ephedra campylocarpa* was found in the control plots.

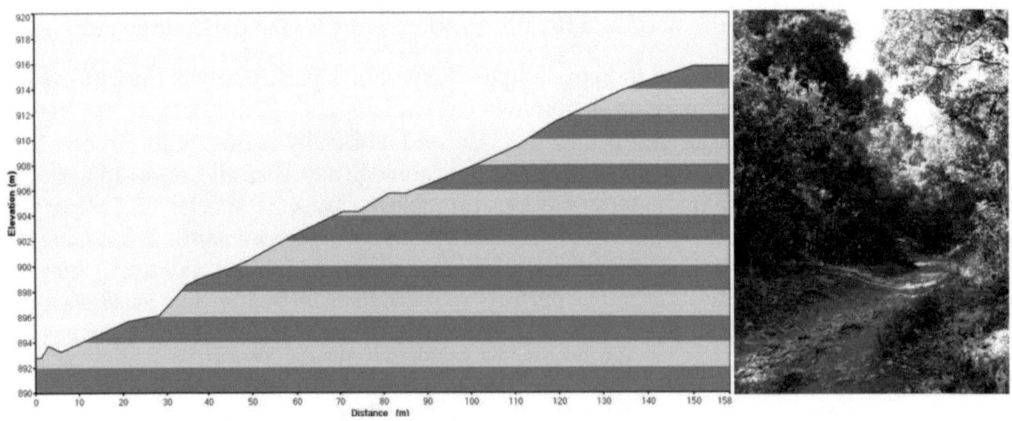

Fig. 8. Elevation profile and a view from Trail 5

Vegetation cover and species richness were supported along the trail by low levels of disturbance and where control plots were characterised mainly by bushes and trees. Unlike other trials in Termessos National Park, there has been regular vegetation clearance on Trail 5 to avoid trail closure by dense vegetation and to allow visitors an easy access. Kobayashi et al. (1997) specified that after vegetation removal there was a rapid recovery leading to more diverse vegetation with many more species lingering under different shade conditions. As an example of very low vegetation disturbance along the trail, Trail 5 can continue to be used as it has been, but must be closely monitored for changes.

5. Conclusion

Impacts of trails and other forms of recreation are concern, because activities like hiking and wildlife-viewing are popular in developing countries. Trails often receive the most intensive use within protected areas (Ouma et al., 2011). In the absence of trail management, resource degradation often occurs along trails and varies greatly in terms of type and severity (Marion & Leung, 2004).

As tourism and recreation in protected areas continue to increase, an accurate knowledge of their ecological impacts and ways to reduce them becomes increasingly important for conservation and visitor management (Buckley & King, 2003). Buckley (2004) stated that the significance of ecological impacts from tourism and recreation has been recognised widely by protected area managers. Site managers need to know the impacts of recreation activities on the natural environment of their units. According to Wimpey & Marion (2010) resource impacts associated with trail use can conflict with natural area resource protection mandates, thereby challenging land managers to implement visitor and resource management actions that avoid or to minimize impacts.

Termessos National Park contains sites of high cultural, ecological and historical values and is therefore an attraction for local, national and international tourists. Because the trails provide opportunities for nature walks and hikes to scenic vistas, most visitors to the park are entirely dependent upon the trails.

Trail impact usually depends on the characterics of the setting in which the trails were established. Talbot et al (2003) remarked that the response of an ecosystem to recreation-related impacts is primarily determined by the ecological characteristics of the biophysical system. The aim of the study was to measure the impacts of trails on the vegetation and to determine the necessary trail management actions that must be taken to avoid further negative vegetation responses.

The impacts of trails and their relative use intensity in Termessos National Park were significantly important based on the assessment of vegetation response in terms of species richness ($P \leq 0,05$), plant height ($P \leq 0,05$) and vegetation cover ($P \leq 0,01$). Generally species richness was higher along trails than in control plots distant from the trails. This was generally a function of opening the typically closed canopy of Mediterranean evergreen sclerophyllus vegetation and competition for light and nutrions. With similar results, Hall & Kuss, (2003) found out that cover and species diversity increased as one moved closer to the trail in eight out of ten cases they studied.

As Hammitt & Cole (1998) noted that the most significiant aspect of vegetation structure is the effect of tree canopy closure on diminishing vegetation amounts. Water and nutrient amelioration and light deprivation along trails become major factors influencing impacts. Similarly Valladares and Gianoli's (2007) confirmed that competition for water tends to accentuate the adverse (to growth) combination of shade and drought among understory plants in Mediterranean forests.

With respect to recreational impacts, site management actions may include relocating use to resistant sites, permanent or temporary site closures, control of the spatial distrubution of use, site hardening and shielding, rehabilitation of closed sites. Some visitor management actions might include use limits, length-of-stay limits, visitor dispersal or concentration, seasonal limitations and low-impact recreation training (Hammitt & Cole, 1998). While knowledge of the role of environmental factors in causing vegetation change can be applied to trail management, trail impact management deals with understanding and managing the role of the visitors (behaviour, timing, frequency or total numbers) in vegetation impacts (Whinam & Chilcott, 2003).

As expressed by Leung & Marion (2000) trail impacts are influenced by a diverse array of use-related and environmental conditions. Analysis of the trail conditions in Termessos National Park has proven that the characteristics of the vegetation, the proximity to cultural ruins, elevation, and slope combine to challenge trail impact management.

Leung & Marion (2000) confirmed that without proper trail management efforts, trail use can alter natural patterns. Visitor use regulations and educational programs can help to reduce impacts associated with trail use. Regulating visitor cycles along Trail 1, particularly when vegetation is most active and/or sensitive, maintaining current levels of use along Trail 2 and Trail 3 in combination with a monitoring program, reducing use of Trail 4, and allowing current levels of hiking along Trail 5 (with new monitoring programs) are the trail management options that would seem to be most in line with reducing vegetation impacts from hiking trails.

Trails management requires management of the type, amount and timing of visitor use and visitors' behaviors to ensure resource protection (Marion & Leung, 2004). Classification of the Termessos National Park trails according to their popularity would have been subjective. Because no direct relationship of vegetation response and use of the trails was apparent (Trail 1 was most used and Trail 5 was the least), an initial use assessment of the trails of the park would provide the basis for understanding the trail-impact threshold of users.

Sayan et al. (2005), in their effort to determine the recreational carrying capacity of Termessos National Park, reported that the number of people visiting the national park each year is nearing the maximum acceptable visitor number. Thus must be seen as an important problem and must be addressed by park management. Termessos National Park has recently been added to the list of UNESCO World Heritage Sites (UNESCO, 2011) and therefore will see an increase in the number of visitors to the estimated 24.000 visitors per year. Therefore, challenges to the protection of the natural environment and cultural heritage of Termessos will become increase with time.

The provision of safe, accessible, and environmentally friendly recreational trails is a key managerial focus for protected areas. The current trail system in Termessos National Park provides opportunities for hiking based on high cultural and natural qualities. The addition of a trail-design programme (supplemented by construction and maintenance) in combination with a visitor management programme would be an effective, long-term site management inititiative.

Trail management for the protection of ecosystems and cultural heritage can provide aesthetically pleasing landscapes and keep recreational uses in harmony with the natural environment. However a complete understanding of the functioning of the Mediterranean environment and trail impacts thresholds is still a far-reaching objective requiring further studies.

6. Acknowledgment

This work was supported by the Administration Unit of Scientific Research Projects of Akdeniz University and partly financed by the Scientific and Technical Research Council of Turkey (TÜBİTAK-TOGTAG-3197). We would like to thank Muzaffer Uysal for his earlier review of our paper and Termessos National Park Service for their support.

7. References

Alçıtepe, E. (1998). *Termessos Milli Parkı (Antalya) Florası Üzerinde Bir Araştırma.* Akdeniz Ü., Fen Ed. Fak., Biyoloji Bölümü Doktora Tezi, 194 sayfa, Antalya.

Altan, T. (1990). *Ekolojik Riziko Analizi ve Turizm Planlaması.* Türkiye Çevre Sorunları Vakfı, Turizm ve Çevre Konferansı, 3-5 Ekim 1990, Ankara.

Andres-Abellan, M.; Lopez-Serrano, F.R.; Morote, F.A.G. & Del Cerro-Barja, A. (2006). Assessment of trampling simulation impacts on native vegetation in Mediterranean sclerophyllous forest. *Environment Monitoring and Assessment,* Vol. 120, pp. 93-107.

Atik, M.; Karagüzel, O. & Ortaçeşme, V. (2005). Türkiye'de Doğa Koruma Yapısı ve Biyolojik Çeşitliliğin Korunması. *Korumada 50 Yıl Sempozyumu,* pp.63-72, Mimar Sinan Güzel Sanatlar Ü., 17-18 Kasım 2005-İstanbul.

Anonymous, (1970). *Termossos Milli Parkı Uzun Devreli Gelişme Planı.* TC Orman Bakanlığı Orman Genel Müdürlüğü Milli Parklar Dairesi-Ankara, 55p.

Aust, M.W., Marion, J.L. & Kyle, K. (2005). Research for the Development of Best Management Practices to Minimize Horse Trail Impacts on the Hoosier National Forest. In: *U.S. Forest Service* Final Research Report, Available from http://www.fs.usda.gov/Internet/FSE_DOCUMENTS/stelprdb5292110.pdf

Bayfield, N.G. (1979). Recovery of four montane heath communities on Cairngorm, Scotland, from disturbance by trampling. *Biological Conservation,* Vol.15, No 3, pp.165-179.

Bayfield, N.G. & Brookes, B.S. (1979). Effects of repeated use of an area of heather *Calluna vulgaris* (L.) *hull* moor at Kindrogan, Scotland, for teaching purposes. *Biological Conservation,* Vol.16, No.1, pp: 31-41.

Bright, J.A. (1986). Hiker impact on herbaceous vegetation along trails in an evergreen woodland of Central Texas. *Biological Conservation,* Vol.36, No.1, pp: 53-69.

Bell, S. (2001). *Designing for Outdoor Recreation.* Published by Spon Press, 11 New Fetter lane, London ISBN 0 419 20 350 8.

Bhuju, D.R. & Ohsawa, M. (1998). Effects of nature trails on ground vegetation and understory colonization of a patchy remnant forest in an urban domain. *Biological Conservation*, Vo.85, No.1-2, pp: 123-135.

Buckley, R. & King, N. (2003). *Visitor-impact Data in a Land-management Context.* In: Nature-based Tourism, Environment and Land Management, CABI Publishing, London.

Buckley, R. (2004). Using Ecological Impact Measurements to Design Visitor Management. In: *Environmental Impacts of Tourism* (ed.R.Buckley), CABI International, Oxfordshire, UK.

Cayuela, L.; Mari, J.; Benayas, R.; Maestre, F.T. & Escudero, A. (2008). Early environments drive diversity and floristic composition in Mediterranean old fields: Insights from a long-term experiment. *Acta Oecologica,* Vol.34, No.3, pp: 311-321.

Cole, D.N. (1978). Estimating the susceptibility of wildland vegetation to trailside alteration. *Journal of Applied Ecology*, Vol.15, pp: 281-286.

Cole, D.N. (1981). Vegetational changes associated with recreational use and fire suppression in the Eagle Cap Wilderness, Oregon: Some management implications. *Biological Conservation,* Vol.20, Vol.4, pp: 247-270.

Cole, D.N. & Spildie, D.R. (1998). Hiker, horse and Ilama trampling effects on native vegetation in Montana, USA. *Journal of Environmental Management,* No.53, Vol.19, pp: 61-71.

De Marinis, A.M. & Masseti, M. (2009). Mammalian Fauna of the Termessos National Park, Turkey. *Zookeys,* Vol.31, pp: 221-228.

Cole, D.N. (2004). Impacts of Hiking and Camping on Soils and Vegetation: In: *Environmental Impacts of Tourism* (ed.R.Buckley), CABI International, Oxfordshire, UK.

Çevre ve Orman Bakanlığı, (2010). Türkiye'deki Koruma Alanları; Milli Parklar, Tabiat Parları, Tabiat Anıtları, Tabiatı Koruma Alanları, Yaban Hayatı Geliştirme Sahaları. In: *Çevre ve Orman Bakanlığı,* Available from http://www.milliparklar.gov.tr

Çevre ve Orman Bakanlığı, (2011). Çevre Durum Raporu. In: *Çevre ve Orman Bakanlığı,* Available from http://www2.cevreorman.gov.tr/durum_rapor/ockkb/ockkb_ild.pdf

Edington, J.M. & Edington, M.A. (1986). *Ecology, Recreation and Tourism.* ISBN 0 521 30646, Published by the Press Syndicate of the University of Cambridge, UK.

Farrel, T.A. & Marion, J.L. (2002). Trail Impact and Trail Impact Management Related to Visitation at Torres del Paine National Park, Chile. *Leisure/Loisir,* Vol.26, No.1-2, pp: 31-59.

Greer, D.L. (2000). Nebraska Rural Trails: Three Studies of Trail Impact. In: *National Park Service,* Available from http://www.unomaha.edu/recadmin/trails/omahatrails.pdf

Hall, C.N. & Kuss, F.R. (1989). Vegetation alteration along trails in Shenandoah National park, Virginia. *Biological Conservation,* Vol.48, No.3, pp: 211-27.

Hall, C.N. & Kuss, F.R. (2003). Vegetation alteration along trails in Shenandoah National Park, Virginia. *Biological Conservation*, Vol.48, Vol.3, pp: 211-227.

Hammitt, W.E. & Cole, D.N. (1998). *Wildland Recreation Ecology and Management.* John Wiley&Sons, Inc. ISBN 0-471-19461-1.

Holden, E. (2000). *Environment and Tourism.* London: Routledge 11 New Fetter Lane, London, ISBN 0-415 20717.

Kirkpatric, E.M. (1980). *Chambers Dictionary.* Printed in Britain by Clay Ltd Bungay, Suffolk, ISBN 0 550 10634 0.

Kobayashi, T.; Hori, Y. & Nomoto, N. (1997). Effects of Trampling and Vegetation Removal on Species Diversity and Micro-Environment under Different Shade Conditions. *Journal of Vegetation Science,* Vol.8, No.6, pp: 873-880.

Kutiel, P.; Zhevelev, H. & Harrison, R. (1999). The effect of recreational impacts on soil and vegetation of stabilised coastal dunes in the Sharon Park, Israel. *Journal of Ocean & Coastal Management,* Vol.42, pp: 1041-1060.

Leung, Y.F. & Marion, J.L. (2000). Recreation Impacts and Management in Wilderness: A State-of-Knowledge Review. *USDA Forest Service Proceedings* RMRS-P-15, Vol 5, pp: 23-48.

Lynn, N.A. & Brown, R.D. (2003). Effects of recreation use impacts on hiking experiences in natural areas. *Landscape and Urban Planning,* Vol.64, pp: 77-87.

Marion, J.L. (1994). *An assessment of trail conditions in Great Smoky Mountains National Park.* US Department of Interior, National Parks Sevice, Southeast Region, Georgia.

Marion, J.L. & Leung, Y.F. (2001). Trail resource impacts and an examination of alternative assessment techniques. *Journal of Park and Recreation Administration,* Vol.19, No.3, pp: 17-37.

Marion, J.L. & Leung, Y.F. (2004). Environmentally Sustainable Trail Management. In: *Environmental Impacts of Tourism* (ed.R.Buckley), CABI International, Oxfordshire, UK.

Marion, J.L.; Leung, Y.F. & Nepal, S.K. (2006). Monitoring Trail Conditions: New Methodological Consideration. *The George Wright Forum,* Vol.23, No.2, pp: 36-48.

McPherson, G. (1990). *Statistics in Scientific Investigation: Its Basis, Application and Interpretation.* In: Fienberg, S., Olken, I. (Eds.), Springer-Verlag, pp. 1-666.

Nepal, S.K. (2003). Trail Impacts in Sagarmatha (Mt.Everest) National Park, Nepal: A Logistic Regression Anaysis. *Environmental Management,* Vol.32, No.3, pp: 312-321.

Nepal, S.K. & Nepal, S.A. (2004). Visitor Impacts on Trails in the Sagarmatha (Mt. Everest) National Park, Nepal. *Ambio,* Vol.33, No.6, pp: 334-340.

Nepal, S.K. & Way, P. (2007). Comparison of vegetation conditions along two backcountry trails in Mount Robson Provincial Park, British Columbia (Canada). Journal of Environmental Management 82 (2): 240-249.

Ouma, K.O., Stadel, C. & Eslamian, S. (2011). Perception of tourist on trail use and management implications for Kakamega Forets, Western Kenya. *Journal of Geography and Regional Planning,* Vol.4, No.4, pp: 243-250, April 2011

Resmi Gazete, (2008). 2008/14355 nolu Bakanlar Kurulu Kararı. In: *Resmi Gazete,* 06.12.2008 tarih ve 72075 sayı, Available from: http://www.resmigazete.gov.tr

Sayan, M.S.; Ortaçeşme, V.; Karagüzel, O.; Atik, M.; Şahin, T.; Yıldırım, E. & Avcı, Ü. (2005). *Termessos (Güllükdağı) Milli parkı'nda Rekreasyonerl Taşıma Kapasitesinin Belirlenmesi.* TÜBİTAK-TOGTAG-3197 Proje Raporu 65 sayfa.

Sert, H. (2000). *Termessos Milli Parkının (Antalya) Avifaunası ve Ekosistemdeki İlişkileri.* Akdeniz Üniversitesi, Fen Edebiyat Fakültesi, Biyoloji Bölümü Yüksek Lisans Tezi, Antalya.

Sinden, J.A. (1976). Carrying capacity as a Planning Concept for National Parks: Available Diserable Capacity? *Landscape Planning,* Vol.2 pp: 243-247.

Sun, D. & Liddle, M.J. (1993). A survey of trampling effects on vegetation and soil min 8 tropical and subtropikal sites. *Environmental Management,* Vol.17, No.4, pp: 497-510.

Talbot, L.M.; Turton, S.M. & Graham, A.W. (2003). Trampling resistance of tropical rainforest soils and vegetation in the wet tropics of North east Australia. *Journal of Environmental Management*, Vol.69, pp: 63-69.

Tufan, Ö., Sümbül, H. & Türk, A.Ö. (2005). The lichen flora of the Termessos National Park in Southwestern Turkey. *MYCOTAXON,* Vol.94, pp: 43-46 P

Turton, S.M. (2005). Managing Environmental Impacts of Recreation and Tourism in Rainforests of the Wet Tropic of Queensland World Heritage Area. *Geographical Research,* Vol.43, No.2, pp: 140-151.

Tzatzanis, M.; Wrbka, T. & Sauberer, N. (2003). Landscape and vegetation responses to human impact in sandy coasts of Western Crete, Greece. *Journal Nature Conservation,* Vol.11, pp: 187–195.

UNESCO, (2011).Güllük Dağı - Termessos National Park. In *UNESCO World Heritage Centre,* Available from: http://whc.unesco.org/en/tentativelists/1412/

Valladares, F. and E. Gianoli. 2007. How much ecology do we need to know to restore Mediterranean ecosystem. *Restoration Ecology,* Vol.15, No.3, pp: 363-368.

Wimpey, J. & Marion J.L. (2010). The influence of use, environmental and managerial factors on the width of recreational trails. *Journal of Environmental Management,* Vol.90, pp: 2028-1037.

Wimpey, J. & Marion J.L. (2011). A spatial exploration of informal trail networks within Great Falls Park, VA. *Journal of Environmental Management,* Vol.92, pp: 2012-1022.

Whinam, J. & Chilcott, N.M. (2003). Impact after four years of Experiemental trampling on alpine/sub-alpine environments in Western Tasmania. *Journal of Environmental Management*, Vol.67, pp: 339-351.

Yalım, B. & Çıplak, B. (2002) Termessos Milli Parkı (Antalya) Orthoptera (Insecta) Faunası: Fauna Elemanlarının Zoocoğrafyaları ve Vejetasyona Göre Dağılışları. *Türk Entomoloji Dergisi*, Vol.26, No.4, pp: 267-276.

Part 5

International Challenges to Contemporary Conservation

11

Applied Landscape Ecology, Future Socioeconomics and Policy-Making in the Neotropics

Arturo Restrepo-Aristizábal[1,2],
Valentina Heggestad[2] and Ian Sajid Acuña-Rodríguez[3]
[1]*Inter-American Biodiversity Information Network (IABIN),*
[2]*Applied Ecology for Tropical Resources Center (EcoTropics),*
[3]*Universidad de La Serena, Laboratorio de Ecología Funcional*
[1,2]*Colombia*
[3]*Chile*

1. Introduction

The colloquial basis of ecology and reality is that "all things interact," in one way (magnitude) or another. All living and non-living entities intertwine constantly over time in amazing forms and complex systems. These entities are linked by higher flows of matter and energy to form what we commonly refer to as nature. Species are a fundamental part of the multifaceted ecological world and act as basic entities in ecosystem-building and evolution (Guisan et al., 2006; Hey et al., 2003) as they constantly change the dynamics of ecological patterns and processes.

Anthropogenic activities have often been responsible for prolonged global and ecological alterations (Halper, 2008; Hobbs, 1997; Vitousek et al., 1997). Extreme levels of population growth, industrial and agricultural consumption affect the ecological processes that drive local and global biogeochemical dynamics. The impact of these alterations, however, is still the topic of scientific debate. Biotic systems have great resilience that for humans has been a double-edged sword; it has given rise to the false belief of an unlimited supply of resources, as in neoclassical economics, and led to a disposable mentality for human consumption and high rates of exploitation. The degradation of ecosystem services has led to several environmental problems that the world now faces even after the "no-return" or "tipping point" has inadvertently been reached (Ives & Carpenter, 2007; Ludwig et al., 1997). Until regional, global and time-series dataset patterns were observed, the negative synergic effects from human activities were not clearly understood (Bøjrnstad and Bryan, 2001; Ducklow et al., 2009; Gaston, 2000; Wiens, 1989). Even with current available knowledge, we still face several ecological data gaps for decision-making. However, a multitude of clues to understanding the structure and function of landscapes in the Neotropics are buried in the data. It takes an entrepreneurial and scientific inquiry mindset to find them.

Over the past three decades, the ecological world and our view of it, has undergone accelerated rate of global changes. Little more than a century ago, pristine ecosystems were still common in Neotropical lands and the favorites of rigorous researchers since Victorian times (Davis, 1996). Now, they are more an ideal that is rarely found in the field, especially in the Neotropics (Hobbs et al., 2006; Vitousek et al., 1997). The ecological science that was based on so-called natural conditions was recently confronted with unnatural contexts and processes: extensive land cover transformations (Fischer & Lindenmayer, 2007), water and nutrient cycles alterations (Smith et al., 1999), and many other disturbances of nature's dynamics have widely been observed in recent years. Conversely, as the disturbed environments are now more common in the world, a new ecology is also arising from the "novel ecosystems" (as defined by Hobbs et al., 2006) formed under such altered conditions. Side-road prairies (Hopwood, 2008), native-exotic mixed forests (Lugo, 2004) and tropical agroforestry and agricultural landscapes (Chazdon et al., 2009) are a few examples of these new communities that surprisingly retain certain ecological services, and even more, these could probably be the best opportunities for maintaining some of the original functions of local ecosystems (Harvey et al., 2008).

Under the mosaic of an altered ecological world, human communities live elsewhere, and decisions about actual and future biotic resources, and ecological services management, are expected. Environmental monitoring and assessment have been recently included in the design of different national and international policies around the world (Ashford & Caldart, 2008; Edwards et al., 2010; Van den Bergh & Grazzi, 2010). Despite how accurate or effective each legal framework could likely be, decisions about environmental issues (conservation areas, biotic exploitation rates, land use and territorial management) are particularly conflictive because they involve interactive processes with components at different spatio-temporal scales and different social actors (Ashford & Caldart, 2008). Laws are usually thought to be general and permanent, but ecological problems are usually very specific and variable depending on the audience.

In a context where landscape ecology could streamline a fast, accurate and integrated assessment of several ecological processes in both time and space. Remote sensing analyses is now a widespread tool in public and private organizations and can be done repeatedly over time and at a range of scales. The synthetic evaluation allowed by landscape analysis is capable of geospatially integrating both environmental and socioeconomic frameworks at the regional scale, but it also identifies hidden political ecologies among people´s empowerment of their territories (Restrepo, 2006). The integration of these two points of view is the basis of multi-scale planning and is clearly fundamental for decision makers in both public and private environmental institutions (Dinerstein et al., 1995).

2. Landscape ecology in the Neotropics

The term Neotropics defines a highly diverse and complex biogeographic region that roughly encompasses Latin America and Caribbean countries and their adjoining seas (Udvardy, 1975; Pielou, 1979). This delimitation, unless based on ecological parameters, also fits a territory with common cultural and socioeconomic realities. Several major ecological threats are undoubtedly associated with regional social problematics (Dinerstein et al., 1995). Despite the great attention that some Neotropical ecosystems (i.e. tropical rain forest, tropical montane cloud forest, and páramos) have received in the past decades, landscape approaches to environmental planning are still scarce in the region. Ninety studies were found in the ISI Web-of-Science database under a search for "landscape" and "Neotropics" (1920-2011). Of them, just 70 were in landscape ecology, the majority related to tropical forest fragmentation and biodiversity conservation. Ninety-two percent of the research published during the last ten years was primarily by scholars at institutions from United States (41%), Mexico (20%), Brazil (18%) and England (11%).

In 1114 the Inca civilization coined the term *"Ispalla pachamama"*; *ispa* means plants, *alla* animals, *mama* mineral and soil attributes and pachamama for the cosmologic insight of the spiritual-natural connector to the land. The Inca civilization had integrated knowledge of their landscape that allowed them to terrace and channel water in the Andean mountain landscapes. This integrated landscape knowledge allowed Inca's civilization to be able to overcome a warming climate over a 400-year period (Chepstow-Lusty, 2009). The word landscape was developed from the Dutch word "landschap" which is originally referred to a patch of cultivated ground in a snapshot through time. Humboldt and Darwin in the 19th century understood the interaction of physical attributes and living nature. In 1884, the term "landscape science" was used to describe physical attributes that assembled processes into geomorphology and physiographic prints. German and Russian geographers in the 20th century envisioned this natural science through socio-cultural and psychological features embedded in physical territory. Then, the term landscape ecology was normalized and applied to the first aerial-photograph interpretation; then to studies of interactions between environmental processes and vegetation (Troll, 1939), and subsequently to its contemporary application as a science by using geographic information systems, remote sensing data and the ecological economics of carbon potential for plant sequestration (Fig. 1) as well as payments (values) for ecosystem services.

Hence, landscape ecology research is primarily focused on the degradation processes of ecosystems, as the result of deforestation, degradation through fragmentation, and the prioritization of selected sites for assertively managing local biodiversity (Guadagnin & Maltchik, 2007; Martinez-Morales, 2005). But the level of degradation and interference that human activities have generated in certain ecosystems needs a different focus: the integration of human dimension (i.e., land use, human demography, socioeconomics) and its effect at landscape scale over ecological dynamics. This would chiefly enhance the theory for this new scenario and facilitate necessary mapping for better planning and management about natural capital. Recently, some studies of ecological processes and interactions in agricultural, urban and sub-urban landscapes are providing insights into this approach (Chacon & Harvey, 2006; MacGregor-Fors, 2010; Sanchez-Clavijo et al., 2008; Suarez-Rubio & Thomlinson, 2009).

Nevertheless, beyond old paintings of landscapes and scientific conceptualizations, the ontology of landscape ecology is synthesized in the Dutch school of thought, which might be described as a big-headed horse riding among geography, ecologies, cultures, and the power of wise agriculture to shape a landscape by revealing geospatial patterns and processes over time (Zonneveld, 1979). As a result, the Dutch applied this knowledge to their landscapes and seascapes while reclaiming land from the sea, and established a coherent policy-making through the "polder model" as a reference to the Dutch term for concession politics, water-boards, constructed wetlands, waterways and pump mills.

Fig. 1. Map distribution with some Neotropical plants at the level of order with keystone potentials for sequestering carbon. Source: Inter American Biodiversity Information Network (IABIN) http://www.iabin.net of Data Integration and Analysis Gateway (DIAG), InBio Costa Rica, Global Biodiversity Information Facility (GBIF).

Landscape ecology in the Neotropics is better applied to biodiversity planning in order to categorize ecosystem services aimed at estimating both functional ecological aspects and effective demand for these services. Since nature conservation is not the exclusive domain of either human condition or the Neotropical societal mindset, biodiversity strategic action planning needs to be better understood philosophically. Instead, the wise use and management of biodiversity is widely assimilated in the Neotropics as nature "elongation" by referring to a co-evolutionary process to extend the breadth and length of wildlife species over interacting cultural landscapes. For example, in past years Nearctic guidelines of non-profit organizations and governments have been aimed at streaming revenue to buy land for protected areas in the Neotropics. They believe that the trespassing legal concept prevails in practice as it occurs over their lands. However, if impoverished communities around buffer areas need to secure their livelihoods, they will most likely use legally protected areas for survival in the Neotropics. All in all, the dominant matrix for a Neotropical landscape is transformed into agrofarming mosaics of socio-ecosystems and remnant patches of protected areas (Fig 2).

Fig. 2. Map of protected areas (pink) in the Neotropics. Source: IABIN DIAG and IABIN protected areas network, World Database on Protected Areas.

Applications of landscape ecology are often assembled in spatial units and time series through overlapped tiles. This analytical approach lays out geology, topography, soil structure, biogeography, vegetation, carbon content, ecosystems, water balance, socioeconomic valuations, road networks, and demographics by using GIS technology. This approach represents a better fit for an applied socio-ecological structure and function, especially in a region characterized as mega-biodiverse, but at the same time socioeconomically mega-poor. This means that communities do not fully comprehend the resource potential around their locality.

Consequently, the region is missing a paramount opportunity compared to Nearctic neighbors, lagging behind on the establishment of effective and timely mechanisms of tax exemption among diverse taxpayers and non-profit organizations in order to bridge the rich and poor gap. A region with hundreds of oil, mining, extensive agriculture and power generation companies should mobilize revenue toward biodiversity research, development and innovation in prioritized landscape units for nature elongation based on a) chorological types (defined as groups of ecosystems that are more spatially related between them than with others); and b) minimum ecosystem targets (defined as the area of an ecosystem that has to be conserved to secure the survival of its species) (Wyngaarden & Fandiño-Lozano, 2005). Ultimately, there is often widespread discussion in the region as to whether or not a coupled socioeconomic and ethnospheric map would be best to describe the degree of geospatial control of truly anthropogenic interaction with its territory; similar to a *fuzzy set* classification that is a class with a continuum of grades of membership (Zadeh, 1987). Maps of this type enable either bottom-up or top-down approaches for delivering effective market mechanisms derived from payment for ecosystem services in accordance with social structures and traditional knowledge throughout the Neotropics.

2.1 The socioeconomic ecologies in the Neotropics

Unplanned urban, road infrastructure and agricultural expansion are the main causes of degradation patterns of ecosystem structures and functions (McKinney, 2006; Morton et al., 2006). The decoupling of ecological and social planning is one of the main causes for the failure of conservation and restoration efforts, the loss of ecosystem services, and in some cases, displacement of large (human) communities (Bennett et al., 2005). Integrating the evaluation of ecological resources and services with human population dynamics allows decision-makers to estimate the ecological costs of a social change or vice-versa; the social "expense" of an ecological change (Mattews & Selmann, 2006). Not only physical elements, such as infrastructure or rates of land transformation, are useful and informative when assessing socio-ecological dynamics; It is important to also assess the capacity of local communities to incorporate the basic changes that are needed to achieve an ecological goal into their daily activities coupled with the knowledge of an impact-threshold at their focal ecosystems. In the next section we will explore this relationship in more detail, focusing on the key factors for effective socio-ecological integration for a better intergenerational justice.

2.1.1 Landscape reasoning applied to small scale problems

Complex ecological problems are easily detected at the landscape scale. Degradation tends to have a pattern that is observable at the regional scale through synthesis indicators. In order to detect and prevent a regional ecological problem, local landscapes, particularly boundary (ecotone) landscapes are monitored and analyzed together to understand regional dynamics. Hence, a landscape approach doesn´t have an associated scale *per se* (Forman, 1995). Small-scale (local) problems are, by definition, undetectable at the landscape (regional) scale, but from the point of view of local communities the problem may be affecting their whole locality. Using the landscape approach, the mosaic of spatial units inside a locality could be analyzed as dynamic entities with their own interactions, fluxes and compositions, leading to their local landscape properties and certainly to better solutions (Zurlini et al., 2006).

For example, a very common situation that restoration or conservation initiatives have to face in implementation is, (despite the techniques and models used for determining the assigned project area), that a designated territory will always encroach onto someone else's property or land use interests. In the simplest case, if there is only one owner (usually a state or public entity) the task may take some time. But if an initiative is expected to work and the project is approved, there is a good chance that it will be incorporated into a permanently established, legally ensured, and environmentally managed area (e.g., a park, reserve, or nature sanctuary). Unfortunately, this is not the typical circumstance. Usually, the cadastral layer is overlapped on ecosystems, and their boundaries rarely coincide. Ecological problems are elsewhere in the Neotropics, but they are primarily associated with human activities (Morton et al., 2006) and historic land tenure transgressions. For this reason, phases of the processes to restore ecological services or manage an ecosystem, there will be a need for a period of exclusion (time without human presence) that could vary between months and permanency. But even locally, the desired exclusion area can be distributed among several owners with diverse and particular socio-ecological demands. The landscape approach suggests that the socio-ecological conditions of all land owners should be characterized and integrated to develop a consistent land-use management program.

Social variables could be related to an owner's income, economic activities, property area or any relevant combination of factors. In the same way, ecological conditions of each property could be described by specific factors such as diversity, focal ecosystem area, or degradation risk. Therefore, a cross-tabulation of this knowledge could reveal the different socio-ecological categories. Through mapping, this information can enable identification of concrete strategies to establish the relevant ecological and communitarian connection to maintain a restoration or conservation project over the time in a shared space. Complementarily, each local landscape could provide comparison points to assess degradation and potential actions across similar localities in a region.

2.1.2 Adapting Neotropical landscapes through voluntary carbon markets for both native afforestation and reduced emissions from forest ecosystems degradation and deforestation (REDD Plus)

There are widespread anthropogenic agents of deforestation and ecosystems degradation in the Neotropics in terms of magnitude and direction: a) linear large-scale agricultural and livestock expansion patterns; b) the linear clearing from cooperative road-building projects by colonists; c) common land-clearing in the fishbone pattern for farm development, and d) the dendritic spread of logging operations. From these processes and the uneven distribution of efforts at afforestation, the Neotropics contains only 40% of its original forest (Fig. 3). Moreover, heuristic accounting for REDD+ using geospatial technologies and ground-truth methodologies for Neotropical deforestation and forest degradation that should be rendered in terms of: a) time-series cumulative deforestation rates of change in accordance with its agents and drivers; either as linear, exponential, logarithmic, potential, logistical or polynomial functions; b) total leaf area; c) water content and quality of the vegetation; d) fragmentation patterns; e) biodiversity estimations; f) socioeconomic welfare indicators; and g) sound metrics about above- and below-ground carbon pools and stocks.

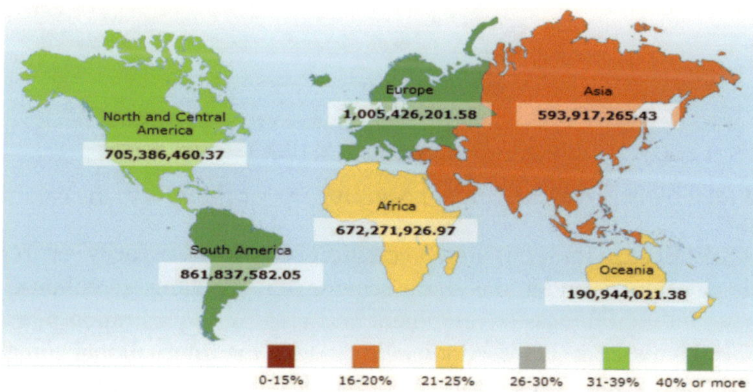

CONTINENT	LAND AREA (ha)	FOREST AREA	FOREST %
AFRICA	2,974,011,000.00	674,419,000.00	22.68
ASIA	3,091,407,000.00	592,512,000.00	19.17
EUROPE	2,214,726,000.00	1,005,001,000.00	45.38
NORTH AND C.A.	2,134,979,000.00	705,393,000.00	33.04
OCEANIA	849,094,000.00	191,384,000.00	22.54
SOUTH AMERICA	1,746,292,000.00	864,351,000.00	29.06

Fig. 3. Global and Neotropical 2011 forest remaining area. Data Source: World Forest Scenario Tool from Global Green Carbon & FAO 2010.

These above-mentioned features are important to determine the health of Neotropical forests, as landscape ecologists have long known that different landscape patterns affect biogeochemical cycles and ecological processes differently. Therefore, these ecological-economics attributes are keystone indicators since the price of voluntary carbon units to be traded into metric tons equivalent ($MtCO_2e$) are driven by these features within the voluntary carbon markets for REDD+. Thus, the potential effect of turning blue areas of the map into yellow or red would stream revenue to the Neotropics, if transparent mechanisms and demand for carbon certificates are effectively supplied (Fig. 4).

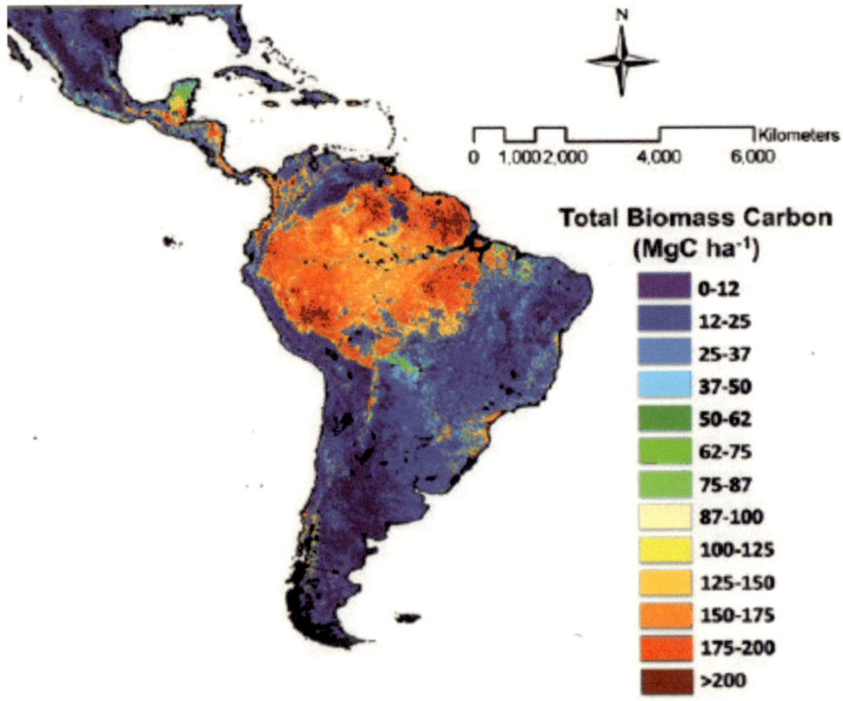

Source: Saatchi S. S. et al., 2011/ PROC. NATL ACAD. SCI. USA.

Fig. 4. Map of total biomass carbon content counting up for the 49% in the Neotropics,

Carbon-offset certificates for emissions reduction, either voluntarily or regulated, are financial tools developed under the trade scheme for reducing greenhouse gas (GHG) emissions by expanding biomass development and making sequestration permanent while phasing out GHG emissions. Transactions take place in the international climate market to ensure proper compensation for climate, communities, biodiversity, and water in order to phase out, avoid, and sequester GHG emissions. This mechanism is based on nature elongation and market strategies to enable better economic incentives for sequestering carbon or avoiding deforestation and ecosystem degradation. Moreover, these markets facilitate the availability of financial resources for implementing ecological actions, diagnosed through ecological informatics software applications over landscapes and cultures dwelling on terrestrial, freshwater and marine ecosystems for footnote blue carbon. These specific actions at a project site, or at national, regional or global scales, should focus on reducing gas emissions by avoiding deforestation.

The carbon-offset certificates are then be standardized in the market for all GHG in $MtCO_2e$ known as voluntary carbon units (VCUs) to be validated through the voluntary carbon standard (VCS) or climate community and biodiversity standard (CCB) and many more as TUV SUD and ISO. Carbon offset certificates are negotiated at a determined price in the voluntary emission reduction (VERs) markets through verified emission reduction purchase agreements. The certificates allow the incorporation of opportunity cost over the damage produced by GHG warming potential in the Neotropics into the world's economy for the first time. The gases can be unnaturally produced by various economic sectors such as energy generation, transportation, agriculture, deforestation, infrastructure construction, ecosystem degradation, and livestock rearing.

Integrating carbon offsets with landscape ecology can improve guidelines for Project Design Documents (PDD) or Project Appraisal Documents (PAD) in order to determine the baseline of carbon stocks accurately and generate verifiable credits. This is what is needed for effective funding in the Neotropics to have baselines and catalyze investment in forest protection. Essential data about the Carbon pools and stocks, additionality or avoidance of emissions, permanence of reduction, and mitigation of leaks demanded by any carbon project to be traded can ensure fair trade compensation for communities near forests or in buffer zones negatively impacted by climate change. Carbon-offset certificates in the Neotropics, therefore, are needed to synergize GHG-reduction with the global economy through applied ecological economics.

These actions can catalyze an adaptive convergence among better climate, community adaptation, biodiversity elongation, and carbon-market conditions. Even if perverse incentives exist, adequate market regulations and communities´ self-regulation will ensure that benefits and advantages will be fairly produced and properly allocated based on sound scientific, technological, legal, and financial frameworks. Ultimately, landscape ecologists that apply and deliver these market-based mechanisms efficiently would be maintaining and improve ecosystem structure and function in the Neotropics, one of the main sources of ecosystem services and resources for global socioeconomic systems. No planet, no markets!

2.2 The policy-making ecologies for forests in the Neotropics

The world's economy, biodiversity, and human population are at stake if nothing is done to mitigate climate-change effects. Climate change is a common challenge to all countries and can be solved only through international cooperation. International initiatives include United Nations Framework Convention on Climate Change (UNFCCC) of 1992 (UNFCCC, 1992), the Kyoto Protocol (KP) of 1997, and the agreements for Reducing Emissions from Deforestation and Degradation, plus Conservation, Sustainable Management of Forests, and Enhancement of Forest Carbon Stocks (REDD+) and Land Use, Land Use Change and Forestry (LULUCF). The carbon market created in the Kyoto Protocol has only proven to be a partial solution to the reduction of the GHG/CO_2 emissions which affect the world's climate, environment and the agricultural economy (De la Torre et al., 2009). However, more committed initiatives will be necessary to effectively reduce climate change and the global "free rider problem" (Ostrom, 1990).

Tropical forests play an essential role in the biogeochemical global carbon cycle and must be at the forefront of the discussion to reduce global GHG emissions and to mitigate climate-change effects. Tropical forests include a variety of ecosystems. These forests, while significant for their biodiversity and other exceptional characteristics, store immense amounts of carbon. Studies show that forest ecosystems store greater amounts of carbon than the total amount of carbon in the atmosphere. When forest areas are cleared or degraded, this carbon is released into the atmosphere. Some calculations from the Intergovernmental Panel on Climate Change (IPCC) show that annual GHG emissions from deforestation and degradation of forests account for 20 percent of global emissions, a figure about equal to the emissions of the global transportation sector. The consensus is that in order to achieve significant GHG reductions worldwide, the international community must focus on halting and degradation of tropical forests (O'Sullivan & Saines, 2009).

The 13th conference of the UNFCCC Committee of the Parties (COP) was held in Bali in 2007. Governments agreed to devote two years to the consideration of "policy approaches and positive incentives" for the reduction of emissions from tropical deforestation. These incentives could become part of an agreement implemented after the expiration of the current obligation under the Kyoto Protocol (KP). In 2009, at COP 15 in Copenhagen, Denmark, parties agreed only REDD+ projects were important and should be financed (UNFCCC, 2009). Currently, UNFCCC member states are considering different policy approaches for REDD+. However, the KP maintains a restricted allocation for forest projects such as REDD+ that can eventually produce certified emissions reductions (CERs). The international alternatives are intended to create mechanisms that will create positive incentives to tackle the reduction of GHG by avoiding deforestation and degradation (O'Sullivan & Saines, 2009). That is why some of the UNFCCC, KP and Bio-Carbon Fund goals are to "improve livelihoods, restore ecosystems, adapt to climate change, remove CO_2 from atmosphere, and provide access to carbon market" (Baroudy, 2009).

Experts recommend market-based mechanisms as one of the most promising ways to provide the considerable financing needed for REDD+ (O'Sullivan & Saines, 2009). Practitioners stress the value of forestry and land-use changes for coping with climate change, thereby improving both livelihoods and environment. They also found that forestry and LULUCF projects are among the few opportunities for impoverished rural communities and indigenous peoples to access the carbon market. However, there are tight restrictions on forest projects under the Kyoto Clean Development Mechanism (CDM) and the international environmental system, including European Union Emissions Trading System (EU ETS). These have concerns with REDD+ and LULUCF projects in all developing countries. These concerns range from CO_2 leakage accumulated in forests projects, uncertain land tenure, and the lack of effective governance capable of monitoring and enforcement in forest areas. The main policy problem is that even though tropical forests provide critical environmental benefits such as wildlife habitat, biodiversity, water and carbon sequestration, they also provide exploitable resources, such as timber, fuel, and agriculture. Since the 1980s, several international organizations have adopted agreements in support of environmental and economic interests contingent on the principles of sustainable forest management. Nevertheless, deforestation in tropical countries continues with alarming speed.

Experts agree that non-market based funds must compliment market approaches if REDD+ policy goals are to be achieved (O'Sullivan & Saines, 2009). Another reality is that REDD+ projects must be sufficiently profitable to compete with other crop or timber businesses (Sandker et al., 2010) in order to meet the interests of local farmers and other investors. Developing countries require consistent funding for REDD+ projects so that they can access national and international carbon markets (O'Sullivan & Saines, 2009).

A parallel objective of this chapter is to present to the international policy advisor for the Neotropics an analysis of environmental, legal, and institutional aspects relating to climate change. This is especially important for developing countries with tropical forests, high biodiversity levels, and human communities whose dependency on the natural resource is fundamental to their well-being. The chapter will be useful to carbon project developers, non-governmental organizations (NGOs), communities, and indigenous peoples. The combination of landscape ecology and offset-payments for ecosystem services will help improve efficiency in GHG emissions reduction. It will also be useful for assessing difficulties and providing recommendations for future climate change regulations in Neotropics. The Neotropical governments will need to obtain international technical cooperation, and create space for GHG emissions reduction projects in the international carbon market created by the Kyoto CDM and the Voluntary Carbon Market. To accomplish these goals, developed countries and the International Climate Change System need to be more aware of the Neotropics' fragile position as most of the countries that comprise it are impoverished and possess tropical forestland (De la Torre et al., 2009).

Unfortunately, opening space for CO_2 reduction projects has proven especially difficult following the COP 15 negotiations in Copenhagen. At these negotiations developed countries were expected to take the lead in developing a carbon mechanism to effectively control climate change, and agreed only to fund REDD+ projects (Figueres & Streck, 2009). Consequently, Neotropical countries' national policies should address agricultural and forestland issues to obtain a more comprehensive and effective approach to climate change. This will be especially important to help reduce leakage in LULUCF and REDD+ projects.

While this recommendation may seem obvious, in countries like Colombia, national policies addressing climate change on forest lands are distinct from the agricultural and rural development law reforms. Many Neotropical developing countries feel uncomfortable addressing both issues jointly because they believe it will negatively affect their agricultural production. However, this is far from the truth; agricultural production is being affected first and most significantly with climate change (De la Torre et al., 2009). Thus, an integrated agricultural and forestland policy is fundamental for Neotropical developing countries. In addition, its absence creates confusion and concern among indigenous people, land tenants and other stakeholders, making any international carbon market or national REDD+ or LULUCF effort unproductive.

This study depicts an integrated analysis of the current situation in some Neotropical countries current situation and potential implementations of CO_2 /GHG emissions reduction mechanisms. This highlights the importance of forest projects for the Neotropics and suggests the importance of holistic analyses that show original and current distribution of Neotropical ecosystems as a tool to prioritize biodiversity conservation plans while mitigating climate change. In the Colombian case this research already exists and should help to establish government policies, to assist project developers and VCM investors in selecting and prioritizing project locations for ecosystem restoration and climate change adaptation (Wyngaarden & Fandiño-Lozano, 2005); the importance of Free Prior and Informed Consent (FPIC) for indigenous peoples and its value for the implementation of REDD+ and LULUCF projects. Moreover, these heuristic studies about carbon property rights of indigenous peoples' reservations and the history of their property rights correlated with the legal framework and tools available in a Neotropical country like Colombia will foster emissions reduction contracts.

Here, we suggest a suite of recommendations for policy strategy, regulations, and mechanisms to channel investment for the effective operation of the carbon trade business in Colombia and other Neotropical countries the project approach, with private sector involvement (Industry, Communities, NGO's & Project Developers) accompanied by results-based contracts such is Colombia's Payment for Environmental Services (PES), as the most effective solution while coordinated command and control regulations are implemented (Richards, et al. 2006 ; Wunder, 2007). This conclusion is based on findings that local efforts initiated by forest communities or those inhabitants that receive benefits from the forest beyond CO_2 emissions reductions have proven successful in Colombia as well as the basis to finance REDD+ initiatives whereas a carbon market is supported by the international and various sectors of the economy.

The need of inclusive forestland, rural development and mining law reforms along with effective use of FPIC for indigenous peoples and other ethnic groups proposes the enforcement of current consumer protection law arguing that while the private sector may present some resistance to GHG emissions reductions, the final consumer also will force environmental responsibility in the business sector. While tax advantages for GHG emissions reduction may seem attractive, Neotropical countries must establish a clear baseline (Passero, 2009) and GHG emissions reduction projects should not be managed and owned solely by the government given the risk that government projects may be targeted by guerrillas, as in the Colombian case. It suggests that at national levels, the government role should be limited to formulating coherent agricultural and forestland policies, contributing advice and partial co-finance for REDD+ and LULUCF projects, aimed at raising awareness and educating about climate change, and facilitating space for private carbon markets to flourish.

2.2.1 International Emissions Trading (IET)

Article 17 of the KP establishes the third mechanism to allow trades or transfers of the emissions rights of assigned amount units (AAU) among Annex I countries. This system has evolved into what is known as international emissions trading (IET). One concern with this type of AAU trading is that it will not be sufficiently green because it is not linked to emissions reduction by either the sellers or buyers. For that reason, a green investment scheme (GrIS) was designed to "earmark AAU revenues for environment-related activities". In this scheme, sellers have the option either to reinvest their proceeds into measures that further reduce their emissions or to support measures with other environmental benefits. Currently, other IETS have been implemented in Australia, the United Kingdom, and Denmark. The IETS and AAU trading became so popular that a Voluntary Carbon Market

(VCM) in emissions reductions has been created and the private sector is becoming the main player (Freestone, 2009; Simonetti & de Witt Wijnen, 2009).

2.2.2 Voluntary Carbon Markets (VCM)

The voluntary GHG emissions reduction market or VCM is a voluntary GHG emissions reduction by one party and the subsequent sale or transfer of these reductions to another party. In exchange for this transfer of reductions, the receiving party provides either financial compensation and obtains the right to claim and use their benefits or to transfer the reductions to another party. Emissions reductions or "offsets" are decreases of GHG emissions by the removal and storage of GHG emissions from the atmosphere (Passero, 2009). Since the GHG emissions voluntary offset began, companies around the world have been investing at an ever-increasing rate in projects that will reduce their footnote at "Carbon footprint"(Hamilton et al, 2008).

There are several VCM strategies, tools and venues such as the Chicago Climate Exchange (CCX), the New York Climate Exchange (NYCX), the Montreal Climate Exchange (MCeX), the Northeastern Climate Exchange (NECX) a manifold of voluntary but legally binding international and U.S.-based cap-and-trade system and "over the counter (OTC) voluntary carbon markets" in which transactions occur on a deal-by-deal basis at a trade level (Bayon, et al., 2006). Some of the sellers in the VCM consist of retailers selling offsets online, project developers or conservation organizations that hope to connect the influence of carbon finance with buyers from the UN Global Compact for example, and potential CDM or JI projects that could neither qualify nor wait for the regulated market. There are also a wide variety of buyers such as international corporations, nonprofit organizations as market catalysts, and individuals. The buyer's incentive rises from their interest in good publicity, good public relations, philanthropy, and the ability to resell and broker credits at a profit. Combining figures from both VCMs, the United States acquired $93 million worth of credits in 2006 (Hamilton et al, 2008; Passero, 2009).

In 2007, $331 million worth of credits were traded and it is expected to increase in the coming years. The VCM's appeal is its innovation, flexibility, and lower transaction costs that benefit both buyers and suppliers, unlike to the CDM or JI regulatory requirements that create cost and time barriers that exclude project participants from accessing the markets (Hamilton et al, 2008; Passero, 2009). The most popular type of GHG emissions reduction in this market comes from renewable energy, energy efficiency, forestry, sustainability benefits, and methane destruction. One of the concerns relating to projects within the VCM is that there is no standard process for monitoring and certifying these projects to avoid leakage or test the results. Thus, baselines are needed for GHG emissions reduction accounting and requirements for registry standards in projects (Passero, 2009). The VCM's flexibility is crucial as a laboratory to be scaled out for the creation, evolution and strengthening on emissions-reducing forestry projects throughout The Americas.

Deforestation alone represents 20 to 25 percent of global carbon emissions. Currently, forest project developers and conservation organizations are interested in financing GHG emissions reduction through reforestation and forest protection projects ("avoided deforestation projects") or REDD+ through the VCM after they discovered that these projects do not fit into the CDM market (Hamilton, 2008). The development of GHG accounting standards and registries have been essential to quantification and verification of GHG emissions reduction. Registries are advantageous because they centralize accounting data, provide transparent monitoring of reductions, and track transactions in order to avoid the double counting of reductions. Registries also provide legal and policy information that ranges from the crafting of GHG emissions-reduction contracts, known as verified emission reduction purchase agreements (VERPAs), to regulatory measures and government policies. Accounting standards for native forest projects included in the VCM are each country's baselines, "additionality" tests, permanence, verifiability, and leakage control. Baselines and additionality tests work hand-in-hand to assess whether additionality exists in a project and requires the establishment of a baseline to reveal "business as usual" or "without a project scenario". Permanence refers to the duration of GHG reductions and the capacity to address any setback in emissions reductions previously verified as a reduction. Permanence is especially significant in soils, grasslands, or forest projects that entail long-term GHG reductions and back-up storage to fulfill VERs. Verifiability requires independent or third party corroboration of the accuracy, permanence and ownership of the GHG reductions overtime. Leakage avoidance aims to reduce or avoid displacement of GHG emissions caused directly by the GHG reduction offset project (Passero, 2009).

The VCM is contributing greatly to climate-change mitigation. Even under the most demanding reduction targets it is doubtful that all GHG emissions-producing activities will be regulated (Hamilton, 2008). Nonetheless, the VCM has obtained GHG emissions reductions, raised awareness and engaged individuals and businesses in emissions reductions. This has also served as the forum for developing important market infrastructure such as purchase/sale contracts, GHG accounting standards protocols and registries (Passero, 2009). The importance of the VCM in LULUCF and REDD+ to climate change cannot be overemphasized. And finally, since the Kyoto CDM not includes REDD, the VCM is the only option to assist developing tropical countries with forest projects.

2.2.3 The Global Environment Facility (GEF)

The Global Environment Facility (GEF) managing system is a permanent and independent secretariathoused within the World Bank. In 2007, the COP 7 in Marrakech asked the GEF to be the financial mechanism of the UNFCCC to start and operate the Special Climate Change Fund (SCCF), the Least Developed Countries Fund (LDCF), and the Adaptation Fund. The GEF has become instrumental in achieving convention goals as it manages Bank funds addressing global environmental issues. Donor countries supply these funds to finance four basic areas: part of ozone depletion not covered by the Montreal Protocol Multilateral Fund, biological diversity, international waters and climate change. In 2002, two new areas were added: land degradation including desertification and deforestation, and persistent organic pollutants (POPs). The latter is the least known in the Neotropics. There is a lack of ecotoxicological research about POPs or pesticide use over the Colombian rainforest targeted to illicit crops, co-funded by the U.S. Department of the State and Colombia's government.

Nonetheless, GEF funds are considered insufficient to cover environmental challenges a recent study found that the GEF role in mitigating climate change is minor. Even though, it plays a key role in "cofinancing and transforming some markets for energy and mobility in developing countries." In accordance with the UNFCCC mandate, the GEF finances "incremental costs," meaning a country contributes the amount it would have supplied to the "least cost" but climate-damaging project and the GEF finances the additional or "incremental cost" of the new climate-friendly technology (Freestone, 2009).

2.2.4 Special Climate Change Fund (SCCF)

The Special Climate Change Fund (SCCF) was established in 2001 to complement other UNFCCC funds that financed projects relating to: adaptation; technology transfer and capacity building; energy, transport, industry, agriculture, forestry, and waste management; and economic diversification (Decision 7/CP.7.) (UNFCCC, 2010).

2.2.5 World bank Climate Investment Fund (CIF)

At the request of G8 donors, in 2008 the World Bank created the Climate Investment Fund (CIF) to invest in developing countries. The funds are distributed as grants, concessional loans, and risk mitigation instruments which are managed through the Multilateral Development Banks (MDB) and the World Bank Group. Two funds are included, the Clean Technology Fund (CTF) and the Strategic Climate Fund (SCF). The CTF finances projects or programs oriented to demonstrate, deploy, and transfer low-carbon technologies with a long-term CO_2 savings. The SCF finances a broader scope of projects and programs to assess creative approaches to climate change mitigation (Freestone, 2009).

2.2.6 World bank biocarbon fund

The BioCarbon Fund, part of the World Bank's Carbon Finance Unit, is a trust fund with public and private financing managed by the World Bank. The fund finances projects intended to absorb or maintain CO_2 in forests, agricultural areas, and other ecosystems. The fund provides financing to developing countries with very few opportunities to participate in the Kyoto CDM as well as to countries with economies in transition through JI. Although still in its evolutionary stages, the BioCarbon Fund is proving how LULUCF projects might generate high quality emissions reduction (ERs), provide benefits for the environment and communities' livelihoods, and persevere investments for the long term, which will allow these projects to be measured, monitored, and certified. Mid-level emitter countries that sign the KP are expected to access this market, but in practice Kyoto's CDM is highly restricted for forest projects. Currently, there are only three pilot projects. REDD+ is preparing forest projects in the VCMs in hopes of accessing Kyoto's CDM or a Post-Kyoto market agreement.

2.2.7 Forest Carbon Partnership Facility (FCPF)

Established in 2008, the Forest Carbon Partnership Facility (FCPF) is a public/private partnership working in conjunction with other programs; the UN-REDD Programme set up by FAO, UNDP and UNEP; and the Forest Investment Program (FIP) (UNFCCC, 2010). REDD+ is divided into three phases: readiness, capacity, and operations. The readiness phase involves diagnosis of the current situation and formulation of a REDD+ strategy and a monitoring system. The capacity reform and investment phase examines the countries' promotion of REDD+, institutional strength, sustainable forest management, forest governance, and investment outside the forest sector. The mission of the FCPF is to assist tropical and subtropical forest countries in the development of systems, policies, and strategies for REDD+; design of monitoring systems; establishment of national management arrangements; and inclusion of all key stakeholders. The fund will also award the countries with performance-based payments for REDD+ (Bosquet & Andrasko, 2010). The World Bank is the trustee for the Readiness Fund and the BioCarbon Fund, provides secretarial services and technical support to REDD+ country participants, and conducts due diligence in fiduciary policies and environmental and social safeguards (UNFCCC, 2010). Presently, the FCPF focus is on REDD+ readiness. This is expected that through the assistance of the BioCarbon Fund, the FCPF will provide payments for VERs from REDD+ programs in countries that have achieved progress towards REDD+ readiness.

2.2.8 Forest Investment Program (FIP)

The Forest Investment Program (FIP), established in 2009, provides financing to developing countries for the policy formulation necessary to reach their REDD+ goals, as well as to assure sustainable forest management (FIP, 2010). The FIP is part of the World Bank under the SCCF and the CIF (FIP, 2010). The program focuses on the public sector's formulation of critical forest policy assuring: (a) regulatory and institutional frameworks to support private and public sector investments; (b) private sector investment in sustainable forests and forest landscape management in reforestation, afforestation and conservation through grants, tax relief, and subsidize loans; and (c) finance for forestry activities that include social and environmental benefits. In essence, the FIP promises "to contribute to multiple benefits such as biodiversity conservation, protection of the rights of indigenous peoples and local communities" (CIF, 2010; BIC, 2010).

2.2.9 Wealth Accounting and Valuation of Ecosystem Services (WAVES) program

Payment for ecosystem services (PES) are important for the Neotropics where communities whose livelihoods depends upon their natural capital exist. As a result, a national accounting PES system unveils winners and losers on supply and demand of services when correlated into a rate of changes over ecosystems.

The World Bank through the footnote WAVES program bolster pilot cases about natural capital for public policy delineations addressed to recover and elongate ecosystems services over time. The welfare accounting of natural capital is measured by revenue of utility margin derived from determining effective demand of functional ecological services. Currently, the World Bank is supporting a readiness plan for Colombia and Costa Rica to become the premier vehicles of evidence based impacts to effectively apply and replicate payment for ecosystem services as integral part of public policies in the Neotropics.

2.2.10 Property rights and carbon property in civil law systems: The case for Colombia

This section will explain general property rights protected by Colombia's civil law system and property rights of indigenous peoples and a myriad of ethnic groups. We focus on the application of the law to owners of carbon property rights in Colombia and those who buy and sell environmental services through the appropriate legal contracts. Article 713 of the second book of Colombia's Civil Code relating to property rights defines accession as "a way to acquire property. The owner of one good passes to be the owner of what that good produces or adheres to it. The products of a good are civil or natural."

Article 714 defines natural fruits as those that "come from nature with the help or not of the human industry". Also, Article 716 states that "the natural fruits of a good belong to its owner" (C. Civ. Col., 1887). Thus, the property owner is required to arrange for carbon storage since, by accession of the natural fruits, the CO_2 created in the biogeochemical cycles adheres to trees and soils through its leaves and roots. The owner is required to arrange for environmental services such as carbon storage. These carbon storage agreements and the environmental services they provide are made with property owners under contractual guidelines such as the payment for environmental services (PES). These agreements are consistent with the civil code and the Colombian legal system.

However, there has been a legal complex gap based on belowground carbon due to the Colombia's and many other Neotropical mining regulation, that states that all minerals below ground deserves royalties for its state countries. This situation is critical because in this gap, inorganic Carbon may be claimed by local governments, losing project site incentives for trading when a baseline has been surveyed.

2.2.11 Indigenous peoples property rights in Colombia

In 1991, the constitution of Colombia established legal recognition of a multicultural society. This recognition was necessary to ensure the permanence of human rights, as well as property rights. Now indigenous peoples are owners of carbon rights and have the authority to allow carbon projects on their property with approval of the indigenous authorities on their reservations. Article 63 of the constitution establishes that "The goods of public use, the natural parks, the communal land of ethnic groups, the land of indigenous reservations, the archaeological patrimony of the Nation, and other goods determined by the law are unalienable, imprescriptibly, and cannot be attached or seized." This measure guarantees that indigenous peoples' property in Colombia cannot be sold or acquired by other people or groups and will not be subject to the statute of limitations. Article 329 and Article 330 numbers 5 and 6 of the constitution also state that indigenous territories are collective property and are governed by their own indigenous council. The council has responsibility to both "5. Oversee the preservation of their natural resources" and "6. Coordinate programs and projects promoted by other communities in their territory" (C. Pol. Col., 1991). These constitutional mandates give indigenous peoples the authority to allow any project on their property, including projects relating to the elongation of natural resources such as A/R, REDD+, and LULUCF. They also have authority to coordinate projects and programs with regional or local institutions and other communities nearby. Thereby, indigenous peoples can exercise their right to give FPIC as well as to take responsibility and ownership of the projects that they decide to approve.

In general, with the exception of some nomad indigenous communities located in the Guaviare-Amazon region (Martinez, 2010), indigenous reservations have two kinds of property: family land and collective land. The collective land is usually forestland that has been typically used sustainably and in ways that do not damage the environment. Indigenous peoples are owners of land that cannot be sold, acquired by prescription, or seized, and this permanence guarantees better participation in carbon storage projects and contracts such as PES. They can also organize themselves through legal entities and sell their carbon storage outputs and other forest by-products.

2.2.12 International support for A/R, REDD+ and LULUCF projects

REDD+ and LULUCF projects present a more sustainable and environmentally consistent alternative to reducing GHG emissions than other methods for the production of biofuels such as soy biodiesel which also slightly reduce GHG emissions. According to several reports soy production for biofuels and other uses have triggered human rights violations in Argentina, Brazil, and Paraguay. The ecotoxicological use of chemical pesticides on soy plantations in South America is affecting the public health of indigenous communities (Ketabi, 2009). Clear-cutting forests for the production of soy also threatens their subsistence (Wallace, 2007). REDD+ and LULUCF projects present a much better alternative for GHG emissions reduction, watershed policy-making, biodiversity use, integrated water resources management and improvement of communities' livelihoods. If the international system favors and prioritizes funding of forest projects in mid-level emitter developing countries in the tropics, it might be achieved through several of the United Nations Millennium Development Goals (UN, 2010).

2.2.13 Participation on International and national carbon markets

The EU ETS, the major market for carbon credits, prohibits forest-based carbon credits (O'Sullivan & Saines, 2009). This is regrettable because Colombia has more potential for participation in the carbon market with GHG emissions reductions through forest projects. High-level GHG emitters like China deliver more GHG reductions in the power sector and, because of this, have greatly benefited from the Kyoto CDM that sells CERs to the EU ETS. Correcting this inequity at the international level has proven difficult. Mid-level emitter countries must assume a more active role in the post-2012 international negotiations to ensure increased market participation for REDD+ and LULUCF projects among developing countries. As mentioned previously, half of the Colombia's tropical ecosystems, and remnants which include islands, are extremely fragile. Because of the numerous islands in the region, experts recommend that Latin America and the Caribbean (LAC) should be "ahead of the pack" in climate-change reductions since the region currently suffers disproportionately from negative climate-change, intensive mining impacts and this will only be worsened.

To solve some of these issues, LAC countries must use a regional approach. The weather insurance market is currently underdeveloped in LAC and experts recommend incentivizing and establishing regulations for this high-risk market. This approach would allow these countries to engage completely in the insurance business and ultimately in the global carbon market. But first committed and continued international technical assistance for basic research is needed (De la Torre et al., 2009). This research provides critical insight needed by the weather-insurance business and UNFCCC to establish a clear baseline for accounting GHG reductions. It is a duty for Colombia and other Neotropical countries to report in detail GHG-CO_2 emissions inventories to establish the countries' baselines. The baseline is imperative to keep an accounting of reductions of

GHG emissions in the UNFCCC, the Kyoto CDM and VCM. These studies will also help collect useful and consistent information required for forest projects, leakage mitigation, forest management, agriculture, and the sustainability of communities dwelling in the forest buffer areas.

Progress was made at the December 2007 COP 13 in Bali where delegates agreed to "invite parties to further strengthen and support ongoing efforts to reduce emissions from deforestation and forest degradation) on a voluntary basis" (BioCarbon Fund, 2009). This decision opened the path for a more comprehensive international approach toward REDD+, LULUCF, and other forest projects. Later, at the 2009 COP 15 in Copenhagen, industrialized countries agreed to only support REDD+ projects (UNFCCC, 2010). This decision provides a small opening for international environmental cooperation on forest projects in developing countries to reduce CO_2 emissions. The pronouncement also recognizes the need to provide support to REDD+ projects in countries with serious concerns about the impacts of climate change. This decision will attract project developers interested in forest projects in the VCM and the Kyoto CDM.

Support and finance in these areas will complement Colombia's efforts at the national and local levels to implement REDD+ and LULUCF projects. Consequently, a team formed from members of the private sector and NGOs in Colombia is creating its own exchange-like market to facilitate the flow of carbon credits and to provide the initial baseline financing for Colombian forest projects that can participate in VCMs with VERs. This is necessary because current CDM projects with CERs restrict space for forest projects. For that reason, through the auspices of the GEF, non-profits, the Colombian Ministry of Environment, and the Inter-American Development Bank (IADB) launched their strategy on "Colombia's underwhelming carbon market presence" in July 2010 (Peters-Stanley, 2010).

2.2.14 National governmental involvement

Some experts recommend that the central government should take ownership of all REDD+ carbon credits (O'Sullivan & Saines, 2009). However, governments' financial resources are scarce and the potential for increased bureaucracy and corruption in monitoring and accounting is significant. In some countries guerrilla groups will most likely target projects if the government is the exclusive owner or demands royalties. Also as a developing country, Colombia has little ability to enforce regulations and the nationalization of forest areas "creates open access resources where limited-access to common-property resources had previously existed" (Ostrom, 1990). In addition, a local community project approach is the most effective (Robledo & Tobon, 2008). Of course there are main "command and control" regulations necessary to maintain a climate-change policy at national and regional levels. Yet, in the end, forest inhabitants and forest-adjacent communities are the ones to either protect or destroy the forests. Thus, they must be included in project formulation and implementation in order for projects to be successful.

2.2.15 Project site approach

The project approach allows local communities and project developers to obtain direct benefits from REDD+ if their projects are doing well. This private investment is required to make a significant dent in GHG emissions reduction from REDD+ and LULUCF projects at a global level. Advisory to the private sector investment will require that policymakers generate long-term demand for REDD+ credits (O'Sullivan & Saines, 2009).

LAC/Neotropical countries must advocate for increased participation in the Kyoto and post-Kyoto regulatory carbon market. Unfortunately, given the current international ambience, this is unlikely unless the countries with major GHG emissions craft more committed goals for climate change. Currently, there is no real market to participate in Kyoto's CDM selling CERs in the EU ETS. For now, in some Neotropical countries, groups like Colombia's NGOs, Ministry of Environment, and those working with the IADB, make efforts to strengthen the carbon market. In July 2010 the Mechanism for Voluntary Mitigation of GHG Emissions in Colombia was launched to encourage participation by the private sector with forest projects in the VCM offering VERs.

2.2.16 Payment for environmental services, legal contracts and price drivers

Colombia's limited technical and financial resources require the implementation of new policy mechanisms. The command-and-control approach is important, yet by itself has not been enough to make the private sector and land owners change their forest management practices. However, the PES introduces an important alternative that complements the command-and-control regulations while catalyzing markets. Many Neotropical countries seek to solve some of these environmental issues applying the PES at the national and local levels (PROFOR, 2004; Richards et al., 2006; Wunder, 2007). In Colombia, experts found that currently the PES is used effectively only at the local level in watershed and CO_2 emissions reduction projects (Blanco, 2010). Many forests produce water for residential consumers and irrigation systems, help to maintain fisheries, and supply hydroelectric power generation which are essential to community livelihoods, business development, and ecological economics growth.

The PES has proven an effective contractual tool to provide ecological services because forests affect the quality and quantities of water flow, carbon sequestration, and biodiversity conservation (PROFOR, 2004). Therefore, a hybrid policy approach of practice-based incentives and results-based incentives or PES can help efficiently implement environmental protection and international climate-change mitigation goals. This will allow the government to coordinate with indigenous peoples and other local communities that are directly involved in the fight against problems like diminishing water supplies, CO_2 emissions and biodiversity conservation (Casas & Martinez, 2008). Because current REDD+ payments are insufficient to achieve UNFCCC, Kyoto, and national goals, these projects must be financed continuously from implementation to completion (Sandker et al., 2010). This will be necessary for the remainder of the financial aid to be delivered depending on the results-based payment for the PES. However, the critical point is governmental monitoring of the projects (Blanco, 2010). This is one of the primary reasons that national-level strategies and governmental programs in forests and climate change based on governments' initiatives have not been successful in Colombia (Watershed Markets, 2010). A consistent national and local monitoring system is needed to establish Colombia's baseline and to evaluate the effectiveness of all PES, REDD+ and

LULUCF programs and projects. Therefore, international financial assistance is needed to improve monitoring systems for use by government employees, project developers, NGOs, and independent companies. If all monitoring responsibility is left to the government, the program will fail because of ever-lasting bureaucratic inefficiency and insufficient training of employees. It will also be important to monitor for leakage and additionality. In practice, the legal instrument used to buy and sell carbon-storage services in Colombia is the PES, which is currently used in other kinds of watersheds protection and forest management projects.

There is a national tendency to use PES in watershed management and biodiversity conservation services (Casas & Martinez, 2008). PES "is a voluntary, conditional agreement between at least one 'seller' and one 'buyer' over a well-defined environmental service—or a land use presumed to produce that service" (Wunder, 2007). Because REDD+ projects require considerable amounts of forestland, the "seller" organizes itself as one legal entity, usually a public, private, and civil society partnership, to represent the community and take responsibility for the project. This entity facilitates the trade of products and environmental services such as watershed provisions or carbon storage (Robledo & Tobon, 2008). Several reforestation and conservation projects with private and public parties are being developed under PES (Blanco, 2010). For example, explain what Chaina is micro-basin project around the Iguaque´s Sanctuary of Fauna and Flora of Colombia, developed in 2005 through PES, solved a historic socio-economic conflict concerning the distribution of natural resources and enforcement of environmental rules (Casas & Martinez, 2008). PES is supported in Colombia by different laws titled here: (1) Articles 58, 79, 80, and 95 of the Colombian Political Constitution; (2) Ley 99 of 1993, Articles 42, 43, and 45; (3) Ley 165 of 1994 and Ley 216 of 2003; (4) Ley 139 of 1994; and (5) Ley 3172 of 2003, Tax Statute Article 158-2.

3. Conclusion

Climate change represents a serious international challenge especially for developing countries in LAC that are mid-level GHG emitters (De la Torre et al., 2009). Neotropical countries are in a fragile position with sensitive tropical forestland, high overall poverty rates, and impoverished communities. Neotropical landscapes are extremely valuable for their great capacity to store carbon dioxide and necessary for the preservation of the world's biodiversity and indigenous peoples. REDD+ and LULUCF projects appear to be one of the few opportunities for impoverished rural communities and indigenous peoples to access the carbon market and get an income source (Carbon Finance Unit & Forestry Team, 2008). The main policy issue is that forests provide critical environmental benefits such as wildlife habitat, biodiversity, and carbon sequestration, while at the same time they are an exploitable resource for timber, fuel, and agricultural production. Experts conclude that a market-based mechanism for REDD+ and LULUCF is the best solution (O'Sullivan & Saines, 2009).

However, in order to create carbon markets mechanisms, continuity in finance from non-market-based funds will be necessary to meet REDD+ policy goals. At the same time, international technical cooperation will be required to assist LAC countries in different

UNFCCC and KP commitments ranging from baseline reports and climate change policy to GHG emissions and Global Warming Potential reduction projects such as REDD.

4. Acknowledgments

Like most chapters, this one has been years in the inception and has many parents. One of them, Richard A. Meganck whose vision of a better Colombia started long time ago when he volunteered as a U.S. Peace Corps in this country. His unfaltering help towards a reinvented cohort of Colombian scientists is now reaping rewards on this writings. We are also grateful to our parents Jose Arturo Restrepo Ospina, Lucero Aristizabal Botero, Nelly Lucia Rodriguez-Vega and Melquisedec Acuña Cuervo for helping us constantly in the long run of our careers. We also want to thank Ana Maria Quintero Caicedo and Liliana A. Zuniga-Venegas for editing and reviewing this document. Also, the Applied Ecology for Tropical Resources - Ecotropics, an international non-profit organization devoted to empower Pantropical communities facing climate change. *Ecotropics.org* applies the latest advancements in ecological informatics, ecohydraulics, renewable energies, integrated carbon and biodiversity offsets to help develop solutions, so local communities can adapt to and mitigate changing climates. This way, Ecotropics develops projects on payments for ecosystem services and natural resources such as water, Carbon stocks, food, fibers, natural medicines and environmental-friendly biofuels based on principles of sustainability and planned biodiversity use. Authors are grateful to Ecotropics donor's base which supported the making of this book chapter. Last but not least, we highlights the use of data and services to make our regional analysis provided by IABIN. As well as Valentina Hegestad's part of this article was submitted to partially satisfy the requirements of a Master of Laws degree at The George Washington University

5. References

Ashford, N. & Caldart, C. (2008). *Environmental law, policy, and economics: Reclaiming the environmental agenda*, M.I.T. Press, ISBN: 978-0-262-01238-6, Cambridge, USA.

Bayon, R; Hawn, A. & Hamilton, K. (2009). *Voluntary Carbon Markets: An International Business Guide to What They Are and How They Work.* 2nd Edition, Earthscan, ISBN: 978-1-84407-561-4 Sterling, USA

Baroudy, E. (July 2009). The BioCarbon Fund & Land Use, Land Use Change and Forestry (LULUCF), In: *Swedish Presidence of the European Union*, Last visited: 28.08.2011, Available from: http://www.se2009.eu/polopoly_fs/1.27673!menu/standard/file/12_dec_Bringing%20_forest_carbon_projects_to%20_the_market_Ellysar_Baroudy.pdf

BIC (September 2010). Forest Investment Program, In: *Bank Information Center (BCI)*, Last visited: 10.07.2011, Available from: http://www.bicusa.org/en/Issue.53.aspx

Bjørnstad, O. & Grenfell, B. (2001). Noisy Clockwork: Time series analysis of population fluctuations in animals. *Science*. Vol. 29, (July, 2001), pp. 638-643, ISSN: 0036-8075

Blanco, J. (2010). La Experiencia Colombiana en Esquemas de Pagos por Servicios Ambientales, In: *Center for International Forest Research (CIFOR)*, Last visited: 18.09.2011, Available form: http://www.cifor.cgiar.org/pes/publications/pdf_files/colombia_experience.pdf

Bosquet, B. & Andrasko, K. (2010). Introduction and Early Lessons - Briefing to Guyana Civil Society , In: *The Forest Carbon Partnership Facility (FCPF)*, Last visited: 02.09.2011, Available from: http://www.forestcarbonpartnership.org/fcp/sites/forestcarbonpartnership.org/files/Documents/FCPF_Intro_Early_Lessons_Guyana_Final%20_04-21-10.pdf

Casas, A. & Martinez, R. (2008). *Marcos legales para el pago por servicios ambientales en America Latina y el Caribe: Análisis de ocho paises*. Departamento de Desarrollo Sostenible - Organización de Estados Americanos (OEA), ISBN: 978-0-8270-5256-7, Washington, DC, USA.

Carbon Finance Unit and Forestry Team (2008). Contributions to global climate deliberations: the experience of the BioCarbon Fund, the Forest Carbon Partnership Facility and the Forest Investment Program, *The World Bank*, Last visited: 04.05.2011, Available from: <http://unfccc.int/resource/docs/2008/smsn/igo/007.pdf

Carbon Finance Unit (2010). BioCarbon Fund, *The World Bank*, Last viewed: 08.04.2011, Available from: < http://wbcarbonfinance.org/Router.cfm?Page=BioCF >

Chacon, L. & Harvey, C. (2006). Live fences and landscape connectivity in a neotropical agricultural landscape. *Agroforestry Systems*, Vol.68, No.1, (September, 2006), pp. 15-26, ISSN: 0167-4366.

Chazdon. R.L.; Harvey, C.A.; Komar O. *et al.* (2009). Beyond reserves: A research agenda for conserving biodiversity in human-modified tropical landscapes. *Biotropica*, Vol.41, No.2 (March, 2009), pp. 142-153, ISSN: 1744-7429.

Chepstow-Lusty, A.J.; Frogley, M.R.; Bauer, B.S. *et al.* (2009). Putting the rise of the Inca Empire within a climatic and land management context. *Climate of the Past*, Vol. 5, (July 2009), pp. 375-388, ISSN: 1814-9332.

CIF (2010). Forest Investment Program: Investment Criteria and Finance Modalities. In: *Climate Investment Funds (CIF)*, Last visited: 13.06.2011, Available from: http://www.climateinvestmentfunds.org/cif/sites/climateinvestmentfunds.org/files/FIP%20Investment%20Criteria%20and%20Financing%20Modalities%204%2021%202010%20FINAL%20CLEAN.pdf

Colombia, Código Cilvil (1887). Articles: 713, 714 and 716.

Colombia, Constitución Política (1991). Article: 330, numerals: 5 and 6.

Davis, W. (1996). One River: Explorations and Discoveries in the Amazon Rain Forest. Simon & Schuster, ISBN 0-684-80886-2, New York, USA.

De la Torre, A.; Fajnzylber, P. & Nash, J. (2009). *Low Carbon-high growth: Latin American responses to climate change, an overview*. The International Bank for Reconstruction and Development / The World Bank, ISBN: 978-0-8213-7619-5, Washington, DC, USA.

Dinerstein, E.; Olson, D.; Graham, D. *et al.* (1995). *A conservation assessment of the terrestrial ecoregions of Latin America and the Caribbean*. The International Bank for

Reconstruction and Development / The World Bank, ISBN: 0-8213-3295, Washington, DC, USA.

Ducklow, H.; Doney, S. & Steinberg D. (2009). Contributions of long-term research and time-series observations to marine ecology and biogeochemistry. *Annual Review of Marine Science*, Vol.1 (January, 2009), pp. 279-302, ISSN:1941-1405.

Edwards, M.; Beaugrand, G.; Hays, G. *et al.* (2010). Multi-decadal oceanic ecological datasets and their application in marine policy and management. *Trends in Ecology and Evolution*, Vol.25, No.10, (October, 2010), pp. 602-610, ISSN: 0169-5347.

Fandiño-Lozano, M. & W. van Wyngaarden, 2005. Prioridades de Conservación Biológica para Colombia. Grupo ARCO, Bogotá. 188 pp.

Figueres, C. & Streck, C. (2009). A post-2012 vision for the clean development mechanism. In: *Legal Aspects of Carbon Trading Kyoto, Copenhagen and beyond*, David Freestone & Charlotte Streck (Eds.), pp. 562-568, Oxford University Press, ISBN: 978-0-1995-6593-1, Oxford, USA.

Fischer, J. & Lindenmayer, D. (2007). Landscape modification and habitat fragmentation: a synthesis. *Global Ecology and Biogeography*, Vol.3 (May, 2007), pp. 265-280, ISSN: 1466-8238.

Forman, R.T. (1995). Some general principles of landscape and regional ecology. *Landscape Ecology*. Vol.10, No.3, (June, 1995), pp. 133-142, ISSN: 1572-9761.

Freestone, D. (2009). The international climate change legal and institutional framework: An overview. In: *Legal Aspects of Carbon Trading Kyoto, Copenhagen and beyond*, David Freestone & Charlotte Streck (Eds.), pp. 3-16, Oxford University Press, ISBN: 978-0-1995-6593-1, Oxford, USA.

Gaston, K. (2000). Global patterns in biodiversity. *Nature*, Vol.405 (May, 2000), pp. 220-227, ISSN: 0028-0836.

Guadagnin, D. & Maltchik, L. (2007). Habitat and landscape factors associated with neotropical waterbird occurrence and richness in wetland fragments. *Biodiversity & Conservation*, Vol.16, No.4, (April, 2007), pp. 1231-1244, ISSN: 0960-3115.

Guisan, A.; Lehmann, A.; Ferrier, S.; *et al.* (2006). Making better biogeographical predictions of species' distributions. *Journal of Applied Ecology*, Vol.43, No.3, (June, 2006), pp. 386-392, ISSN: 0021-8901.

Halpern, B.; Walbridge, S.; Selkoe K.A.; *et al.* (2008). A global map of human impact on marine ecosystems. *Science*, Vol.319, (February, 2008), pp. 948-952, ISSN: 0036-8075.

Hamilton, K.; Bayon, R. & Hawn, A. (2008). Carving a niche for forests in the voluntary carbon markets. In: *Climate change and forests: emerging policy and market opportunities*, Charolette Streck, *et al.* (Eds.) pp. 282-292, Chatham House / Brookings Institution Press, ISBN: 0-8157-8192-X, Washington, DC, USA.

Harvey, C.A.; Komar, O.; Chazdon, R.; *et al.* (2008). Integrating agricultural landscapes with biodiversity conservation in the mesoamerican hotspot. *Conservation Biology*, Vol.22, No.1, (February, 2008), pp. 8-15, ISSN: 0888-8892

Hey, J.; Waples, R.; Arnold, M.; *et al.* (2003). Understanding and confronting species uncertainty in biology and conservation. *Trends in Ecology and Evolution*, Vol.18, No.11, (November, 2003), pp. 597-603, ISSN: 0169-5347.

Hobbs, R.J.; Arico, S.; Aronson, J.; et al. (2006). Novel ecosystems: theoretical and management aspects of the new ecological world order. *Global Ecology and Biogeography*, Vol.15, No.1, (January, 2006), pp. 1-7, ISSN: 1466-8238.

Hobbs, R. (1997). Future landscapes and the future of landscape ecology. *Landscape Urban Planning*, Vol.37, No.1, (June, 1997), pp. 1-9, ISSN: 0169-2046.

Hopwood, J. L. (2008). The contribution of roadside grassland restorations to native bee conservation. *Biological Conservation*, Vol.141, No.10, (October, 2008), pp. 2632-2640, ISSN: 0006-3207.

International Institute for Environment and Development (2007). Case study: Colombia, Plan Verde. In: *Watershed Markets*, Last viewed: 01.30.2011, available from: <http://www.watershedmarkets.org/casestudies/Colombia_Plan_Verde_E.html>

Ives, A. & Carpenter, S. (2007). Stability and diversity of ecosystems. *Science*, Vol.317, (July, 2007), pp. 58-62. ISSN: 0036-8075.

Ketabi, N. (2009). The blood of going green: Using environmental initiatives to account for the human rights volations of the green movement, *Fordham International Law Journal*, Vol.32, No.6, (November, 2008), pp. 1911-1946, ISSN: 0747-9395.

Ludwig, D.; Walker, B. & Holling, C.S. (1997). Sustainability. stability. and resilience. *Conservation Ecology*, Vol.1, No.1, (June, 2007), ISSN: 1708-3087.

Lugo, A. E. (2004). The outcome of alien tree invasions in Puerto Rico. *Frontiers of Ecology and the Environment*, Vol.2, No.5, (June, 2004), pp. 265–273, ISSN: 1540-9295.

MacGregor-Fors, I. (2010). How to measure the urban-wildland ecotone: redefining 'peri-urban' areas. *Ecological Research*, Vol. 25, No.4, (July, 2010), pp. 883-887, ISSN: 1440-1703.

Martinez, A.G. (2010). Nukak Macu. In: *Toda Colombia*, Last visited: 22.08.2011, Available from: http://www.todacolombia.com/etnias/gruposindigenas/nukak.html

Martinez-Morales, M. A. (2005). Landscape patterns influencing bird assemblages in a fragmented neotropical cloud forest. *Biological Conservation*, Vol.121, No.1, (January, 2005), pp. 117-126, ISSN: 0006-3207.

Matthews, R. & Selman, P. (2006). Landscape as a focus for integrating human and environmental processes. *Journal of Agricultural Economics*, Vol.57, No.2, (July, 2006), pp. 199-212, ISSN: 1477-9552.

McKinney, M. L. (2006). Urbanization as a major cause of biotic homogenization. *Biological Conservation*. Vol.127, No.3, (January, 2006), pp. 247-260, ISSN: 0006-3207.

Morton, D.; DeFries, R.; Shimabukuro, Y.; et al. (2006). Cropland expansion changes deforestation dynamics in the Southern Brazilian Amazon. *Proceedings of the National Academy of Sciences of the United States of America*. Vol.103, No.39, (), pp. 14637-14641 ISSN: 0027-8424.

Ostrom, E. (1990). *Governing the Commons: the evolution of institutions for collective action*. Cambridge University Press, ISBN: 0-521-40599-8, Cambridge, UK.

O'Sullivan, R. & Saines, R. (2009). International Market Solutions to Protect Tropical Rainforest. In: *Legal Aspects of Carbon Trading Kyoto, Copenhagen and beyond*, David Freestone & Charlotte Streck (Eds.), pp. 583-584, Oxford University Press, ISBN: 978-0-1995-6593-1, Oxford, USA.

Passero, M. (2009). The voluntary carbon market: Its contributions and potential legal and policy issues. In: *Legal Aspects of Carbon Trading Kyoto, Copenhagen and beyond*, David Freestone & Charlotte Streck (Eds.), pp. 517-521, Oxford University Press, ISBN: 978-0-1995-6593-1, Oxford, USA.

Peters-Stanley, M. (2010). New voluntary carbon platform rooted in Colombian forests. In: *Forest Carbon Portal: Tracking terrestrial Carbon*, Last visited: 15.06.2011, Available from: http://www.forestcarbonportal.com/content/colombia-launching-national-ecosystem-marketplace,

Pielou, E.C. (1979). *Biogeography*. John Wiley & Sons, ISBN: 0-471-05845-9, New York, USA.

PROFOR (2004). Economic incentives for SFM and landscape restoration, a workshop summary. In: *The World Bank Program on Forest (PROFOR) Vol.1, No.2*, Last visited: 21.07.2011, Available from:
http://www.profor.info/profor/Documents/pdf/PES_English.pdf

Restrepo-Aristizabal, J.A. (2006) Landscape dynamics, forest fragmentation and their relation to socioeconomic history and biophysical attributes in the Colombian highlands. In: Schumann, A., Pahlow, M. Reducing the Vulnerability of Scoieties to Water Related Risks at the Basin Scale. pp. 60-65. IAHS Publication 317. ISSN: 01447815

Richards, K.R.; Simpson, N. & Brown, S. (2006). Agricultural and forestlands for U.S. carbon policy strategies. In: *The Pew Center on Global Climate Change*, Last visited: 11.08.2011, Available from:
http://www.pewclimate.org/publications/report/agricultural-and-forestlands-us-carbon-policy

Robledo, C. & Tobon, P. (2008). Case Study: The San Nicolas project in Colombia. In: *Climate change and forests: emerging policy and market opportunities*, Charolette Streck, *et al.* (Eds.) pp. 122-124, Chatham House / Brookings Institution Press, ISBN: 0-8157-8192-X, Washington, DC, USA.

Sanchez-Clavijo, L.M.; Arbelaez-Alvarado, D. & Renjifo, L.M. (2008). Bird research and conservation in neotropical rural landscapes. *Ornitologia Neotropical*, Vol.19, pp. 549-565, ISSN: 1075-4377.

Sandker, M.; Nyame, S.K.; Forster, J. *et al.* (2010). REDD Payments as Incentives for Reducing Forests Loss. *Conservation Letters*, Vol.3, pp. 114-121, ISSN: 1755-263X.

Saatchi, S.; Harris N.; Brown, S. *et al.* (2011). Benchmark map of forest carbon stocks in tropical regions across three continents. *Proc. Natl. Acad. Sci. USA* Vol. 108, No. 24, pp.1-6, DOI: 10.1073.

S. & De Witt Wijnen, R. (2009). International Emissions Trading and Investment Schemes. In: *Legal Aspects of Carbon Trading Kyoto, Copenhagen and beyond*, David Freestone & Charlotte Streck (Eds.), pp. 157-158, Oxford University Press, ISBN: 978-0-1995-6593-1, Oxford, USA.

Smith, V. Tilman, G. & Nekola, J. (1999). Eutrophication: impacts of excess nutrient inputs on freshwater, marine. and terrestrial ecosystems. *Environmental Pollution*, Vol.100, No.1-3, pp. 179-19, ISSN: 0269-7491.

Suarez-Rubio, M. & Thomlinson, J.R. (2009). Landscape and patch-level factors influence bird communities in an urbanized tropical island. *Biological Conservation*, Vol.142, No.7, (July-2009), pp. 1311–1321, ISSN: 0006-3207.

Troll, C. (1939). Luftbildplan und ökologische Bodenforschung. *Zeitschrift der Gesellschaft Für Erdkunde zu Berlin*, No.7-8, pp. 1-56, ISSN: 0013-9998.

Udvardy, M.D.F. (1975). A classification of the biogeographical provinces of the world. *IUCN Occasional Papers*, No.18, International Union of Conservation of Nature and Natural Resources (IUCN). Retrived from: <http://cmsdata.iucn.org/downloads/udvardy.pdf>

UN (1998). Kyoto Protocol to the United Nations Framework Convention on Climate Change (UNFCCC). *United Nations*, Last visited: 19.04.2011, Available from: <http://unfccc.int/resource/docs/convkp/kpeng.html>

UNFCCC (1992). Full text of the United Nations Framework Convention on Climate Change. In: *United Nations Framework Convention on Climate Change*, Last visited: 25.02.2011, Available from: <http://unfccc.int/essential_background/convention/background/items/1349.php>

UNFCCC (2010). Report of the Conference of the Parties on its 15th session, held in Copenhagen from 7 to 19 December 2009 Addendum Part Two: Action taken by the Conference of the Parties at its 15th session: Decisions adopted by the Conference of the Parties. *UN/UNFCCC*, Last visited: 01.09.2011, Available from: <http://unfccc.int/resource/docs/2009/cop15/eng/11a01.pdf#page=3>

UNFCCC (2010) Demonstration activities: Forest Carbon Partnership Facility (FCPF). *United Nations Framework Convention on Climate Change*, Last viewed: 12.05.2011, Available from:<http://unfccc.int/methods_science/redd/demonstration_activities/items/4536.php>

Van den Bergh, J. & Grazi, F. (2010). On the policy relevance of ecological footprints. *Environmental Science & Technology*, Vol.44, No.13, (May, 2010), pp. 4843-4844, ISSN: 0013-936X.

Vitousek,P.; Mooney, H.; Lubchenco, J. & Melillo, J. (1997). Human domination of earth's ecosystems. *Science*. Vol.177, (July, 1997), pp. 494-499, ISSN: 0036-8075.

Wallace, S. (2007). Last of the Amazon. In: *National Geographic*, Last viewed: 12.12.2010, Available from: <http://ngm.nationalgeographic.com/2007/01/amazon-rain-forest/wallace-text>

Wiens, J.A. (1989). Spatial Scaling in Ecology. *Functional Ecology*. Vol.3, No.4, pp. 385-397, ISSN: 1365-2435.

Wunder, S. (2007). The efficiency of payments for environmental services in tropical conservation. *Conservation Biology*, Vol.21, No.1, (February, 2007), pp. 48-50, ISSN: 0888-8892.

Wyngaarden, W. & Fandiño-Lozano, M. (2005). Mapping the actual and original distribution of the ecosystems and the chorological types for conservation planning in Colombia, *Diversity & Distribution*, Vol.11, No.5, (September, 2005), pp. 461-73, ISSN: 1472-4642.

Zadeh, L. (1987). Fuzzy Sets. In: Yager, R.R, Ovchinnikov, S., Tong, R.M., Nguyen, H.T. *Fuzzy sets and applications*. John Wiley &Sons. ISBN:0-471-85710-6.

Zonneveld, I.S. (1979). Land Evaluation and Landscape Science. ITC, Enschede.

Zonneveld, I.S. (1979). Land evaluation and land (scape) science. In: *ITC Textbook VII.4* (2nd ed.), ITC Enschede. pp. 134

Zurlini, G.; Riitters, K.; Zaccarelli, N.; *et al.* (2006). Disturbance patterns in a socio-ecological system at multiple scales. *Ecological Complexity*, Vol.3, No.2, (June, 2006), pp. 119–128, ISSN: 1476-945X.